# HEAT AND FLUID FLOW IN WATER REACTOR SAFETY

# HEAT AND FLUID FLOW IN WATER REACTOR SAFETY

## I Mech E CONFERENCE PUBLICATIONS 1977–8

Conference sponsored by the
Thermodynamics and Fluid Mechanics Group of
The Institution of Mechanical Engineers

Manchester, 13–15 September 1977



Published by
Mechanical Engineering Publications Limited for
The Institution of Mechanical Engineers
LONDON

First published 1977



This volume is complete in itself.

There is no supplementary discussion volume.

ISBN 0 85298 379 4

Printed and bound at The Burlington Press, Foxton, Royston, Hertfordshire

# CONTENTS

# Corrigenda

Page

69 The author's name and qualifications should read:
S. FABIC, MMechE, MS, PhD

145 W.B. Hall has provided the following correction for paper C212/77:
The example quoted in Section 5 of the paper contains an arithmetical error. The table at the end of the section should read:

| $\frac{Ar}{Ao}$ | 0.384 | 0.311 | 0.249 | 0.187 | 0.124 |
|---|---|---|---|---|---|
| % Blockage | 61.6 | 68.9 | 75.1 | 81.3 | 87.6 |
| $\Delta T_c(^\circ C)$ | 79 | 115.7 | 166.8 | 237.9 | 464.3 |
| $\Delta T_s(^\circ C)$ | 79 | 137.4 | 215.2 | 347.3 | 627.4 |

169 The author's inform us that in para 21, the equations should read:

$$j_s^* = j_s \rho_s^{\frac{1}{2}} \left[g D_H (\rho_w - \rho_s)\right]^{-\frac{1}{2}}$$

and
$$j_w^* = j_w \rho_w^{\frac{1}{2}} \left[g D_H (\rho_w - \rho_s)\right]^{-\frac{1}{2}}$$

where $j_s = Q_s/A$ and $j_w = Q_w/A$ and the characteristic dimension is taken as the hydraulic diameter $D_H = 4A/P$.

C191/77

# THE GENERAL PHILOSOPHY OF RISK

F.R. FARMER, OBE, HonFSE, BSc, FInstP
Safety Adviser to the Authority, UKAEA, Culcheth, Warrington

SYNOPSIS As the Conference is concerned with heat and fluid flow in Water Reactor Safety, this paper will consider some of the issues which influence safety research. If research were determined only by cost of potential error, then most effort would be devoted to accident related studies.

Perhaps too much money is already invested in the improbable event, and the theme developed endeavours to relate effort to a probability based potential loss.

The study of transient events is costly and it is difficult to define the field of interest sufficiently well to convince the critic. It is impossible to prove that all accidents have been considered and that all models represent real situations. There is a recognition of risk of injury or death in work and play, not so clear a recognition that risk must be accepted in terms of uncertainty in data in accident analysis and in our perception. There is some attempt to match, at least qualitatively, the degree of uncertainty with the probability weighted penalty of error.

## COSTS VERSUS BENEFITS

1. In some industries and for some projects there is an attempt to relate the cost of research to the possible benefits accruing. In safety research in the nuclear industry there has been a tendency to argue that risk must be removed at any cost. This is obviously impossible and in general the programmes are determined by pressure groups, or by the degree of technical challenge or as a continuation of existing research.

2. Although this meeting is concerned with heat and fluid flow - I wish to widen the basis of argument to illustrate some of the difficulties in directing safety programmes. Several estimates have been made as to the cost of a serious reactor accident, they lie in the range of $\$10^8$ to $\$10^9$ The cost to an operator will be limited by his resources, or insurance cover. Resources are unlikely to exceed $\$10^8$, so that in practice a company might ask what investment today might be worth covering a long-term risk of this order.

3. If the risk is of the order of $10^{-5}$ per year, then it can hardly be worth as much as $1,000 per year to eliminate some or all of the penalty. Even if we consider 20-40 installations over a ten year programme, it can still be worth no more than $10,000-$20,000 per year to investigate the problem. Fundamental in any consideration of this type lies the uncertainty in obtaining any solution; we cannot say that a problem is worth $20,000 per year, therefore spend up to this sum to solve it.

4. In the early 1960's, how could one assess how much should be spent to investigate the possible failure of pressure vessels? Even then, it might reasonably be said that the failure rate of well constructed vessels was less than $10^{-5}$ per vessel per year. It should be shown that thick-walled vessels could be regarded as falling within the family of past experience. It was claimed that this had been achieved by pressure vessel hydraulic tests and through the temperature failure criteria determined by the Robertson machine. Hence, thick-walled vessels could not fail catastrophically if maintained at a temperature of NDT (nil ductility temperature) + 60°F. If this had been universally accepted there would have been no logical basis for further work. However, research into crack propagation continued - perhaps to show that pressure vessel failure could still be supported at $10^{-5}$ per year, as against an alternative proposition that thick structures might show some new problems and could have a higher failure rate, even $10^{-3}$ per year. Any supporting argument for research is not as clear cut as implied in the foregoing. What has stemmed from pressure vessel research?

- a better understanding of crack growth rates
- a closer control of pressure and temperature cycles in operation
- better inspection techniques
- better design in detail to eliminate stress concentration and to ease inspection

## THE FALLACY OF ABSOLUTE SAFETY

5. However, underlying this work (in pressure vessels) as also work in many aspects of safety, is first the search for a perfect solution, or one deemed adequate within some assumed risk rate - "it is good enough" - "we may proceed without undue risk."

6. The pressure vessel criteria, NDT + 60°F fall into this framework, as also in earlier days the belief in containment - a reactor would be safe if contained. A desire to believe that

safety is achieved - or achievable - in an absolute sense may divert attention from necessary research in other directions and has done so in the past.

7. The pressure vessel argument delayed work into more sophisticated studies of crack propagation and inspection - the containment argument delayed more detailed examination of depressurisation phenomena - including transient heat and fluid flow. Through the years of 1970-1974 the examination of emergency cooling criteria might have appeared against a background that the right series of experiments would enable definitive criteria to evolve, such that conformity with these criteria would ensure safety.

8. Various limitations have been considered - burn-out or dry-out on limited surfaces - deformation of cladding - zirconium-water reactions - hydrogen explosion etc. In all of these studies, there is need to consider the range of the variables, the distribution of flow, of power, of surface dimensions, of burn-up, and of cladding or fuel properties. Nowadays reasonably sophisticated ways of combining distributions are used - but always in a way which precludes some extreme distributions. It is accepted now that no criterion is free from some small residual risk of failure. Implicit in this is the uncertainty in the pattern of the accident initially imposed; only a finite range can be analysed and envelope assumptions are seldom satisfactory or tenable.

ADDITIONAL BENEFITS

9. Whereas some research is directed solely to a safety issue, it is more likely that the results of many reactor programmes show benefit in improved performance, a lower cost, or a lower risk of failure or accident. Safety alone is unlikely to justify on a logical basis the extent of effort devoted to it. This would be particularly true if we extended the basis for safety decisions beyond the field of nuclear risks. Some of us have for years maintained that non-nuclear risks were as great in their potential damage, had a higher likelihood of occurrence and were not receiving a fraction of the effort needed to control the hazard. However, this lies beyond the scope of the present meeting.

THE PURPOSE OF SAFETY EXPERIMENTS

10. Even if nuclear safety now receives excessively generous support, it is important that the programme should be critically reviewed. Many of us - including friends in the USA - feel that the LOFT programme was ill-conceived. I remember in 1965 the plea from industry to set up a demonstration programme to show that a pipe fracture would not lead to a significant release of fission products beyond the containment. What has followed has been a research programme exploring in detail the transient pressures and transient heat and fluid flow for a finite range of accident initiating events.

11. During any prolonged series of experimental programmes with associated modelling and theoretical interpretation and extrapolation there is earnest debate about the uncertainty of the data, the model and its relevance to some other condition. This is still true in current work associated with emergency cooling, and even more so in the understanding of the thermal interaction of hot fuel and coolant. Too often experts in these fields of transient heat and flow phenomena view their problems in isolation, and also believe it is their responsibility to continue to work to remove residual uncertainties.

12. In most safety related programmes there will always remain uncertainties. It was not fashionable to admit this in the period 1965/1970 when there was an apparent intent to design for zero escape or to license only those plants which carried no credible risk of harm to people.

RISK ASSESSMENT DIRECTING RESEARCH

13. More recently there has been open recognition that there is some small risk of a major accident - the incredible US Class 9. The Rasmussen report (WASH-1400) examines most accident patterns and although his work has been challenged in detail, there are many who feel that the overall conclusion is reasonable - there is a risk in the range of $10^{-4}$ to $10^{-6}$ per year that any US water reactor might suffer a severe accident. The consequences then span a range of many decades - as would be the case for many non-nuclear accidents (Ref.1)

14. I suggest then, that if many types of accident fall within some roughly discernible framework, it is not unreasonable to assess the current state of art in relation to the accident frequency and consequences which may be linked with specific phenomena.

15. I regard the Marshall report on pressure vessels as doing just this, throwing light on the areas of pressure vessel design construction and inspection which have a sensitive bearing on the continued safety of the vessel (Ref.2)

16. The Rasmussen report provides a basis on which the adequacy of knowledge in other directions might be tested. If an initiating event such as gross melting of the core is thought to carry a risk of $10^{-5}$ per reactor per year, then I would not spend much money investigating metal-water reactions. If, however, it was felt that localised melting could lead to a form of propagation then this would need a more careful study.

THE NEED FOR A BALANCED APPROACH

17. An early introduction to probability criteria arose in relation to the operating temperature of magnox reactor fuel (Ref.3):

> "The problem of specifying a 'safe' temperature distribution is a statistical one for which a number of alternative approaches can be adopted. The one chosen specifies that the distribution of fuel element temperatures in the reactor under normal operating conditions shall be such that the initial transient following a bottom duct failure, the probability of can temperature exceeding the ignition point of magnox, shall be less than 0.1 at a 90% confidence level".

In this reactor we had a uniquely definable initiator - magnox fire which would soon spread at least within a channel.

18. I see the difficulty in many safety analyses of deciding the size in area, volume, number of assemblies etc and the extent of deviation from

some specified limit in temperature, or flow - which jointly leads to an irretrievable situation. Any such analysis must also take into account the uncertainty in timing and functioning of the remedies (such as emergency cooling).

19. I do not propose any easy answer, it is more important than ever in the pursuit of knowledge to eliminate low risk situations, that the uncertainties in any research programme should be weighed against the uncertainties of the accident model proposed, the uncertainties in engineering or operational aspects which affect the initiating of the accident or its subsequent control. Finally some risk will remain; it is not unreasonable to admit uncertainty in analysis, in engineering and in perception at the level of $10^{-5}/10^{-6}$ per year - as quoted recently by Vinck (Ref.4):

> "I do not believe target value lower than $10^{-6}$ events/reactor-year are practical, nor are they necessary, taking account for instance of the severity with which the nuclear activities are handled."

REFERENCES

1. Reactor Safety Study - An Assessment of Accident Risks in US Commercial Nuclear Power Plants. WASH-1400, USNRC, Washington DC, October 1975.

2. An Assessment of the Integrity of PWR Pressure Vessels. Report by a Study Group under the chairmanship of Dr W Marshall. UKAEA Report, 1 October 1976, HMSO.

3. HUGHES H A, and HORSLEY R M. Application of Safety Limitation against Depressurisation to the Calder and Chapelcross Reactors, BNES Symposium on the Safety of Magnox Reactors, London 11 November 1964, pp 198-203.

4. VINGK W, and REIJEN G VAN, Quantitative Risk Assessment; the promised but not the sacred, ANS/ENS International Conference on World Nuclear Power, Washington DC, Session on 'Implications of WASH-1400 - the Reactor Safety Study' November 14-19, 1976.

C192/77

# INTERFACIAL PASSAGE FREQUENCY FOR TWO-PHASE, GAS-LIQUID FLOWS IN NARROW RECTANGULAR DUCTS

O. C. JONES, Jr, BS, MS, PhD
Brookhaven National Laboratory, Upton, NY, USA
N. ZUBER, BS, MS, PhD
US Nuclear Regulatory Commission, Washington, DC, USA

SYNOPSIS Hydrodynamic and thermal transfer and relaxation phenomena in two-phase gas-liquid flows are strongly dependent on departure from equilibrium and the interfacial area density available for phase interactions. Such interactions include mass, momentum, and energy exchange for which the transfer coefficient must be known accurately as well as the area available for interactions to occur.

The direct measurement of interaction interfacial area density is an exceedingly difficult task in view of today's knowledge. On the other hand, knowledge of the interfacial passage frequencies coupled with information regarding the flow structure would enable a model to predict the area densities reasonably well. Such information has been obtained using a sub-miniature, hot-film anemometer in air-water flows at velocities up to approximately 25 m/s. These data include bubbly, slug, annular, and annular mist flows. This report describes the method, presents the data, and shows how certain conclusions regarding the flow structure can be drawn, particularly regarding the nature of roll waves in narrow ducts.

## INTRODUCTION

1. Hydrodynamic and thermal transfer and relaxation phenomena in two-phase, gas-liquid flows are strongly dependent on departure from equilibrium, and on the interfacial area density available for phase interactions. Ishii (1, 2) has shown the importance of the interfacial passage frequency in his discussions of short-term averaging of the field equations at a point in a two-phase flowing mixture. Jones (3) used Ishii's method rewritten to obtain the time averaged mixture balance as

$$\frac{\partial \overline{\psi}}{\partial t} + \hat{\nabla} \cdot \overline{\psi \hat{v}} = \overline{s} - \hat{\nabla} \cdot \overline{\hat{\phi}}$$
$$+ f_i(\hat{x},t)\ \overline{\{\{\psi\} - \{\psi\hat{v} + \hat{\phi}\} \cdot \nabla\Delta_k\}}$$

where the impulsive jump in $\psi$ from $(t + \Delta_k^+)$ to $(t + \Delta_k^-)$ is denoted by $\{\ \}$. Note that $\psi$ is any volumetrically intensive property, $\hat{v}$ is the material velocity, s the volumetric source of $\psi$, $\hat{\phi}$ the flux of $\psi$, and $\Delta_k$ is the time delay between the current time, t, and the time of passage of the $k^{th}$ interface, $\eta_k$: i.e., $\Delta_k = \eta_k - t$. It is readily seen that the strength of the additional effect of interfacial passage on short term averaged, (overbar), system behavior at a point in space, $\vec{x}$, is directly proportional to the interfacial passage frequency $f_i(\vec{x}, t)$. The need to know the interfacial passage frequencies to describe the point behavior of the two-phase mixture based on the time average is directly analogous to the need to know the interfacial area densities for the volume-averaged case, in order to describe interactions such as non-equilibrium mass exchange and momentum transfer.

2. The direct measurement of interfacial area density is an exceedingly difficult task in view of today's knowledge. On the other hand, knowledge of the interfacial passage frequencies coupled with information regarding the flow structure would enable a model to predict the area densities reasonably well. It is the purpose of this paper to present local measurements of such data obtained in profiles across the small dimension of a narrow rectangular duct, and to show how certain conclusions regarding the flow structure can be drawn, particularly regarding the nature of roll waves in these ducts.

## NOTATION

3. English

A flow area

d diameter

f interfacial passage frequency

j volume flux (of the mixture if no subscript)

n number

Q volume flow rate

s source

t time

v velocity

V volume

x position

4. Greek

$\alpha$ void fraction

$\Delta$ interface time delay

$\phi$ flux

ψ volumetric intensive property

5. Subscript

b bubble

f liquid

g gas

i interface

k interface-k

p probe factor

EXPERIMENTAL EQUIPMENT

6. Details of the experimental equipment and procedures are given in References (3), (4), and (5), so only a brief outline shall be given herein.

7. Data were obtained by traversing a 50-μm diameter, hot-film anemometer, oriented perpendicular to the flow, in small, accurate (~± 0.5 μm), increments across the 0.51-cm dimension of a calibrated, 0.498 x 6.35 x 300 cm rectangular duct, 12.3 cm from the exit in vertical, upward, air-water flow. The Thermo-Systems, Inc. (TSI), subminiature probes, bent 90° to the probe shaft projecting upstream, were traversed to within 25-50 μm from the opposing wall as determined by electrical contact with a flush metal insert. Special circuitry was used in conjunction with the TSI 1050 anemometer system to yield pulses due to passage of interfaces causing sharp changes in the system signal beyond the normal range of those associated with turbulent velocity fluctuations. Counting of these pulses over a 100-sec period yielded the desired frequencies at one pulse per interface.

8. Except for the bubbly flow data where shear bubbles were obtained by injecting air through the 6.35 cm square porous plates 1.22 m from the channel exit, filtered air was injected into the inlet to an accuracy of 1% of reading while filtered, demineralized water was injected through the plates 2.44 m from the exit with an accuracy of 1/2% of reading. The 100-sec interfacial frequency counts were obtained simultaneously with the Reference 4 data and under similar, repeat conditions as the x-ray void fluctuation statistics of Reference 5. These data were found to be reproducable within ± 3% and are estimated to be accurate within 10%.

RESULTS AND DISCUSSION

9. The detailed profile data obtained are shown in Reference 3, corresponding directly to the data shown in Figures 9, 11, and 13 of Reference 4. While corrections were made to the void residence time fraction data previously reported (4) due to voids not "seen" by the sensor, none were made to these data due to the lack of a suitable reference. Thus, the data shown herein are undoubtedly biased somewhat low. Chord-averaged values for the passage frequencies are given in Table 1. From these values some gross observations can be made using flow regime as a descriminating parameter.

| Run No. | Liquid Flux ft/sec. | Gas Flux ft/sec. | Flow Pattern | Void Fraction *** | Interface Frequency Hz |
|---|---|---|---|---|---|
| A3 | 0.0 | 0.26 | S | 0.156 | 129 |
| A4 | 0.0 | 1.23 | S | 0.442 | 197 |
| A5 | 0.0 | 4.31 | S | 0.701 | 101 |
| A6 | 0.0 | 11.4 | A | 0.793 | 64 |
| A9 | 0.0 | 0.09 | B | 0.073 | 2.1 |
| A11 | 0.5 | 0.05 | B | 0.032 | 15 |
| A12 | 0.5 | 0.23 | B | 0.133 | 103 |
| A13 | 0.5 | 0.12 | S | 0.059 | 36 |
| A14 | 0.5 | 0.43 | S | 0.174 | 109 |
| A15 | 0.5 | 1.62 | S | 0.422 | 122 |
| A16 | 0.5 | 6.25 | S | 0.695 | 125 |
| A17 | 0.5 | 20.8 | A | 0.803 | 179 |
| A20 | 1.0 | 0.28 | B | 0.115 | --- |
| A21 | 1.0 | 0.42 | S | 0.135 | 118 |
| A22 | 1.0 | 2.66 | S | 0.461 | 198 |
| A24 | 1.0 | 5.32 | S | 0.594 | 177 |
| A25 | 1.0 | 14.5 | A | 0.754 | 247 |
| A25R | 1.0 | 12.6 | A | 0.741 | 225 |
| A26 | 1.0 | 26.3 | A | 0.798 | 359 |
| A27 | 1.0 | 45.5 | A | 0.823 | 1149 |
| A30 | 2.0 | 0.18 | B | 0.048 | 130 |
| A31 | 2.0 | 0.57 | B | 0.140 | 229 |
| A32 | 2.0 | 1.22 | B | 0.252 | 403 |
| A33 | 2.0 | 0.59 | S | 0.129 | 100 |
| A34 | 2.0 | 1.34 | S | 0.247 | 176 |
| A35 | 2.0 | 3.57 | S | 0.440 | 187 |
| A36 | 2.0 | 8.62 | S | 0.608 | 230 |
| A37 | 2.0 | 32.1 | A | 0.783 | 708 |
| A40 | 5.0 | 0.64 | B | 0.079 | 397 |
| A41 | 5.0 | 2.30 | B | 0.230 | 1377 |
| A42 | 5.0 | 1.66 | S | 0.167 | 153 |
| A43 | 5.0 | 4.78 | S | 0.352 | 281 |
| A44 | 5.0 | 14.7 | S | 0.574 | 495 |
| A45 | 5.0 | 45.9 | A | 0.745 | 1512 |
| A46 | 5.0 | 78.2 | A | 0.796 | 3443 |
| A47 | 10.0 | 1.04 | B | 0.068 | 963 |
| A48 | 10.0 | 2.17 | B | 0.131 | 2260 |
| A49 | 10.0 | 2.64 | S | 0.146 | 543 |
| A50 | 10.0 | 7.8 | S | 0.309 | 1730 |

Table 1. Average interfacial passage frequencies

Bubbly Flow

10. The bubbly flow data are shown as functions of void fraction in Figure 1. For bubble chains, Silberman (6) assumed that a gas jet issuing into a liquid stream would be quickly accelerated to the liquid velocity and would then break up due to Rayleigh instability. The characteristic wave length would be 6.48 times the jet diameter. Assuming each cylinder of diameter $d_j$ and length 6.48 $d_j$ resulted in a spherical bubble, he was able to calculate the characteristic bubble diameter and frequency as

$$d_b = 2.4\sqrt{\frac{Q_g}{j_f}} \; ; \qquad f = \frac{Q_g}{V_b} = 0.137\, j_f\sqrt{\frac{j_f}{Q_{gb}}} \qquad (2)$$

Since the bubbly flow velocity profile is reasonably flat, (Reference 4), the length between bubble centers in the stream for this single chain is $v_f/f_b$ which gives the void fraction due to n jets in chains as

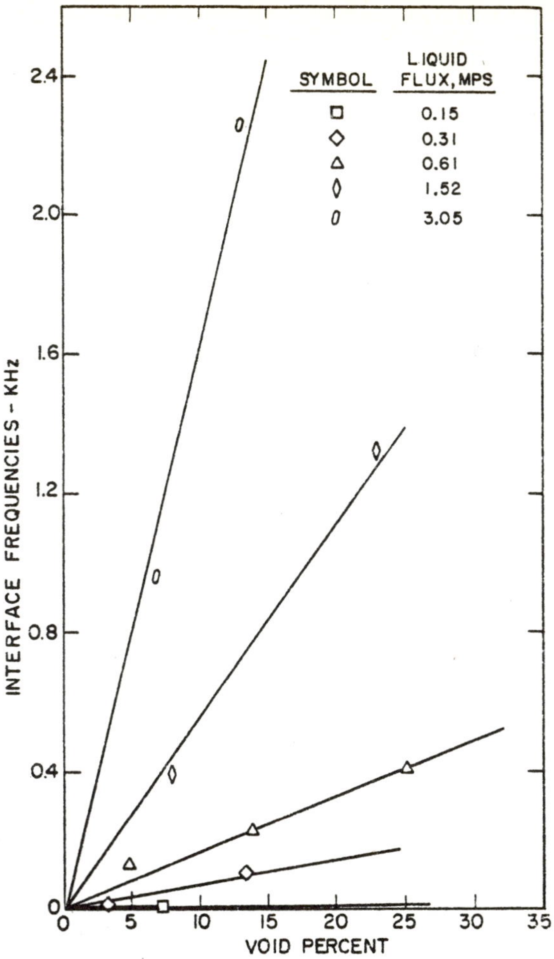


Fig. 1: Variations of chord-averaged interface passage frequency with void fraction at constant liquid flux for bubbly flow

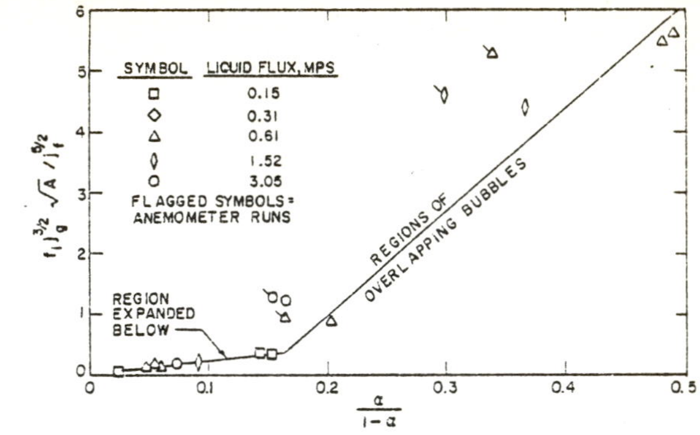


(a) All bubbly flow data

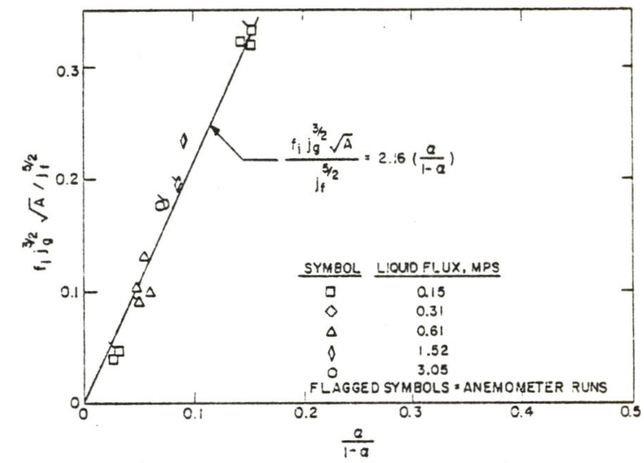


(b) Region of no overlapping bubbles

Fig. 2: Correlation of bubbly flow interface frequencies

$$\alpha = \frac{n f_b V_b}{A v_f} = \frac{j_g}{j_f}(1 - \alpha) \qquad (3)$$

indicating unity slip ratio or zero drift has been tacitly assumed. Solving for the bubble frequency in terms of the void fraction and fluxes yields a dimensionless interfacial frequency of

$$\frac{f_i \sqrt{A}\, j_g^{3/2}}{j_f^{5/2}} = 0.273 \sqrt{n} \left( \frac{\alpha}{1 - \alpha} \right) \qquad (4)$$

Note that it is the single-phase velocity, $j_f$, that is responsible for bubble formation, and that there are two interfaces per bubble.

11. The data for bubbly flow are plotted in Figure 2 and seem to fall into two regimes according to whether they are isolated or interacting. (The anemometer data, that is, the data where the mean channel void fraction are calculated only rather than measured with an x-ray densitometer, are flagged in this figure.) When the bubbles become closely packed and overlap the behavior is seen to change and does not yield good reproducibility on these coordinates. The coefficient is found to be 2.16 for the isolated bubbles yielding n ~ 60 or about 0.7 per square centimeter on the porous plates, in agreement with our qualitative observations.

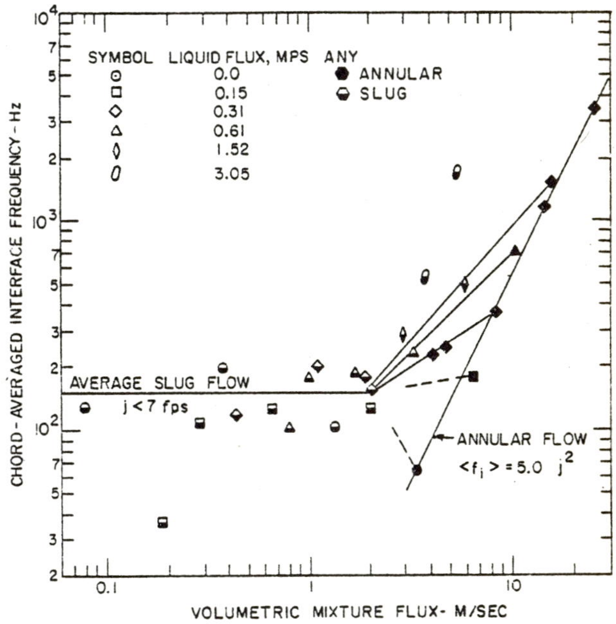


Fig. 3: Variations of chord-averaged interface passage frequencies with mixture volume flux in slug and annular flows

## Slug and Annular Flow*

12. There appear to be three major regimes connected by two transition regimes in the data shown in Figure 3:

(a) Isolated spherical caps or Taylor bubbles where the interface frequency is twice the slug passage frequency (Reference 5);

(b) Slug flow where $j < 2.1$ m/sec and where the average interfacial passage frequency seems constant at 150 Hz $\pm$ 50 Hz;

(c) Annular flow where the interfacial passage frequency seems to be proportional to the square of the volume flux.

Fully Developed Slug Flow: $j < 2.1$ m/s. Slug passage frequency is relatively insensitive to the gas flux (5).

13. In order for the interface passage frequency to be constant, the interface number density must decrease as $1/j_g$ to offset the interface transport effect of increasing volume flux. As shown in Reference 5, the slug length goes as $1/j_g$, but the slug frequency does not change much with $j_g$. Thus, the number density of interfaces in the slug must remain constant while their total numbers decrease due to decreasing slug length only. In addition, the existence of interfaces within slugs seems due predominantly from trailing edge extrusion effects due to flow of liquid around the major bubbles. Any small bubbles generated in this manner must then be offset by agglomeration with the following major bubbles, leading to diffusion of gas downward with respect to the motion of the major bubbles. As a result the velocity of the major bubbles, $v_b$, must be greater than the mass averaged gas velocity, $j_g/\alpha$, while the velocity of the gas in the slugs must be less.

14. Fully Developed Annular Flow. In this case, the terms "fully developed" refer to the apparent fact that there can be some annular flow with behavior similar to the transitional slug flow for mixture fluxes above 2.1 m/s. Thus, two runs, A25 and A25R at $j_f = 0.3$ m/s, do not follow the same behavior as the other annular flow data. Otherwise, it seems as though the average interfacial frequency is given by $\langle f_i \rangle = 5\, j^2$, (j in m/s).

15. Examination of the various wave models in the literature (Hall-Taylor and Nedderman (7), Hewitt and Lovegrove (8), Hewitt and Hall-Taylor (9), Phillips (10), Miles or Brooke-Benjamin (10, 11), Levich (12)) yielded none which agreed with the observations shown in Figure 3. It may be noticed that if the wave celerity being measured moves with the gas velocity (waves appear out to the centerplane) wave lengths of 0.3 - 2.5 m are found for the lower frequencies while wave lengths as low as 0.5 cm are obtained for the higher frequencies. The waves observed by Craik (13) are clearly inapplicable with film Reynolds numbers below 80 whereas, except for the static liquid cases, those tested herein are greater than 850. It appears there is no generally available theory which explains the present data.

16. Examination of photographs of the annular flow data (Ref. 4) shows both regularly and irregularly shaped waves. At higher flows the wave structure appears diffuse and short wave length roll waves (5-7 cm) become evident. Irregularity and three-dimensionality seem generally applicable to these waves. If the flow structure remains fixed the interface frequency must vary directly with j. It is thus judged that interfaces possibly form in direct proportion to j as well, in order to give the observed behavior. Two additional possibilities present themselves regarding wave formation:

(a) Wave length decreases as $1/j$ causing the interface density to increase directly with the mixture flux;

(b) The degree of amorphousness in roll waves increases so that the roll waves have more interfaces at higher fluxes, at the same time the wave length for these waves decreases with increasing mixture flux.

17. The onset of significant entrainment seems to result in a decrease in slip ratio as confirmed in Figure 4. The dashed line in this figure corresponds to the onset of entrainment given by Zuber (14). Except for Run A46, and possibly A27, all the annular flow data herein are below this inception point. If there is truly no significant amount of entrainment in these other runs, the frequencies measured must be due to wave passages, supporting the latter of the two hypotheses above.

18. Without roll waves, interface frequencies would increase sharply to a maximum and then decrease to zero away from the wall, similar to the behavior of Run A17. On the other hand, without roll waves or entrainment, the profile data in Ref. 3 indicate waves growing rapidly to the duct midplane. There is only one way this could happen. The waves would have to grow due to instability, rapidly spurt out to the midplane region, and then be swept back into the film by the high speed gas. While this may happen occasionally, it is not expected that this is the predominant mode of behavior.

19. A seemingly more reasonable explanation takes into account the existence of roll waves. These must be relatively homogeneous mixtures of gas and liquid which form when a burst of liquid forms on the film. This burst of larger than ordinary liquid content of a normal wave, is picked up by the gas, curled up, and projected down the channel. It is judged that these are isolated regions of strong vorticity, the axes of which are perpendicular to the axis of the duct and parallel to the large dimension of the channel. Due to the small spacing, these roll waves are probably near in size to the half-channel spacing. The formation of a roll wave should be due to the same forces which form wind-driven breakers in shallow water of constant depth.

*Note that the three dominant flow regimes specified in Reference 5 are utilized herein. Hence, churn turbulent is treated as a subset of slug flow while wispy annular is simply considered annular flow.

20. Wiegel (11) presents a large amount of data for wave periods as functions of wind speed and fetch, (development length downwind), on small lakes and reservoirs. It appears that for a reasonably large fetch, the wave frequencies vary predominantly as the square of the wind speed, in substantive agreement with the data in Figure 3. In this case, however, gravity is not present to pull the liquid back into the film so the liquid remains in the gas core subject to the drag on the interface, and the drag due to the gas velocity near the center of the gas core. The waves then just roll up and remain in the core to be given both linear and angular acceleration. Drag on the surface of the liquid film, coupled with the higher speed of the gas in the center of the duct, act to spin the wave as it moves down the channel. Being amorphous, the interface content of these waves is uniform with distance from the wall. The effect is to add a constant value to the interface frequency profile associated with the plane annular wave pattern itself, and make the core frequency profile appear as in Reference 3.

21. It is considered possible that it is the centripetal action of the spinning fluid which is responsible for the entrainment associated with roll waves (8). The wakes of several impinging drops are seen on the liquid film in the photographs for Run S25 and Run S37. It is thus concluded that the interface frequencies measured in the core for the annular flow runs are due predominantly to roll waves, giving a measure of the internal structure of these waves. Thus, if roll waves moving at 35 Hz produce frequencies of 3500 Hz, there must be about 100 interfaces along a chord of each wave on the average. Since each wave is observed to be on the order of 2-3 cm long, a very fine structure indeed is indicated in these waves.

22. For the two mist flow runs, Runs A27 and A46, the interface frequency rises toward the channel centerplane (Ref. 3), rather than remaining constant. If the break in the slope of frequency versus location is taken as the point indicating the total number of interfaces due to waves, the balance of the frequency measurement must be due to core-borne liquid. This can be as much as 1800 interfaces per second as seen for Run A46.

23. <u>Transition Slug Flow: $j > 2.1$ ms</u>. In the transition to annular flow there have to be at least two competing mechanisms. The first is the disappearance of bridging and the resulting decrease of core interfaces tending to decrease the chord-averaged interface frequency. The second mechanism involves the production of waves on the film adjacent to the major bubbles in slug flow along with the beginning of roll waves, this latter mechanism tending to increase the frequencies. The amplitude of regular waves increases and the frequency of these waves and of roll waves increases with relative velocity. While transition to ideal, plane annular flow would cause a drop to near zero frequency, the increase in wave density and amplitude along with the appearance of roll waves causes the frequency to increase.

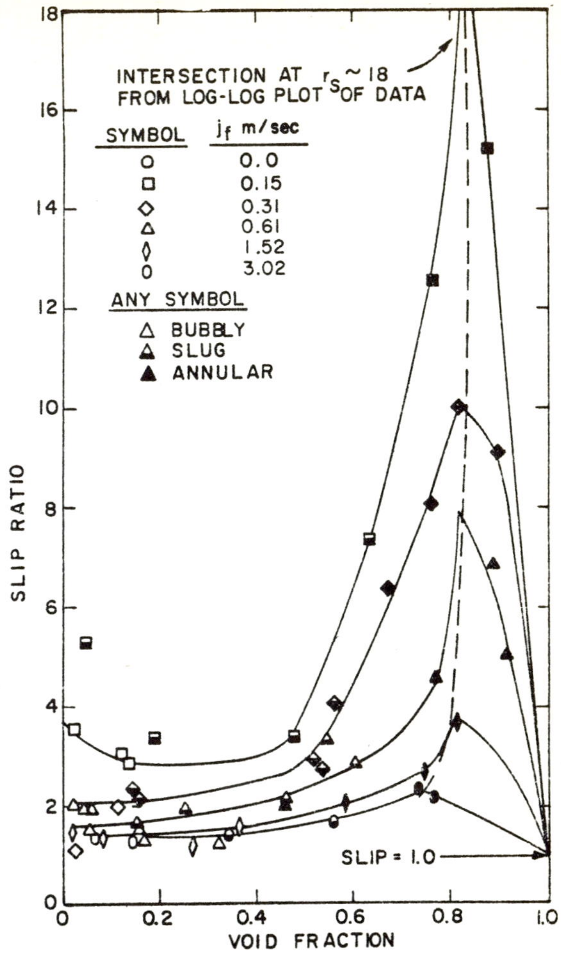


Fig. 4: Variation of slip ratio with void fraction at constant liquid volume flux

## CONCLUSIONS

24. Data have been presented for interfacial passage frequency profiles in cocurrent, upwards, air-water flow in a narrow rectangular duct. It is shown that Rayleigh instability concepts can be used to correlate these frequencies in isolated bubbly flow. On the other hand, slug flows having a volume flux less than 2.1 m/s appear to have constant passage frequencies of 150 Hz $\pm$ 50 Hz. Fully developed annular flow appears to have interfacial frequencies correlated with $j^2$ while the competing effects of the disappearance of liquid bridging and production of waves on the films adjacent to the major bubbles in slug flow interact in the transition region.

25. Regarding slug flow, it was concluded that the number density of voids in the liquid slugs remains constant and that these voids arise predominatly from trailing edge extrusion of gas from the major bubbles leading to a net downstream diffusion of the major bubbles with respect to the gas as a whole. For annular flow, no current theory was found which predicted the variation of the interfacial passage frequency with volume flux. On the other hand, Wiegel's data (11) for wind driven waves in shallow

lakes and reservoirs indicated similar behavior to that reported herein. Interfacial frequencies were more than two orders larger than dominant frequencies obtained from void fluctuation spectra (5) and observed by Hewitt and Lovegrove (8). It was judged that these frequencies were due to extremely fine dispersion of liquid and gas in roll waves, these waves being directly analogous to wind-driven breakers except for the restorative gravity forces, absent in this case.

## ACKNOWLEDGEMENT

26. This work was supported by the U.S. Atomic Energy Commission.

## REFERENCES

1. ISHII, M., 'Thermally Induced Flow Instabilities in Two-Phase Mixtures in Thermal Equilibrium,' PhD thesis, Georgia Institute of Technology, Atlanta Georgia, June, 1971.

2. ISHII, M. _Thermo-Fluid Dynamic Theory of Two-Phase Flow_, Collection de la Direction des Études of Recherches d'Electricité de France, Eyrolles, Paris, (1975).

3. JONES, O. C., _Statistical Considerations in Heterogeneous Two-Phase Flowing Systems_, PhD Thesis, Rensselaer Polytechnic Inst. Troy, New York, (1973).

4. JONES, O. C. and N. ZUBER, 'Use of a Cylindrical Hot-Film Anemometer for Measurement of Two-Phase Void and Volume Flux Profiles in a Narrow Rectangular Channel,' AIChE paper No. 45, presented at the 1975 National Heat Transfer Conference, San Francisco, California, August 11-13, (1975).

5. JONES, O. C., and N. ZUBER, 'The Interrelation Between Void Fraction Fluctuations and Flow Patterns in Two-Phase Flow,' _Int. J. Multiphase Flow_, _2_, 273-306, (1975).

6. SILBERMAN, E., 'Production of Bubbles by Disintegration of Gas Jets in Liquids,' St. Anthony Falls Hydraulic Laboratory Technical Report No. 12 Series A, Reprinted in the _Fifth Midwestern Conference on Fluid Mechanics_, (1957).

7. HALL-TAYLOR, N.S., and R. M. NEDDERMAN, 'The Coalescense of Disturbance Waves in Annular Two-Phase Flow, ' _Chem. Eng. Sci._, _23_, 551, (1968).

8. HEWITT, G. F., and P. C. LOVEGROVE, 'Frequency and Velocity Measurements of Disturbance Waves in Annular, Two-Phase Flow,' AERE-R 4304, April, (1969).

9. HEWITT, G. F., and N. S. HALL-TAYLOR, _Annular Two-Phase Flow_, Pergamon Press Oxford, (1970).

10. KINSMAN, B., _Wind Waves - Their Generation and Propagation on the Ocean Surface_, Prentice Hall, Englewood Cliffs, New Jersey, (1965).

11. WIEGEL, R. L., _Oceanographic Engineering_, Prentice Hall, Englewood Cliffs, New Jersey, (1964).

12. LEVICH, V. G., _Physiochemical Hydrodynamics_, Prentice Hall, Englewood Cliffs, New Jersey, (1962).

13. CRAIK, A.D.D., 'Wind-Generated Waves in Thin Liquid Films,' _J. Fluid Mech._, _26_, 2, 369, (1966).

14. ZUBER, N., 'On the Atomization and Entrainment of Liquid Films in Shear Flow', General Electric Report No. 62GL153, Sept., (1962).

C193/77

# CRITICAL CONDITIONS DURING THE FLOW OF TWO-PHASE MIXTURES THROUGH PIPES

D. CHISHOLM, BSc, PhD, ARCST, CEng, MIMechE
National Engineering Laboratory, East Kilbride

SYNOPSIS The equations for the prediction of the critical or choking mass velocity during two-phase flow are transformed to the most suitable form for numerical calculations. The available evidence for gas-liquid choked flow is examined and deductions made regarding the behaviour of the relative movement of the phases at the point of choking. Data for steam-water mixtures under choked flow conditions are then compared with homogeneous-equilibrium and slip-frozen models. With values of the velocity ratio approximating to available data the latter model gives good agreement with much of the available data for the choking mass velocity.

NOTATION

| | |
|---|---|
| B | Coefficient defined by equation 7 |
| b | Coefficient defined by equation 16 |
| $G_C$ | Choking mass velocity |
| $G_{CG}$ | Choking mass velocity in two-phase region, x = 1.0 |
| $G_{CGS}$ | Choking mass velocity for single-phase gas or vapour |
| $G_{CHE}$ | Choking mass velocity for homogeneous-equilibrium theory |
| $G_{CL}$ | Choking mass velocity in two-phase region, x = 0 |
| $G_{CLS}$ | Choking mass velocity for single-phase liquid |
| $G_{CSF}$ | Choking mass velocity for slip-frozen theory |
| K | The ratio of the gas velocity to the liquid velocity |
| $n_s$ | The isentropic expansion exponent |
| p | Pressure |
| s | Entropy |
| $s_G$ | Entropy of vapour |
| $s_L$ | Entropy of liquid |
| $T_{sat}$ | Saturation temperature ($^oC$) |
| $U_{CGS}$ | Single-phase choking velocity of gas or vapour |
| $U_{CLS}$ | Single-phase choking velocity of liquid |
| v | Specific volume |
| $v_G$ | Specific volume of gas or vapour |
| $v_L$ | Specific volume of liquid |
| x | Mass dryness fraction |
| $x_m$ | Mass dryness fraction assuming non-equilibrium |
| $\Lambda^2$ | Physical property coefficient: single-phase regions (equation 8) |
| $\Lambda_E^2$ | Physical property coefficient: two-phase region (equation 26). |

INTRODUCTION

1. This paper is concerned with the evaluation of the critical or choking flowrate from a pipe during two-phase flow, where the critical pressure at pipe exit is known. The critical conditions occur when the pressure gradient to accelerate the fluid equals the static pressure gradient, and for a given exit pressure is associated with the maximum possible flowrate.

2. Critical flow conditions where gas-liquid mixtures flow are first considered and the equations for the choking flowrate transformed to the most convenient form for numerical calculations. Following this, equations are developed for the case of vapour-liquid mixtures.

CRITICAL FLOW OF GAS-LIQUID MIXTURES

3. The critical mass flowrate of a fluid is given by

$$-\frac{1}{G_C{}^2} = \left(\frac{\partial v}{\partial p}\right)_\phi , \tag{1}$$

where $G_C$, v, and p are respectively the choking mass velocity, specific volume, and pressure. The subscript $\phi$ indicates the derivative is evaluated for an isentropic expansion.

4. In a two-phase flow, care is required when the above expression is used to define the specific volume. For a separated flow model the specific volume is appropriately defined as

$$v = \left[x + \frac{1 - x}{K}\right]\{xv_G + K(1 - x)v_L\}, \quad (2)$$

where K is the ratio of the phase velocities. This specific volume differs from that based on the actual content of the phases in the mixture (Ref. 1).

For the flow of gas-liquid mixture there is no change of phase hence

$$\left(\frac{\partial x}{\partial p}\right)_\phi = 0. \quad (3)$$

5. From equations 1 to 3

$$-\frac{1}{G_C{}^2} = \left\{\frac{1}{K}x(1 - x) + x^2\right\}\left(\frac{\partial v_G}{\partial p}\right)_\phi + $$

$$+ \left\{(1 - x)(1 - x + Kx)\right\}\left(\frac{\partial v_L}{\partial p}\right)_\phi +$$

$$+ \left\{x(1 - x)\left(v_L - \frac{v_G}{K^2}\right)\right\}\left(\frac{\partial K}{\partial p}\right)_\phi. \quad (4)$$

6. From equation 1 it is apparent that the choking mass velocity for liquid in single-phase flow is given by

$$-\frac{1}{G_{CLS}^2} = \left(\frac{\partial v_L}{\partial p}\right)_\phi \quad (5)$$

and for the gas in single-phase flow

$$-\frac{1}{G_{CGS}^2} = \left(\frac{\partial v_G}{\partial p}\right)_\phi. \quad (6)$$

7. From equations 4 to 6

$$\left(\frac{G_{CLS}}{G_C}\right)^2 = 1 + (\Lambda^2 - 1)\{Bx(1 - x) + x^2\}, \quad (7)$$

where

$$\Lambda^2 = \left(\frac{G_{CLS}}{G_{CGS}}\right)^2 = \left(\frac{\partial v_G}{\partial p}\right)_\phi \left(\frac{\partial p}{\partial v_L}\right)_\phi, \quad (8)$$

and

$$B = \frac{\left\{\frac{\Lambda^2}{K} - 2 + K + \left(v_L - \frac{v_G}{K^2}\right)\left(\frac{\partial K}{\partial v_G}\right)_\phi \Lambda^2\right\}}{(\Lambda^2 - 1)}. \quad (9)$$

The group $\Lambda^2$ is analogous to the Baroczy index (Ref. 2) or the 'physical property coefficient' (Ref. 3) used in the correlation of pressure gradients. It should be noticed that the coefficient B is not explicitly a function of the dryness fraction x.

8. From equations 7 and 8

$$\left(\frac{G_{CGS}}{G_C}\right)^2 = \frac{1}{\Lambda^2} + \left(1 - \frac{1}{\Lambda^2}\right)\{Bx(1 - x) + x^2\}. \quad (10)$$

For most gas-liquid mixtures $\Lambda^2$ is much greater than unity; for an air-water mixture at ambient conditions, $\Lambda^2$ is approximately $10^7$. Except at very small values of $x(x \sim 1/\Lambda^2)$ there is little error in numerical calculations in making the approximation

$$\left(\frac{G_{CGS}}{G_C}\right)^2 = Bx(1 - x) + x^2. \quad (11)$$

9. Inspection of equation 9 indicates that with homogeneous flow (K = 1) the coefficient B is unity, hence from equation 11

$$\left(\frac{G_{CGS}}{G_C}\right)^2 = x. \quad (12)$$

In equations 11 and 12 the single-phase choking mass velocity of the gas can be evaluated from

$$G_{CGS}^2 = \frac{n_s p}{v_G}, \quad (13)$$

where $n_s$ is the isentropic expansion exponent.

EVALUATION OF THE COEFFICIENT B

10. Henry (Ref. 4) assumed that at the critical condition the velocity ratio remained constant $\{(\delta K/\delta p)_\phi = 0\}$, hence equation 8 reduces to

$$B = \left(\frac{\Lambda^2}{K} - 2 + K\right)\Big/(\Lambda^2 - 1) \quad (14)$$

and where $\Lambda^2$ is large, as is the case with air-water mixtures at ambient pressure, equation 14 can be approximated to

$$B = \frac{1}{K}. \quad (15)$$

11. Chisholm (Ref. 1), on the other hand, assumed that along an isentrope

$$K = 1 + b\frac{v_G}{v_L}, \quad (16)$$

where b is a coefficient independent of the specific volumes from which

$$\left(\frac{\partial K}{\partial v_G}\right)_\phi = \frac{K - 1}{v_G}. \quad (17)$$

Substituting equation 17 in 9 gives

$$B = \left\{\frac{\Lambda^2}{K^2} - 2 + K + (K - 1)\frac{v_L}{v_G}\Lambda^2\right\}\Big/(\Lambda^2 - 1). \quad (18)$$

If $\Lambda^2$ and $v_G/v_L$ are large, equation 18 reduces to

$$B = \frac{1}{K^2}. \quad (19)$$

12. Table 1 tabulates the air-water critical flow data of Fauske obtained with air-water data at a critical pressure of 1.17 bar (Ref. 5). Figure 1 compares the measured values of the choking mass velocity with values predicted by equations 11 and 12 with B equal to unity (homogeneous flow) and with B given by equations 15 and 19, using the experimental values of K, respectively.

Table 1

Critical flow data: Fauske[(5)] 3.12 x 25.4 mm channel. Critical pressure, 1.17 bar

| x | Predicted K | Mass velocity $kg/m^2$ s |
|---|---|---|
| 0.000 60 | 1.173 | 22 000 |
| 0.000 87 | 1.255 | 19 500 |
| 0.001 26 | 1.341 | 16 100 |
| 0.001 73 | 1.420 | 15 100 |
| 0.002 35 | 1.500 | 13 700 |
| 0.002 76 | 1.544 | 12 700 |
| 0.005 68 | 1.759 | 10 200 |
| 0.008 85 | 1.904 | 8 800 |
| 0.013 0 | 2.041 | 7 800 |
| 0.017 7 | 2.158 | 7 080 |
| 0.033 4 | 2.390 | 5 370 |
| 0.082 7 | 2.390 | 3 120 |

13. The best agreement is obtained when equation 19 is used. The good agreement is surprising as there is experimental evidence that, because of the effects of entrainment, the simple separated flow model, from which equation 2 is obtained, does not give satisfactory agreement with momentum flux measurement (Refs 6,7). Entrainment might also be expected to be a factor in choking.

HOMOGENEOUS-EQUILIBRIUM FLOW OF VAPOUR-LIQUID MIXTURES

14. For homogeneous flow from equations 1 and 2

$$-\frac{1}{G_C} = \left[\frac{\partial}{\partial p}\{xv_G + (1-x)v_L\}\right]_\phi \qquad (20)$$

which can be expanded to give

$$-\frac{1}{G^2_{CHE}} = x\frac{dv_G}{dp} + (1-x)\frac{dv_L}{dp} - (v_L - v_G)\left(\frac{\partial x}{\partial p}\right)_\phi \qquad (21)$$

where the subscripts HE indicate homogeneous-equilibrium; equilibrium implies that the expansion is in thermal equilibrium.

15. The partial derivations of the mass dryness fraction can be expressed

$$\left(\frac{\partial x}{\partial p}\right)_\phi = -\left\{x\frac{ds_G}{dp} + (1-x)\frac{ds_L}{dp}\right\}\frac{1}{s_G - s_L}. \qquad (22)$$

From equations 21 and 22 it is evident that as x approaches zero the mass velocity approaches $G_{CL}$ given by

$$-\frac{1}{G^2_{CL}} = \frac{dv_L}{dp} - \frac{v_G - v_L}{s_G - s_L}\frac{ds_L}{dp} \qquad (23)$$

and as x approaches unity the mass velocity approaches $G_{CG}$ given by

$$-\frac{1}{G^2_{CG}} = \frac{dv_G}{dp} - \frac{v_G - v_L}{s_G - s_L}\frac{ds_G}{dp}. \qquad (24)$$

It should be noted $G_{CL}$ and $G_{CG}$ differ from $G_{CLS}$ and $G_{CGS}$, the values associated with the single-phase regions; there are discontinuities in choking mass velocity as a function of mass dryness fraction (or enthalpy) at the boundaries of the two-phase region as shall be discussed later.

16. From equations 21 to 24

$$\frac{1}{G^2_{CHE}} = \frac{x}{G^2_{CG}} + \frac{1-x}{G^2_{CL}}. \qquad (25)$$

If $\Lambda_E{}^2$ is defined as

$$\Lambda_E{}^2 = \left(\frac{G_{CL}}{G_{CG}}\right)^2 \qquad (26)$$

then from equations 25 and 26

$$\left(\frac{G_{CL}}{G_{CHE}}\right)^2 = 1 + (\Lambda_E{}^2 - 1)x \qquad (27)$$

and

$$\left(\frac{G_{CG}}{G_{CHE}}\right)^2 = \frac{1}{\Lambda_E{}^2} + \left(1 - \frac{1}{\Lambda_E{}^2}\right)x. \qquad (28)$$

THE GROUPS $\Lambda_E{}^2$ AND $\Lambda^2$

17. The term $\Lambda_E{}^2$ for steam-water mixtures is plotted in Figure 2 to a base of pressure; its values are order of magnitudes smaller than the $\Lambda^2$ which can be approximately evaluated for steam and water immediately outwith the two-phase region from the following formulae. Equation 8 can be expressed as

$$\Lambda^2 = \left(\frac{U_{CLS}}{U_{CGS}}\frac{v_G}{v_L}\right)^2 \qquad (29)$$

and the ratio of the single-phase velocities for the saturated fluids can be obtained in terms of the saturation temperature (in $^oC$) of water from, for $T_{sat} < 320^oC$,

$$\frac{U_{CLS}}{U_{CGS}} = 1 + (440 - T_{sat})0.007 \qquad (30)$$

and for $T_{sat} > 320^oC$,

$$\frac{U_{CLS}}{U_{CGS}} = 1 + (373.7 - T_{sat})0.0156. \qquad (31)$$

18. At atmospheric pressure equations 30 and 31 give $\Lambda^2 = 29 \times 10^6$ as compared with $\Lambda_E{}^2 = 17$ from Figure 2. The data used in obtaining these equations are reported in References 8 and 9.

19. $n_2$ can be defined in a manner analogous to equation 13,

$$n_2 = G^2_{CG}\frac{v_G}{p}. \qquad (32)$$

20. Values of $n_2$ are given in Figure 3 for steam-water mixtures as a function of pressure. At lower values of pressure, $n_2$ has a value of about 1.1 compared with a value for the isentropic expansion exponent ($n_s$) of 1.3 for dry steam. Thus the discontinuities in mass velocity at the two-phase boundary (x = 1) is approximately 10 per cent, the mass velocity with dry steam being the greater. It follows

therefore that at the phase boundary, where x = 0, with water at atmospheric pressure,

$$\frac{G_{CLS}}{G_{CL}} = \frac{G_{CLS}}{G_{CGS}}\frac{G_{CGS}}{G_{CG}}\frac{G_{CG}}{G_{CL}}$$

$$= (29 \times 10^6)^{\frac{1}{2}}(1.1)\frac{1}{(17)^{\frac{1}{2}}}$$

$$= 1.4 \times 10^3.$$

There is therefore a major discontinuity in choking mass velocity at this boundary.

SLIP-FROZEN FLOW OF VAPOUR-LIQUID MIXTURES

21. A slip-frozen model assumes that the vapour and liquid move at different velocities and no change of phase occurs at the point of choking (equation 3 applies); the mixture behaves as a gas-liquid mixture at the point of choking. In that case equations 11, 13, and 19 can be assumed to apply (Ref. 10) giving

$$\left(\frac{G_{CSF}}{G_{CGS}}\right)^2 = \frac{1}{\left\{\frac{1}{K^2}x(1-x)+x^2\right\}} \qquad (33)$$

or

$$G^2_{CSF} = \frac{n_s p}{v_G\left\{\frac{1}{K^2}x(1-x)+x^2\right\}}. \qquad (34)$$

22. Chisholm obtained the following equations for K from examination of Burnell data for choking mass velocities at low critical pressures with steam-water mixtures (Ref. 10).

$x \leqslant 0.122$ and $v_G/v_L \leqslant 720$

$$K = 1 + 0.0139\frac{1}{1-x}\frac{v_G}{v_L} \qquad (35)$$

$x > 0.122$ and $v_G/v_L \leqslant 720$

$$K = 1 + 0.001\,93\frac{v_G}{v_L} \qquad (36)$$

$x \leqslant 0.122$ and $v_G/v_L > 720$

$$K = 1 + 10\frac{x}{1-x} \qquad (37)$$

$x > 0.122$ and $v_G/v_L > 720$

$$K = 2.39. \qquad (38)$$

These equations give velocity ratios of similar magnitude (Ref. 1) to the measurements of Fauske (Ref. 5) for the velocity ratio under choked flow conditions with air-water mixtures.

23. Table 2 compares Burnell's experimental values of $G_C$ (Ref. 11) with $C_{CSF}$ from equations 34 to 38. The agreement is within ±15 per cent; the procedure can be considered as one of finding an empirical equation for the coefficient B.

24. Combining equations 28 and 33 gives

$$\frac{G_{CSF}}{G_{CHE}} = \left\{\frac{\frac{1}{\Lambda_E^2}+\left(1-\frac{1}{\Lambda_E^2}\right)x}{\frac{1}{K^2}x(1-x)+x^2}\right\}^{\frac{1}{2}}\frac{G_{CGS}}{G_{CG}} \qquad (39)$$

or using equations 13 and 32

$$\frac{G_{CSF}}{G_{CHE}} = \left\{\frac{\frac{1}{\Lambda_E^2}+\left(1-\frac{1}{\Lambda_E^2}\right)x}{\frac{1}{K^2}x(1-x)+x^2}\right\}^{\frac{1}{2}}\left(\frac{n_s}{n_2}\right)^{\frac{1}{2}}. \qquad (40)$$

Table 2

Comparison of Burnell's data with prediction

| Pressure bar | x | Mass velocity kg/m$^2$ s | |
|---|---|---|---|
| | | Experiment | Equations 34 to 38 |
| 1.02 | 0.025 8 | 2010 | 2210 |
| 0.70 | 0.018 32 | 1730 | 1680 |
| 0.68 | 0.046 5 | 1150 | 1290 |
| 0.60 | 0.024 4 | 1530 | 1350 |
| 0.48 | 0.035 1 | 1100 | 990 |
| 0.43 | 0.040 2 | 858 | 840 |
| 0.35 | 0.025 | 930 | 790 |
| 0.34 | 0.048 | 676 | 660 |
| 0.29 | 0.056 25 | 550 | 540 |
| 0.23 | 0.037 5 | 564 | 480 |
| 0.20 | 0.069 1 | 353 | 372 |

25. Figure 4 plots experimental data of Zaloudek (Ref. 12) obtained with 13.2 mm bore pipe and steam-water mixtures and makes comparison with values predicted by equations 34 to 38 and 40 for dryness fractions in excess of 10 per cent. The experimental values are up to 20 per cent in excess of those predicted; in addition a predicted decrease in the ratio $G_C/G_{CHE}$ with pressure is not discernable in the data.

26. Figure 5 is a similar plot to Figure 4, but now for dryness fraction below 15 per cent. This figure also includes the data of Henry with a 120° included angle divergence at the exit plane of an 8 mm bore pipe (Ref. 4): Zaloudek's data were obtained with 180° angle divergence Henry also examined the influence of smaller divergence angles and found the mass velocity decreased with divergence angle; he attributed this to the two-dimensional nature of the flow (Ref. 4).

27. At a pressure of 3.45 bar the values of $G_C/G_{CHE}$ are predicted to within ±15 per cent. Three points at 7.45 bar are however over-predicted by 25 per cent.

28. Where x is less than 2 per cent the prediction is in agreement with the experimental observations that $G_C/G_{CHE}$ increases rapidly with decreasing x value. At 3.45 bar, for example, $v_G/v_L$ is 493, hence from equation 38 the velocity ratio K is 1.137 with x = 0.02. It follows that the empirical equations for K do not play a significant role in predictions at these low dryness fractions.

29. The 'frozen' model therefore predicts trends at low dryness fractions satisfactorily; however it can be seen that for x < 0.025 some of the data tend to lie above the 'frozen' model. This is presumably due to an assumption not so

far discussed: it has been assumed that the mixture is at its equilibrium dryness fraction, whereas because of the rapidity of depressurization the dryness fraction at the point of choking may be less than the equilibrium value. Henry and Fauske in their analytic treatment allow for this effect (Refs 13,14). Define

$$N = \frac{x_m}{x} \tag{41}$$

where $x_m$ is the dryness fraction assuming non-equilibrium and x the dryness fraction assuming an isentropic expansion. Assuming the liquid is incompressible it can be shown using equation 25 that

$$\left(\frac{G_C}{G_{CHE}}\right)^2 = \frac{1}{N}. \tag{42}$$

At low dryness fractions ($x \rightarrow x^2$) equation 40 simplifies to

$$\left(\frac{G_{CSF}}{G_{CHE}}\right)^2 = \left(\frac{1}{\Lambda_E^{\ 2}x} + 1 - \frac{1}{\Lambda_E^{\ 2}}\right)K^2\left(\frac{n_s}{n_2}\right). \tag{43}$$

30. The two approaches will give identical values ($G_C = G_{CSF}$) when

$$N = \left(\frac{\Lambda_E^{\ 2}x}{1 + \Lambda_E^{\ 2}x - x}\right)\frac{1}{K^2}\frac{n_2}{n_s}. \tag{44}$$

This compares with Henry's empirical formula (Ref. 4)

$$N = 20x. \tag{45}$$

31. Equation 44 gives values below equation 45 except at low pressures (<0.2 bar), thus the present method will give greater predicted mass velocities over most of the pressure range of industrial interest.

## SUMMARY AND CONCLUSIONS

32. For gas-liquid choking flow the following equation is derived:

$$\left(\frac{G_{CGS}}{G_C}\right)^2 = \frac{1}{\Lambda^2} + \left(1 - \frac{1}{\Lambda^2}\right)\{Bx(1 - x) + x^2\} \tag{46}$$

where $\Lambda^2$ is a physical property coefficient defined by equation 8, and the coefficient B is defined by equation 9; B is primarily a function of the velocity ratio K.

33. The coefficient B is shown to be sensitive to the derivative of K with respect to pressure; the assumption sometimes made that the derivative is zero does not appear justified by the analysis presented here.

34. For vapour-liquid mixtures, homogeneous-equilibrium theory can be expressed as

$$\left(\frac{G_{CG}}{G_{CHE}}\right)^2 = \frac{1}{\Lambda_E^{\ 2}} + \left(1 - \frac{1}{\Lambda_E^{\ 2}}\right)x, \tag{47}$$

where $\Lambda_E^{\ 2}$ is a physical property coefficient defined by equations 23, 24, and 26.

35. Equation 10 can also be developed for the 'slip-frozen' model for vapour-liquid flow and in that form has been used to analyse the steam-water data of Burnell, Henry, and Zaloudek. Coefficient B when evaluated from equations 19 and 35 to 38 correlates the data to ±25 per cent with this model.

36. At dryness fractions less than 2 per cent the assumption that the dryness fraction at the point of choking is the equilibrium value may be the cause of the data tending to lie up to 15 per cent above predictions.

ACKNOWLEDGEMENT

37. This paper is published by permission of the Director, National Engineering Laboratory, Department of Industry. It is Crown copyright.

## REFERENCES

1. CHISHOLM, D. Critical conditions during flow of two-phase mixtures through nozzles. Proc. Instn mech. Engrs, 1967-68, 182(Pt 3H), 145-151.

2. BAROCZY, C.J. A systematic correlation of two-phase pressure drop. Chem. Engng Prog. Symp. Ser. 1966, 62(44), 232-249.

3. CHISHOLM, D. and SUTHERLAND, L.A. Prediction of pressure changes in pipeline systems during two-phase flow. Proc. Instn mech. Engrs, 1969-70, 184(Pt 3C), 24-32.

4. HENRY, R.E. A study of one- and two-component two-phase critical flow at low qualities. Report No ANL-7430. Argonne, Illinois: Argonne National Laboratory, 1968.

5. FAUSKE, H.K. Two-phase and one-component critical flow. In: Symposium on Two-phase Flow, Exeter, 21-23 June 1965, Vol. 1, University of Exeter, Dept Chem. Engng, G101-G114.

6. ANDEEN, G.H. and GRIFFITH, P. Momentum flux in two-phase flow. J. Heat Transfer, 1967, Paper No 67-Ht-32.

7. WIAFE, F. Two-phase flow in rough tubes. PhD Thesis, University of Glasgow, 1970.

8. AVDONIN, V.I. Measurements of the velocity of propagation of sound waves in saturated water vapour. Soviet Physic, Tech. Physics, 1961, 5(10), 1180-1185.

9. LANDOLT-BÖRNSTEIN. Numerical data and functional relationships in science and technology. Vol. 5. Molecular Acoustics. Berlin: Springer, 1967.

10. CHISHOLM, D. Calculation of 'choking' flowrates during steam-water flow through pipes. NEL Report 377. East Kilbride, Glasgow: National Engineering Laboratory, 1968.

11. BURNELL, J.G. Flow of boiling water through nozzles, orifice, and pipes. Engineering, London, 1947, 164(4272), 572-576.

12. ŻALOUDEK, F.R. The low pressure critical discharge of steam-water mixtures from pipes. USAEC Report HW-68934(Rev). Contract (AT4S-1)-1350. Rechland, Washington: General Electric Company, Hanford Atomic Products Operation, 1961. Washington, 25, D.C.: Office of Technical Services, US Dept of Commerce, 1961.

13. HENRY, R.E. The two-phase critical discharge of initially saturated or subcooled liquid. Nuc. Sc. Engng, 1970, 41, 336-342.

14. HENRY, R.E. and FAUSKE, H.K. The two-phase critical flow of one-component mixtures in nozzles, orifices, and short tubes. J. Heat Transfer, 1971, 93, 179-187.

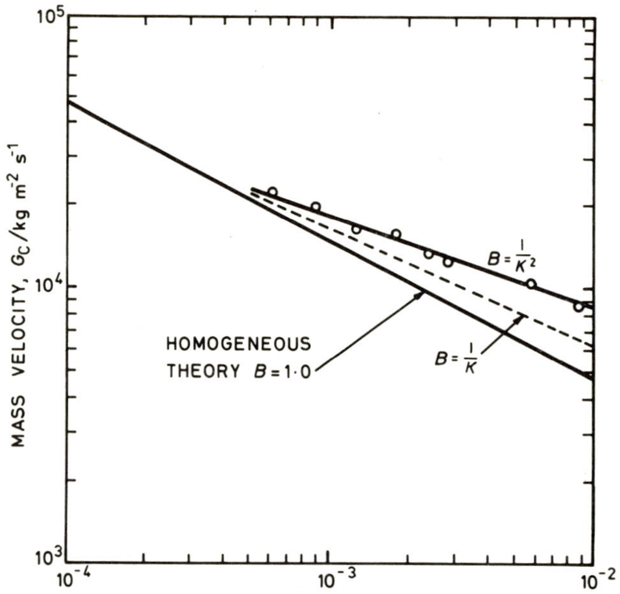


Fig. 1: Choked mass velocity to a base of dryness fraction x for air-water mixtures at 1.17 bar. Data of Fauske (Ref. 5)

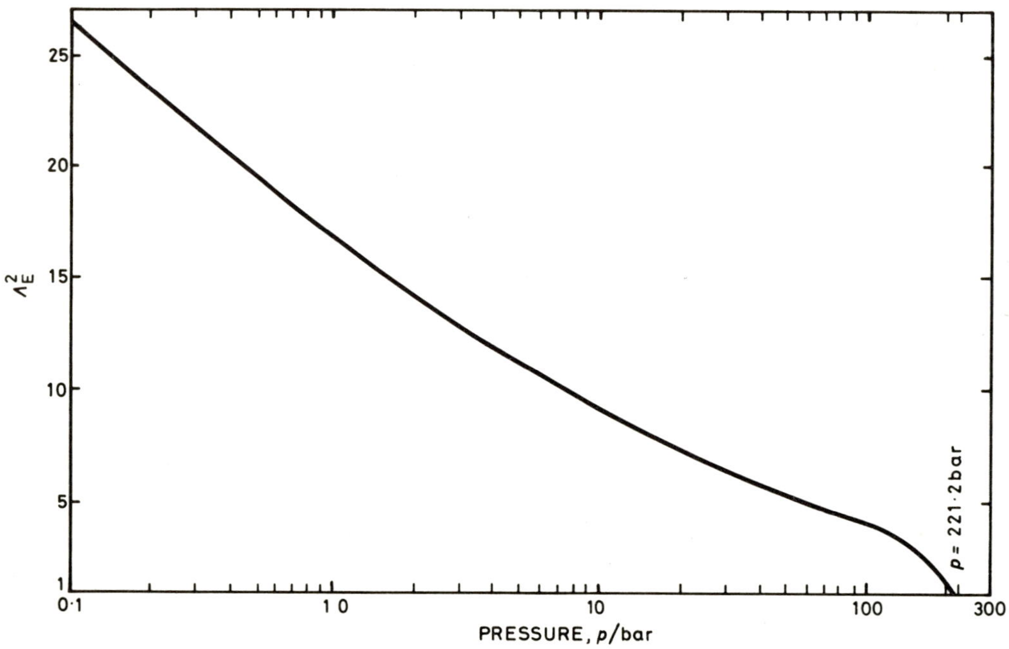


Fig. 2: $\Lambda_E^2$ to a base of pressure for steam-water mixtures

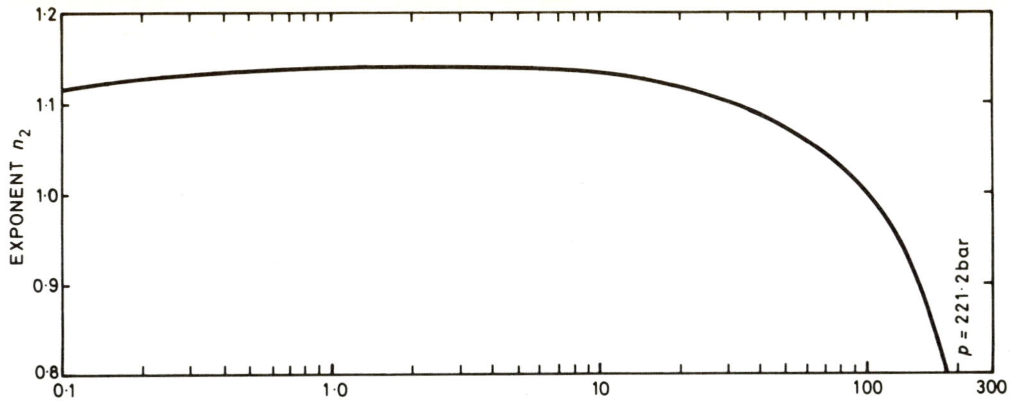


Fig. 3: The exponent $n_2$ to a base of pressure for steam-water mixtures

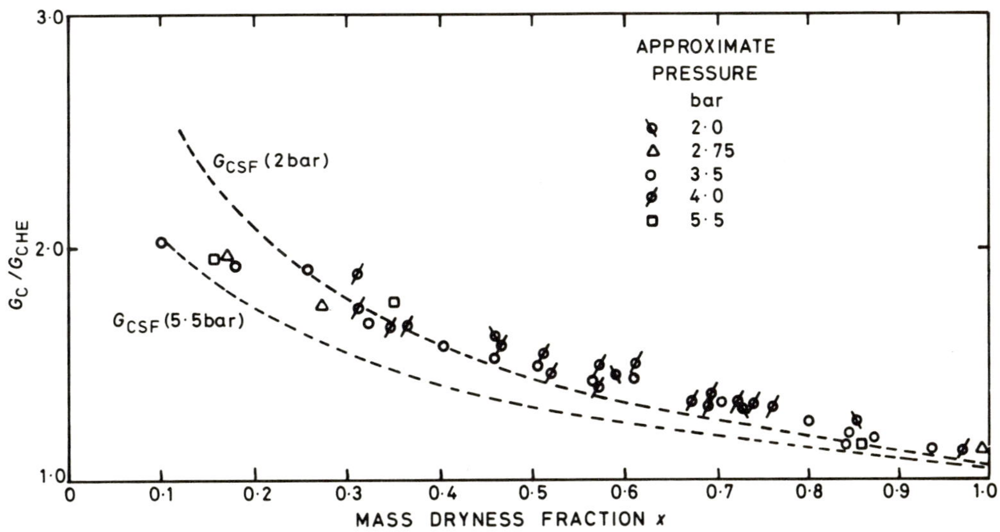


Fig. 4: $G_C/G_{CHE}$ to a base of mass dryness fraction x for steam-water mixtures. Data of Zaloudek (Ref. 12) for $x > 0.1$

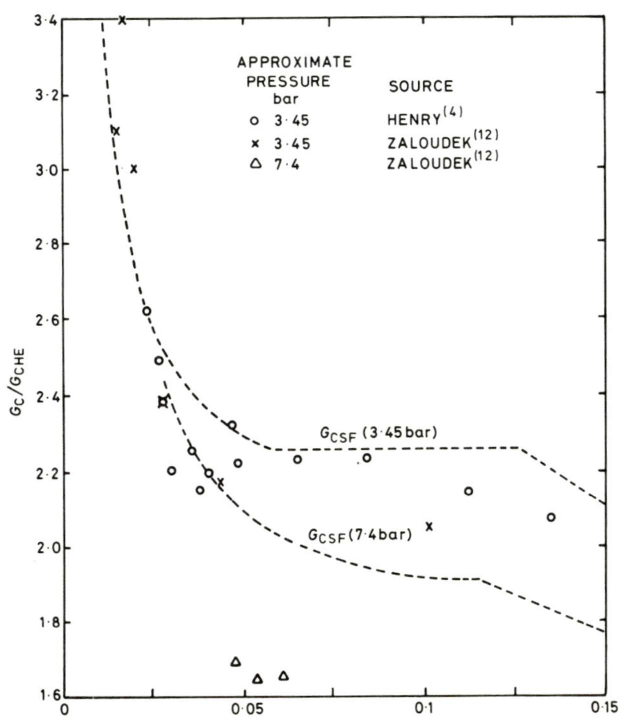


Fig. 5: $G_C/G_{CHE}$ to a base of mass dryness fraction x for steam-water mixtures. Data for $x < 0.15$

C194/77

# NON-EQUILIBRIUM EFFECTS IN TRANSIENT TWO-PHASE FLOW

H.C. SIMPSON, BSc, SM, ScD, FIMechE, FInstP, FRSE
D.H. ROONEY, BSc, PhD, ARCST, CEng, MIMechE, MIED
M.N.A. HAWLADER, BSc, MSc
University of Strathclyde, Glasgow G1 1XJ

SYNOPSIS Depressurisation tests were carried out on Refrigerant 113 liquid flowing in a horizontal pipeline, under conditions where vapour was formed by flash evaporation. The tests covered a range of initial velocities up to 2.1 m/s in a 51 mm diameter glass pipeline at starting pressures around 1.5 bar and with varying rates of depressurisation. Measurements were made of local pressure and temperature, circulation rate, pressure difference and void fraction variation over a test section length of 2 m. The local pressure and temperature measurements give a direct indication of non-equilibrium effects. The vapour formed during the flash evaporation process quickly formed a stratified type flow and a theoretical model was developed on this basis. The model includes the transient two phase flow conservation equations allied to a heat transfer equation. Satisfactory agreement between theoretical predictions and experimental results was obtained.

## INTRODUCTION

1. In the computer modelling of Loss of Coolant Accidents in water cooled nuclear reactors, several simplifying assumptions are made regarding the nature of the transient two phase flow that takes place. For example, the combined phase flow equations are normally used with the assumption of thermodynamic equilibrium between the phases, and slip between the phases is often included in terms of steady state empirical correlations of doubtful validity in the transient case. Experimental studies at Strathclyde, carried out on a circulating loop with vertical, glass test sections and using Refrigerant 113 to model steam water mixtures at high pressures, showed that when a pressure reduction was imposed on the loop, a bubble front advanced downwards into the downcomer tube from the vapour drum in a complex manner, with some evidence of non-equilibrium effects (1) (2). In the subsequent analysis, it was found difficult to separate the slip and non-equilibrium effects.

2. It was decided therefore to carry out a similar experiment with R-113 in a horizontal rig where gravitational effects, and hence phase slip, would be less important. In this way it was hoped to isolate the non-equilibrium effects during the transient.

## TEST RIG

3. A diagram of the test rig is shown in Figure 1. Made of borosilicate glass, the rig was designed to operate at pressures up to 1.5 bar. The R-113 passed from a vapour/liquid drum, through a glass pump, to the straight, horizontal test section and then to another horizontal section heated externally by a 16 kW electric radiant furnace. Connected to the vapour space of the drum was an integral condenser and a separate external condenser.

4. The system pressure was controlled by setting the heat input to the furnace and varying the water circulation rate through the integral condenser; the circulation rate was controlled by varying the speed of the pump using a variable speed gear box. After attaining the desired operating pressure and circulation rate, depressurisation was effected either by increasing the water flow rate through the integral condenser or by blowing off vapour through the external condenser. The rig was so designed that when the pressure was reduced flashing took place preferentially in the test section.

5. The test section tube had a nominal bore of 50 mm and a length of 2000 mm. It was fitted with pressure and temperature tapping points 80 mm from the inlet end and a second pressure tapping point 75 mm from the exit end. Measurements taken during the transient were the pressure and temperature near the entrance to the test section, the pressure difference across the test section, the circulation rate immediately after the pump in a horizontal section where no flashing took place, and the full bore void fraction at three locations in the test section. All these instruments were connected to a U.V. recorder. In addition, the pressure in the vapour drum was continuously observed.

6. Pressures and pressure differences were measured using transducers, and the test section temperature by a copper/constantan thermocouple mounted diametrically opposite the inlet pressure tapping. The flow rate was measured using a pitot static tube and differential pressure transducer,

and the voidage at the three stations (150, 1010 and 1890 mm from the inlet) on the test section by the $\gamma$-ray attenuation, single beam technique using $C_s$ 137 sources and scintillation counters in the horizontal plane.

## EXPERIMENTAL RESULTS

7. Starting with a steady state pressure of 136, 146 or 156 $kN/m^2$, a series of test runs was carried out with circulating velocities between 0.29 and 2.1 m/s and various depressurisation rates which produced voidages up to 70%. A typical set of experimental results which include test section pressure and pressure differenc., circulation velocity and test section temperature as a function of time into the transient are shown in Figure 2. Also shown are the void fractions measured at the three stations in the test section and the temperature difference $\Delta T_L$ between the measured liquid temperature and the saturation temperature corresponding to the local pressure.

8. It can be seen from Figure 2 that, as the pressure in the test section dropped, the circulation velocity decreased and the void fraction increased. Voidage appeared almost simultaneously at the three observation stations, producing voidage/time curves of similar shape and only slightly displaced from one another in time. In addition, the non-equilibrium temperature difference $\Delta T_L$ built up during the initial part of the transient until enough liquid superheat was available to initiate flashing.

9. In most cases this initial superheat was of the order 1 to 2°K. Finally it should be noted in Figure 2 that the pressure drop across the test section decreased in step with the flow rate but was always small.

10. The most striking feature of the experiment was the fact that at all flow rates a separated or stratified flow pattern was observed, with gravity pulling the liquid to the bottom of the tube. In no case was annular or bubble flow observed, as in the vertical rig, hence relative movement between the phases could be neglected. This emphasizes the difficulty of modelling the transient two phase flow in a complicated reactor where different flow regimes may occur in different parts of the coolant circuit.

## THEROETICAL ANALYSIS

11. In the theoretical analysis presented here, two of the above observations are exploited

(i) the variation of pressure and voidage with distance along the test section is small so that terms such as $U\frac{\partial p}{\partial z} \ll \frac{\partial p}{\partial t}$

(ii) the flow is separated, horizontal flow with the relative velocity between the phases negligible Thus the general equations for transient two phase flow in a horizontal duct of constant cross-section, with the additional assumptions of negligible heat loss and energy dissipation due to friction in the test section, become

Mass $$\frac{\partial G}{\partial z} + \frac{\partial \rho_m}{\partial t} = 0 \tag{1}$$

Momentum $$\frac{\partial G}{\partial t} + \frac{\partial}{\partial z}\left[\frac{G^2}{\rho_m}\right] + \frac{\partial p}{\partial z} = F_w \tag{2}$$

Energy $$\frac{\partial}{\partial t}\left[\alpha \rho_g h_g^* + (1-\alpha)\rho_L h_L^*\right] = \frac{\partial p}{\partial t} \tag{3}$$

where $$h^* = h + \tfrac{1}{2}U^2$$

If, in addition, density variations in the liquid and vapour phases are neglected compared with the effect of voidage change on mean density $\rho_m$ so that $\frac{\partial \rho_f}{\partial t} \approx \frac{\partial \rho_L}{\partial t} \approx 0$ and $\frac{\partial \rho_m}{\partial t} \approx -(\rho_f - \rho_g)\frac{\partial \alpha}{\partial t}$ and if equation (2) is used to eliminate the kinetic energy terms, equation (3) becomes

$$\left[\frac{\rho_f \rho_g h_{Lg}}{\rho_m}\right]\frac{d\alpha}{dt} = \left[1 - \alpha\rho_g \frac{dh_g}{dp} - (1-\alpha)\rho_f \frac{dh_L}{dp}\right]\frac{dp}{dt} \tag{4}$$

where $h_{Lg}$ , the difference between the vapour enthalpy and the liquid enthalpy, is given by,

$$h_{Lg} = h_{fg} - C_{p_L}\Delta T_L \tag{5}$$

$h_{fg}$ being the enthalpy of evaporation and $\Delta T_L$ the liquid superheat. Substituting equation (5) in (4) gives

$$\frac{d(\Delta T_L)}{dt} - \left[\frac{\rho_g \frac{d\alpha}{dt}}{(1-\alpha)\rho_m}\right]\Delta T_L = -\left[\frac{\rho_g h_{fg}}{(1-\alpha)\rho_m C_{p_L}}\cdot\frac{d\alpha}{dt} + \frac{dT_g}{dt}\right] \tag{6}$$

a first order differential equation relating the liquid superheat $\Delta T_L$ at time t to the voidage $\alpha$ Integrating equation (6) from t = 0 where $\alpha$ = 0 and $\Delta T_L = \Delta T_0$, just before voidage formation, gives

$$\frac{\Delta T_L}{\Delta T_0} = \frac{\rho_m}{(1-\alpha)\rho_f} - \frac{\rho_m}{(1-\alpha)}\left[\frac{N\alpha}{\rho_m} + \frac{1}{\Delta T_0}\int_0^t \frac{1-\alpha}{\rho_m}\cdot\frac{dT_g}{dt}dt\right] \tag{7}$$

or if $\rho_f >> \rho_g$

$$1 - \frac{\Delta T_L}{\Delta T_0} = \frac{N\alpha}{1-\alpha} + \frac{T_g - T_{go}}{\Delta T_0} \tag{8}$$

where $$N = \frac{\rho_g h_{fg}}{\rho_f C_{p_L} \Delta T_0}$$

Equation (1) shows that $\frac{\partial G}{\partial z} \neq 0$ and equation (2) enables the mass circulation rate to be estimated. It is worth noting that equation (8) implies that, prior to voidage formation, the liquid temperature $T_L$ remains constant. However, since we are primarily interested here in the non-equilibrium effects, attention is focussed on equations (7) and (8) relating $\Delta T_L$ to the dimensionless group N, the pressure (or saturation temperature) variation given by $\frac{dT_g}{dt}$ , the initial superheat $\Delta T_0$ required to start flashing, and the voidage

$\alpha$ at time t. It is desirable to have another relationship between $\Delta T_L$ and $\alpha$ .

12. For this we exploit the fact that the flow was always separated with the vapour layer lying on top of the liquid, as shown in Figure 3. In no case was internal nucleation of the liquid layer observed after stratification. It is assumed that non-equilibrium is caused by the finite rate of evaporation at the liquid surface during the transient, that this rate is determined by the heat transfer between the liquid and vapour, and that temperature gradients in the liquid are negligible. Thus, calling the relevant heat transfer coefficient h and neglecting as before, the variation of $\rho g$ with pressure, a heat transfer balance across the surface gives,

$$\frac{h Y \Delta T_L}{A} = h_{fg} \rho_g \frac{d\alpha}{dt} \qquad (9)$$

where A is the area of the test duct and Y and $\alpha$ are related to the angle $\theta$, shown in Figure 3, by

$$Y = D \sin\frac{\theta}{2} \quad , \quad \alpha = \frac{1}{2\pi}\left[\theta - \sin\theta\right] \qquad (10)$$

Equation (9) was used to evaluate h from the experimental values of $\Delta T_L$ and $\alpha$ .

13. Equations (10) and (8) can be combined to eliminate $\Delta T_L$ or t. In the former case,

$$\frac{\pi}{4}\frac{D}{Y}\frac{d\alpha}{d\tau} + \frac{\alpha}{1-\alpha} + \frac{1}{N}\left[\frac{1}{\Delta T_o}\left(\frac{d\bar{T_g}}{d\tau}\right)\tau - 1\right] = 0 \qquad (11)$$

In the latter case,

$$d\left[\frac{\Delta T_L}{\Delta T_o}\right] + \left[\frac{N\frac{\pi}{4}\frac{D}{Y}R}{\Delta T_L/\Delta T_o} + \frac{N}{(1-\alpha)^2}\right]d\alpha = 0 \qquad (12)$$

where $\tau = \frac{h t}{C_{pL} \rho_f D}$ and $R = \frac{1}{\Delta T_o}\left(\frac{d\bar{T_g}}{d\tau}\right)$ a

a measure of the mean rate of pressure drop.

## COMPARISON OF THEORY AND EXPERIMENT

14. Using the experimental values of void fraction $\alpha$ and saturation temperature $T_g$ in equation (8), it is possible to evaluate $\Delta T_L$ and compare with the corresponding experimental values. Typical comparisons are shown in Figure 4 of (i) $\Delta T_L$ versus $\alpha$ and (ii) $\Delta T_L$ versus time measured from the start of the transient and from the onset of voidage formation. In item (ii), the predicted values of $\Delta T_L$ using equation (8) only apply after the onset of voidage formation.

15. The curves of $\Delta T_L$ versus time show the build up of liquid superheat from an initially subcooled condition to a superheated condition prior to voidage formation. Thereafter, the values predicted from equation (8) can be compared with experiment and show reasonably good qualitative agreement, with the predicted values usually slightly higher than experimental. This difference, although not excessive, may be due to the assumptions made of no heat loss and no temperature gradient throughout the liquid. It should be pointed out that these comparisons include imperfections associated with using experimental void fraction values and irregular rates of pressure drop. Similar comments can be applied to the curves of $\Delta T_L$ versus $\alpha$ where, again, reasonable agreement between experiment and prediction is obtained.

16. However, it is desirable to predict $\alpha$ and $\Delta T_L$ with less empirical input than experimental void fraction data, consequently equation (9) was fitted to all of the experimental data of $\Delta T_L$ versus $\alpha$ and a mean value for the heat transfer coefficient h deduced. A value of $h = 4$ kW/m$^2$K was found to fit the data best and this was used in subsequent analysis in order to facilitate calculation within the current state of knowledge. A suitable value of $\Delta T_o$ was also deduced from the experimental data and taken as 2 K.

17. Using the above values of h (to define $\tau$ ) and $\Delta T_o$, equation (11) was integrated numerically to give $\alpha$ as a function of $\tau$ for a particular mean rate of saturation temperature change $d\bar{T_g}/d\tau$. Typical comparisons with experimental results are shown in Figure 5 and indicate fairly good agreement.

18. Similar comparisons can be made by integrating equation (12) numerically and these are plotted on the $\Delta T_L$ versus $\alpha$ curves also shown in Figure 5. As can be seen, $\Delta T_L$ remains constant or slightly increases during the early part of the transient and then decreases fairly rapidly. This may be due to the increased surface area available for evaporation as the voidage increases, at least up to $\alpha = 0.5$. The agreement between theory and experiment in test run B1, with a low initial liquid velocity, is fairly good; a similar comparison with test run C9 shows much greater scatter of the experimental data, possibly due to the higher initial liquid velocity.

19. The value of $h = 4$ kW/m$^2$K could only be compared with the known work of Addlesee and Dickson (Reference 3). They used a value for h of between 11 and 16 kW/m$^2$K for the heat transfer between steam and water at sub-atmospheric pressures in a multi-flash plant model flash chamber with a flow regime closer to bubbly flow. Based on similar phase density ratios, their corresponding figure for R-113 would be about 10 kW/m$^2$K, although allowance for density ratio was arbitrarily eliminated in the Addlesee-Dickson correlation. Consequently the value of $h = 4$ kW/m$^2$K used seems reasonable in view of the relatively quiescent stratified type flow observed here compared to the more turbulent, bubbly flow experienced in a flash evaporator.

## CONCLUSIONS

20. (i) The experimental results showed that the flow in the horizontal test section was separated during the transient and that non-equilibrium existed between the liquid and vapour.
(ii) Roughly 2°K of superheat was required with the R-113 in the glass rig before flashing occurred.
(iii) The energy equation in which spatial variations were neglected in comparison with temporal variations of properties appeared to describe the flashing phenomena well, relating $\Delta T_L$ to $\alpha$ at a given time into the transient.
(iv) A simple theoretical model, which postulated that the non-equilibrium was controlled by heat transfer across the liquid/vapour interface, described the experimental facts well and produced a reasonable value for the heat transfer coefficient of 4.0 $kW/m^2K$.

## REFERENCES


1. SIMPSON H.C., ROONEY D.H., CALLANDER T,M.S. Transient Effects in Two-Phase Bubbly Downflow. European Two Phase Flow Meeting, Harwell, June 3-7, 1974.

2. SIMPSON H.C., ROONEY D.H., CALLANDER T.M.S., KENDOUSH A.A. Depressurisation Effects on a Freon 113 Circulating Loop. Paper A3, European Two Phase Flow Meeting, Erlangen, May 31 - June 4, 1976.

3. ADDLESEE A.J., DICKSON A.N. Equilibration and Flow Studies with a Model Flash-Chamber. I.Mech.E. and I.Chem.E. Symposium on Multiphase Systems, I.Chem.E. Symposium Series No 38, Paper G5, April 2-4, 1974, University of Strathclyde, Glasgow


## NOMENCLATURE

$A$ Cross sectional area
$C_{pL}$ Specific heat of liquid at constant pressure
$D$ Diameter
$F_w$ Wall friction force
$G$ Mass velocity
$h$ Heat transfer coefficient
$h$ Specific enthalpy
$h_f$ Specific enthalpy of saturated liquid
$h_g$ Specific enthalpy of dry saturated vapour
$h_{fg}$ Enthalpy of evaporation $h_g - h_f$
$h_L$ Specific enthalpy of liquid
$h_{Lg}$ $h_g - h_L$
$N$ Dimensionless group = $\frac{\rho_g h_{fg}}{\rho_f C_{pL} \Delta T_o}$
$p$ Pressure
$R$ Dimensionless group = $\frac{1}{\Delta T_o}\left[\frac{d\overline{T_g}}{d\tau}\right]$
$t$ Time
$T_g$ Saturation temperature of vapour
$T_{go}$ Vapour temperature at onset of flashing
$d\overline{T_g}$ Mean change in vapour temperature
$T_L$ Liquid temperature
$\Delta T_L$ Liquid superheat = $T_L - T_g$
$\Delta T_o$ Liquid superheat necessary to initiate flashing
$U$ Velocity
$Y$ Surface length defined in Fig. 3
$z$ Length
$\alpha$ Void Fraction
$\theta$ Angle defined in Fig. 3
$\rho_f$ Density of saturated liquid
$\rho_g$ Density of dry saturated vapour
$\rho_L$ Density of liquid
$\rho_m$ Mean density = $\alpha \rho_g + (1-\alpha)\rho_f$
$\tau$ Dimensionless group = $\frac{ht}{C_{pL}\rho_f D}$

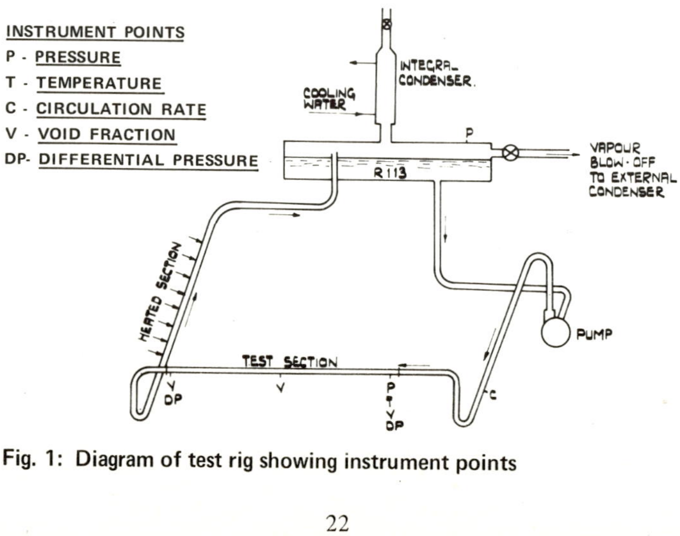


Fig. 1: Diagram of test rig showing instrument points

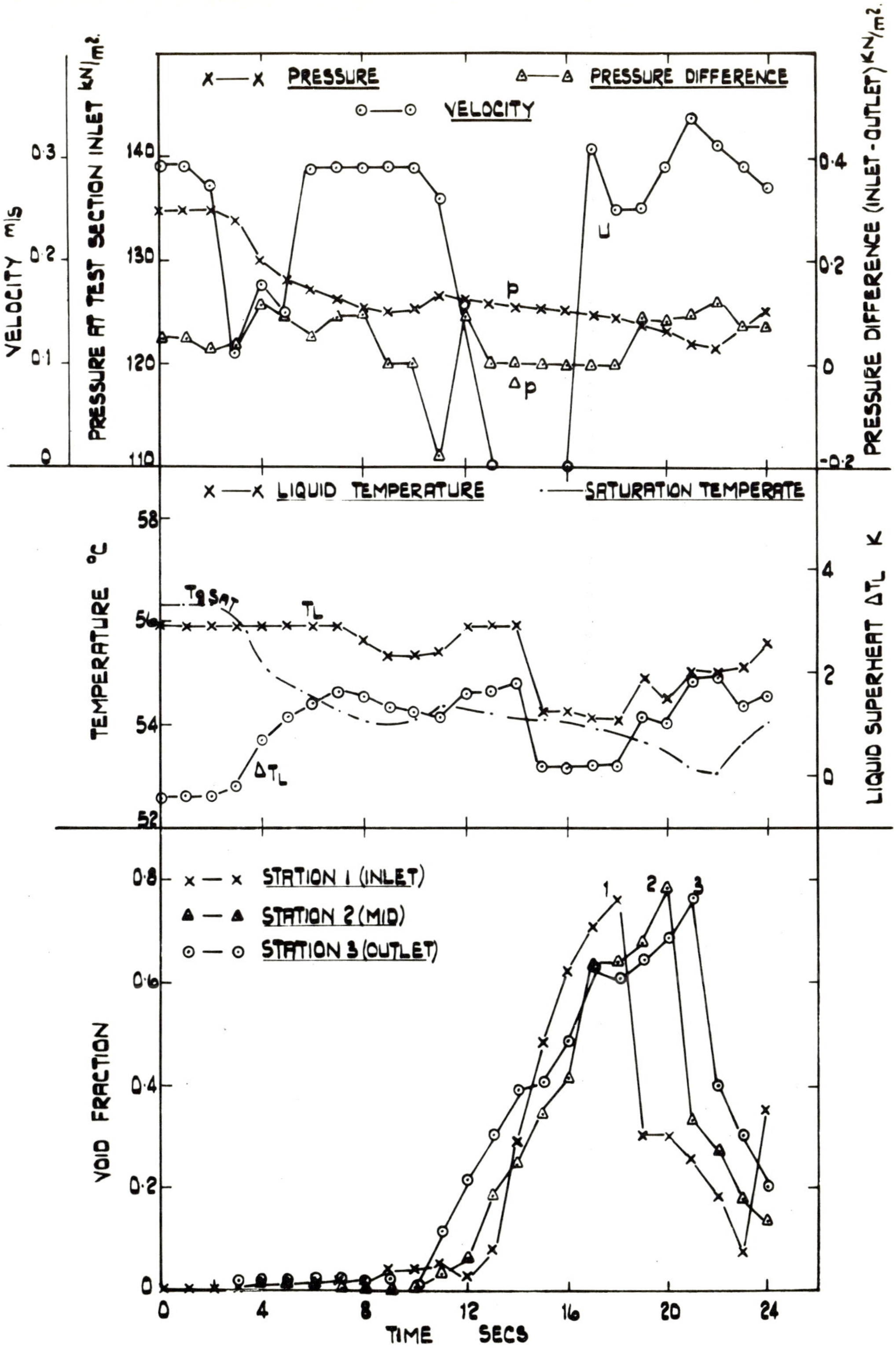


Fig. 2: Specimen set of experimental data (test run A1)

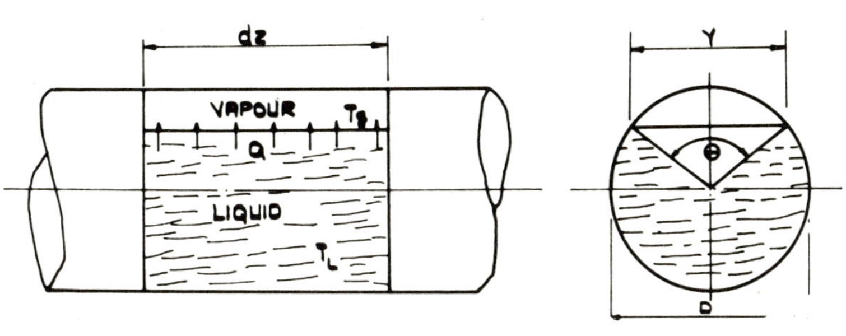


Fig. 3: Separated flow heat transfer model

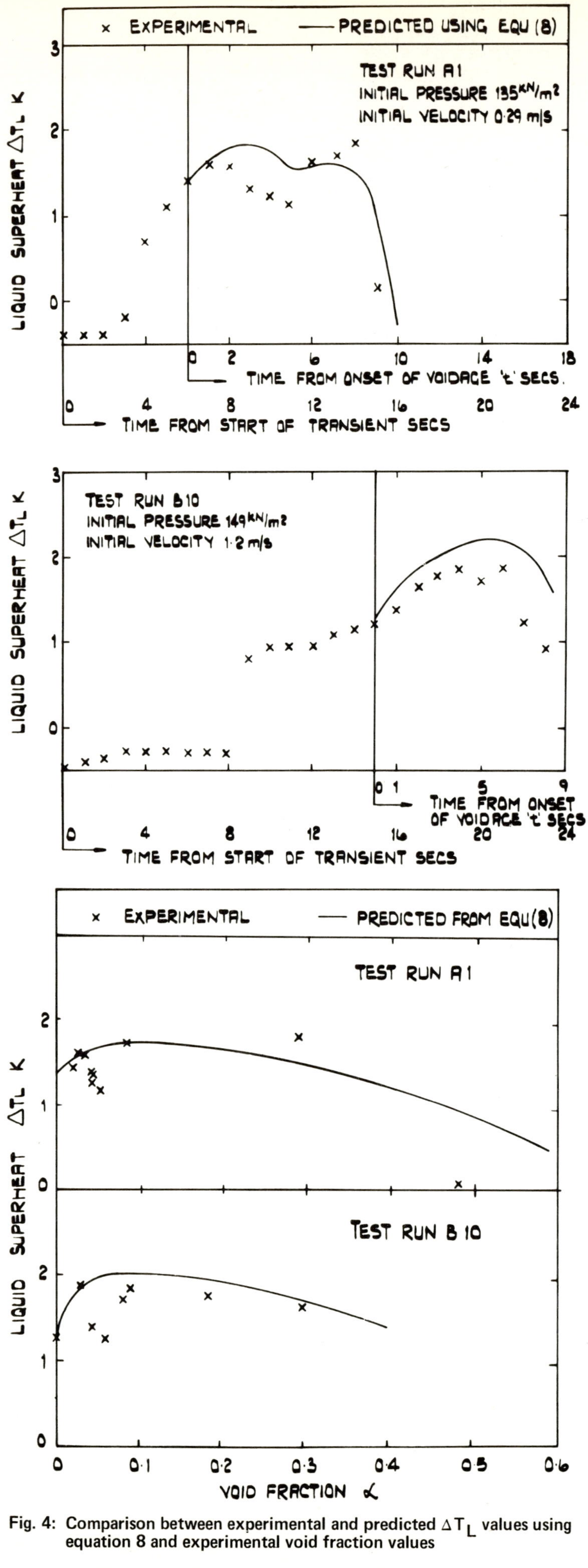


Fig. 4: Comparison between experimental and predicted $\Delta T_L$ values using equation 8 and experimental void fraction values

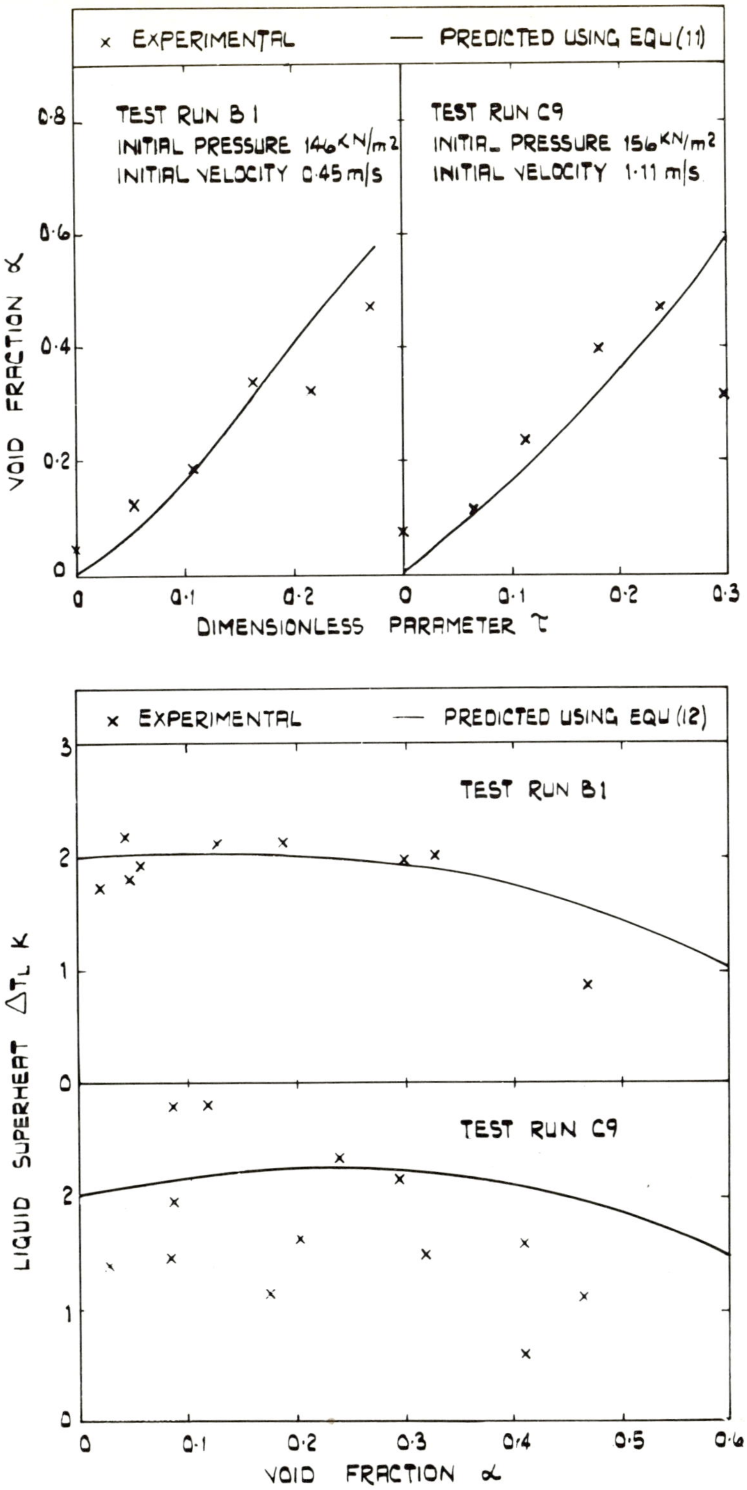


Fig. 5: Typical comparisons between theory and experiment

C195/77

# SOME LIMITATIONS ON THE USE OF THE ASSUMPTION OF THERMAL EQUILIBRIUM BETWEEN PHASES: A COMPARISON BETWEEN THEORY AND EXPERIMENTS ON SIMPLE WATER FILLED SYSTEMS

A. R. EDWARDS, BSc, D. J. MATHER, BSc, PhD,
S. FLANAGAN, BSc, PhD, and D. MARTIN, BSc, PhD
Safety and Reliability Directorate, UKAEA, Culcheth, Warrington

SYNOPSIS Calculations based on the alternative assumptions of equal or unequal phase temperatures are compared with measurements from depressurization experiments on two geometrically simple systems. Water conditions approximated to those in typical water cooled reactors. In general, a thermal non-equilibrium model was required to get agreement with the early transient conditions, when local void fractions were close to zero. But the simple equilibrium assumption proved to be adequate for the longer term calculation of pressure history, however, the transient two-phase frictional losses appear to be underestimated.

## INTRODUCTION

1. Over the past few years increasing attention has been paid to the details of flow conditions during depressurization of water cooled reactor systems. The simple assumption of thermal equilibrium, widely used in the development of steady flow correlations and critical flow models, has been shown to be in error in short small diameter (6 to 10mm id) test pipes, particularly at low voidage conditions (ref. 1); also in long systems where vapourization starts close to the choking plane (ref. 2). In many postulated reactor accident situations the flow is far from steady and the system pressure may also be changing rapidly. A major question is whether meaningful calculations can be carried out using simple assumptions, or is it necessary to develop more elaborate flow models.

2. In this paper comparisons are presented between theoretical predictions from computer programs using the alternative assumptions of equal or unequal phase temperatures and experimental measurements from two simple pipe systems. The experiments are part of the basic studies being carried out for Safety and Reliability Directorate (SRD) at the Ministry of Defence Establishment at Foulness.

## EXPERIMENTAL PROGRAMME AND TEST MEASUREMENTS

3. The first configuration was a simple constant diameter pipe, 4.1m long, internal diameters of 73mm and 32mm (Phase 1) and 206mm (Phase 3) were tested. The second configuration used the 32mm id, 4.1m pipe with a reservoir 132mm id, 2.2m long (Phase 2). Main dimensions of the two systems are given in Fig 1, further details may be obtained from refs. 3, 4 and 5. The initial pressures and temperatures for the five Phase 3 and Phase 2 tests referred to in this paper are given in Table 1. Further results for Phase 3 tests are in the course of publication (refs. 6 and 7).

4. Transient pressure measurements were taken at 8 positions along the 4.1m pipe, and at 4 positions in the reservoir. The fast response piezo electric pressure transducers were basically calibrated in situ by assuming the initial and final signal levels corresponded to the initial system pressure (measured on twin Bourdon gauges) and to atmospheric pressure respectively. Overall accuracy of the final results after digitization and computer plotting is believed to be $\pm 0.5$ bar at the centre of the pressure range and $\pm 0.15$ at the ends.

5. To provide additional information on the state of the fluid in the pipe, the attenuation of a beam of X-rays was used to deduce void fraction. A single 10mm dia beam, across the horizontal diameter, was used on the 32mm id pipe; but a fan of 5 beams was used on the 206mm id pipe (ref. 3). The general level of accuracy for void fraction estimated at $\pm 0.03$ over the range 0-1. The void fraction data, presented here for the 206mm id pipe, are area averaged values. The records from the 5 beam system indicated a general pattern of bubbly flow with a vapour annulus at the pipe wall.

Table 1

Initial conditions for experiments

| Test No. | Pressure MPa | Temperature °C |
|---|---|---|
| 141 | 5.82 | $244^{+14}_{-10}$ |
| 142 | 5.45 | $242^{+14}_{-9}$ |
| 143 | 5.65 | $243^{+15}_{-14}$ |
| 144 | 5.52 | $247^{+23}_{-15}$ |
| 145 | 5.52 | $245^{+14}_{-13}$ |
| 115 | 8.9 | $265^{+14}_{-10}$ |
| 130 | 8.6 | $263^{+15}_{-9}$ |

 

## THEORETICAL CALCULATIONS

### Thermal Non-Equilibrium Code Assumptions

6. A transient flow model, based on a thermal conduction controlled bubble growth concept has been developed and a computer code HUBBLE-BUBBLE written to solve the equation set. The code works with a spatial mesh in a Lagrangian co-ordinate system and tracks a representative bubble or drop for each mesh point. In the first version a detailed temperature distribution around a bubble was computed at each mesh point. A second version has recently been developed, where bubble growth is based on the model developed by Hunt (ref. 8), which uses a single representative temperature for the liquid phase. Nucleation is assumed to start when the pressure falls below the local saturation value. The number of nuclei formed is related to the third power of the superheat, and is fixed when the maximum temperature difference has been computed.

7. The relative velocity between the phases is computed from the pressure gradient and bubble drag, and used to evaluate heat transfer effects in the liquid. This velocity difference is ignored in the hydrodynamic equations as the velocity ratio is less than 1.2. The subroutine handling the exit plane, checks the local fluid velocity against an estimate of the local sonic velocity derived from the basic equations (ref. 9). This sonic velocity relationship is similar in form to that obtained from the assumption of no heat transfer between phases. In order to obtain an accurate estimate of the approach to choking, a fine mesh spacing must be used near the open end which, with the current explicit solution technique, can severely restrict the permissible time step. An inertia term, representing the glass bursting disc, controls the initial fluid accelerations.

8. As the bubble cloud grows, there is an increasing probability of bubbles coalescing. This effect was studied using a table of random numbers to position the centres of a typical number of bubbles in a unit volume. For an assumed void fraction, the average bubble radius was then evaluated. If bubbles touched or intersected, they were treated as one unit and the average radius adjusted to restore the void fraction. The total surface area was then estimated and compared with that for no coalescense. An exponential factor was found to give a reasonable fit, and this has been incorporated in the calculation of total bubble surface area as a first attempt at deriving a suitable compensating factor.

9. For the constant diameter pipe system, 10 mesh points were used over a 60mm length at the open end, with 77 mesh points spread over the remaining length. For the two-pipe system, a second fine mesh of 20 points was needed in the area change region spread over a 225mm length. A total number of 99 mesh points was used for the whole system.

### Non-Equilibrium Code Results

10. The results obtained for the simple constant diameter pipe (Fig. 2) illustrate the main characteristics of a thermal non-equilibrium code using a fine mesh. The initial depressurization wave is well reproduced, both in shape and time of travel, with the pressure falling significantly below the local saturation value. A rapid pressure recovery follows the formation of sufficient bubbles to give a significant heat transfer area. This effect is most marked at the closed end of the system, where the wave strength has been enhanced by reflection effects.

11. The current bubble growth model, using a mean liquid temperature and allowing for the effects of relative motion between the phases on heat transfer, gives good comparisons with the short-term experimental results. The calculated pressure plateau at 10-30ms was not particularly sensitive to the nucleation coefficient assumed. Hunt (ref. 8) recommended a value of 5000, which gave a pressure of 3.30 MPa halfway along the pipe. The calculations in Fig 2 used a value of 100, which gave a plateau value of 3.16 MPa, with better agreement between calculation and experiment. Even so, up to 5 x $10^5$ nuclei/kg were triggered by the peak superheat value, the theoretical model is being studied to try and resolve the problem of needing so many bubble nuclei to get the required order of heat transfer.

12. As the calculated blowdown proceeds, mass and energy are discharged too quickly, so that the predicted pressure falls too soon. In general, the bubble model does not provide sufficient heat transfer at high void fractions. A physically more realistic droplet model, which should result in a larger heat transfer area, is under development but results are not yet available.

13. Again, for the two pipe system (Fig. 3), the HUBBLE-BUBBLE calculations trace the initial depressurization wave accurately, including the recompression effects due to the wave reflection from the reservoir. A bubble nucleation coefficient of 100 has also been used in these calculations. Because of the need to use a very short time step whilst the system contained subcooled water, these calculations were terminated at 24ms.

14. The simple pipe calculation demanded 80 mins CPU time on an IBM 370, and the two pipe calculation used 70 mins in tracing the first 24ms of the depressurization.

### Thermal Equilibrium Code Assumptions

15. The calculations, based on the assumption of thermal equilibrium between the two phases, have been carried out using the computer code RELAP-UK. This code was based on the American code RELAP 3 (ref. 10), but contains features from RELAP 4 (ref. 11) and certain additions, such as a slip flow model and a wide choice of choking flow options, which have been introduced by AEE Winfrith (ref. 12). For the calculations presented here, special routines were introduced by SRD using two-phase slip and friction loss correlations developed by CISE (ref. 13). This slip correlation gives values tending to one at high mass velocities and was considered to be the most appropriate available. Choking flow has been based on the assumption of a slip ratio of one, ie the homogeneous equilibrium model, and hence is compatible with the slip model.

16. The calculation for the simple constant diameter pipe used 13 nodes, 8 of which were spaced to be central on the gauge stations (Fig. 1a). For the two-pipe system 17 nodes were used (Fig. 1b). To start the calculation, the leak was assumed to open linearly over the first 5ms,

which is rather slower than observed in the experiments. However, the code is slow running with sudden changes of conditions and this compromise is normally used in depressurization calculations.

Equilibrium Code Results

17. The results obtained for the simple constant diameter pipe (Fig. 2) illustrate the main characteristics of a typical thermal equilibrium nodal type code. The initial depressurization wave has been spread out by the 5ms opening time assumed for the break. The pressure falls rapidly as far as the local saturation value and then levels off, hence the calculated pressure is above the measured values for the first 50-100ms.

18. Very good agreement between calculated and measured void fractions is given by the slip ratio assumption used, and the generally good agreement on overall blowdown time confirms the choice of critical flow model.

19. A check calculation, using a break opening time of 0.5ms, gave very good agreement for propagation of the initial wave front down the length of the pipe. The use of 13 nodes, with an approximate length of 750mm for 4 of them, was sufficient to avoid any significant wave spreading in the initial 4ms needed for the subcooled decompression wave to reach the closed end of the system. A nodal code, such as RELAP-UK can therefore be made to track strong pressure waves provided sufficient nodes are used to maintain detail. A minimum time step of $10^{-6}$ sec was specified for this calculation and was used by the integration routine a few times in the first 5ms of depressurization. This calculation used 25sec of CPU time for the first 10ms period compared to 2.8 secs for the calculation with the slower break opening assumption. The whole of the standard calculation took 35 sec CPU time on the IBM 370.

20. Several computer runs, using a variety of assumptions have been carried out for the two-pipe system. Variations in the slip correlation, used internally in the pipe, did not make very significant differences, but the use of slip in the discharge model significantly shortened the predicted blowdonw time. So to avoid incompatibilities in the flow modelling as the discharge point was approached, the CISE slip correlation was preferred for the final calculations.

21. The results for the two pipe system are shown in Fig. 3. The assumption of a 5ms break opening time has had some influence on the short term results. Halfway along the small diameter pipe (2621 station), the pressure has not quite fallen to the saturation value before recompression effects from the reservoir are felt. The computed pressure at 2-3ms is not as low as the observations, and the pressure recovery at 6-7ms is not as high. Except for the station close to the open end of the small diameter pipe, the calculated and experimental pressures are in good agreement from 20ms onwards, even though the void fractions are less than 0.01. The generally slower rate of depressurization (about 1MPa/sec) induced by the reservoir, allows near equilibrium conditions to be attained, except close to the discharge where local acceleration effects maintain significant phase temperature differences. A typical calculation for the two pipe system used 200 secs CPU time on an IBM 370.

22. As the long term calculated pressures, using the CISE two-phase friction pressure drop correlation, fell somewhat too quickly, additional runs were carried out with two-phase friction effects enhanced by factors of 3 and 9. The results are shown in Fig. 3, and it will be seen that best agreement with the reservoir pressure measurements was obtained using a factor 9 on friction pressure drop. But the calculation somewhat overpredicts the pressure drop along the discharge pipe between the reservoir and the gauge station at 2621mm. A similar change was tried for the simple pipe, but friction was found to be unimportant, because of the large diameter and short blowdown duration of this test. It is not clear whether transient two-phase flow losses are really higher than corresponding steady state conditions, due possibly to changes in the flow patterns, or whether the computer model is inadequate in some respect.

23. In the present series of RELAP-UK calculations no allowance has been made for vapour separation in the vertical reservoir. The measurements indicate that some phase separation takes place, leading to the "S" shape of the pressure history at 3.5 secs. Further calculations are in progress, but the simple bubble rise model in RELAP-UK does not fit the observations. It is hoped that the new drift flux model under development at Winfrith will give a better comparison.

CONCLUSIONS

24. The main conclusions to be drawn from this study are:

(1) Effects due to a lack of thermal equilibrium between phases can be seen in geometrically simple systems during the initial stages of depressurization. But when vapourization limits the rate of depressurization to less than 1MPa/sec non-equilibrium effects are negligible.

(2) Persistent local thermal non-equilibrium effects can however be seen close to the open end of a pipe, where acceleration effects cause the fluid to experience transient conditions.

(3) A bubble growth model can be made to give good agreement with the short term pressure measurements; but more work is required to arrive at soundly based heat transfer correlations, which do not demand the assumption of excessive numbers of bubble nuclei.

(4) A model, which gives a large heat transfer area, is required to give suitable conditions very close to the open end of a pipe, to improve the agreement on discharge flows.

(5) Best agreement, between calculations based on a thermal equilibrium assumption and the experiments, was obtained with the use of a slip correlation which tends to one at high mass flows and the homogeneous equilibrium choking flow model.

(6) There appears to be a need to increase the two-phase friction pressure drop during a transient situation by a factor of nearly ten times the smooth pipe steady state losses; whether this is a real effect or due to deficiencies in the computer model needs to be investigated further.

(7) A nodal code, such as RELAP-UK, can reproduce the initial depressurization wave travel, provided the node lengths and calculation time step are properly chosen.

REFERENCES

1. HENRY R.E. and FAUSKE H.K. The Two-Phase Critical Flow of One Component Mixtures in Nozzles, Orifices, and Short Tubes. Journal of Heat Transfer, Trans, ASME Vol. 93 Series C, No. 2, May 1971, p. 179-187.

2. REOCREUX M.L. Experimental Study of Steam-Water Choked Flow. CSNI Specialist Meeting on Transient Two-Phase Flow, Toronto (Canada), August 1976, Pergamon Press (in publication).

3. SMITH A.V. A Fast Response Multi-Beam X-ray Absorption Technique for Identifying Phase Distributions during Steam-Water Blowdowns. J Br Nuclear Energy Soc., 1975 14 July, No. 3, 227-235.

4. EDWARDS A.R. and O'BRIEN T.P. Studies of Phenomena connected with the Depressurization of Water Reactors. Journal BNES, Vol. 9, p. 125-135, April 1970.

5. EDWARDS A.R. and JONES C. An Analysis of Phase IIA Blowdown Tests - The Discharge of High Enthalpy Water from a Simple Volume into a Containment Volume, UKAEA Report SRD R27 (1974).

6. BORGARTZ B.O., GOODMAN R.M.E., O'BRIEN T.P., RAWLINGSON M. and EDWARDS A.R. Depressurization Studies. Phase 3. Results of Tests 142 and 143. UKAEA Report SRD R76, 1977.

7. BORGARTZ B.O., GOODMAN R.M.E., O'BRIEN T.P., RAWLINGSON M. and EDWARDS A.R. Depressurization Studies. Phase 3. Results of Tests 144 and 145. UKAEA Report SRD R77, 1977

8. HUNT D.L.M. The Effect of Delayed Bubble Growth on the Depressurization of Vessels Containing High Temperature Water. UKAEA Report AHSB(S) R189, 1970.

9. EDWARDS A.R. Conduction Controlled Flashing of a Fluid, and the Prediction of Critical Flow Rates in a One Dimensional System. UKAEA Report AHSB(S) R147.

10. RETTIG W.H. and others. RELAP-3 A Computer Program for Reactor Blowdown Analysis. USAEC Report IN 1321, June 1970.

11. MOORE K.V. and RETTIG W.H. RELAP-4 A Computer Program for Transient Thermal Hydraulic Analysis. USAEC Report ANCR 1127.

12. BRITTAIN I. et al. The Status of RELAP-UK Mk II, May 1975. UKAEA Report AEEW M1357.

13. LOMBARDI C. and PEDROCCHI E. A Pressure Drop Correlation in Two-Phase Flow. Energia Nucleare, Vol. 19, No. 2, February 1972, p. 91-99.

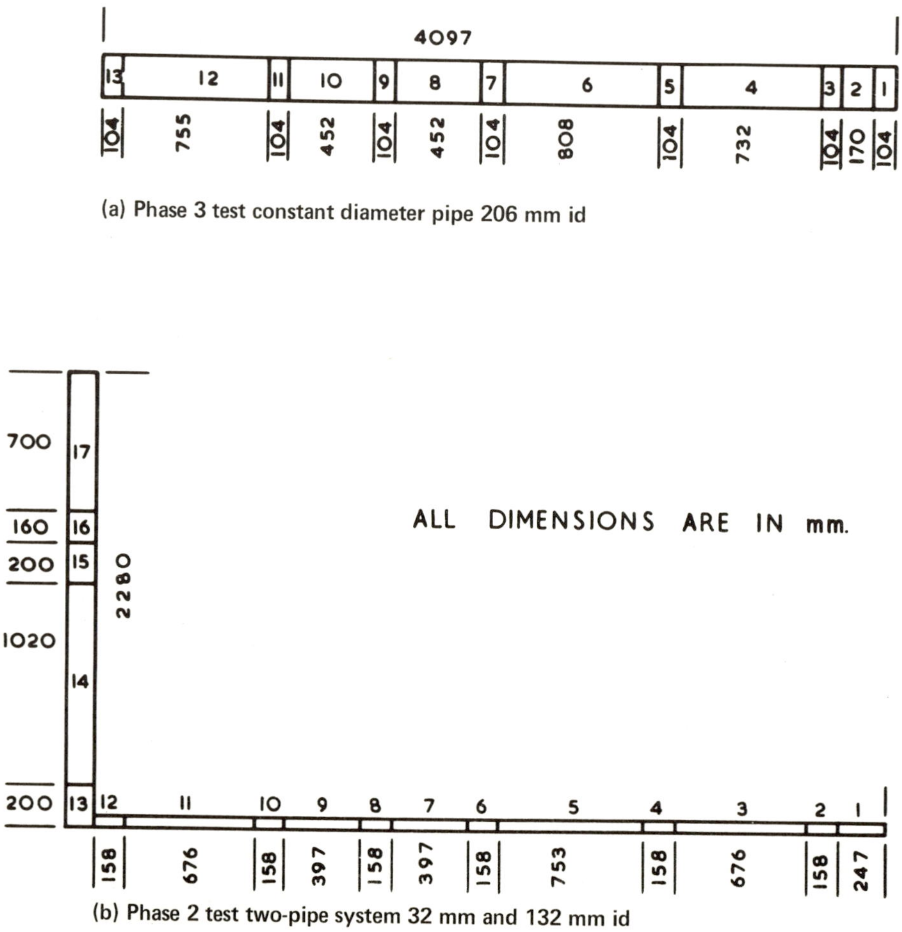


(a) Phase 3 test constant diameter pipe 206 mm id



(b) Phase 2 test two-pipe system 32 mm and 132 mm id

Fig. 1: Node distribution used for RELAP-UK calculations

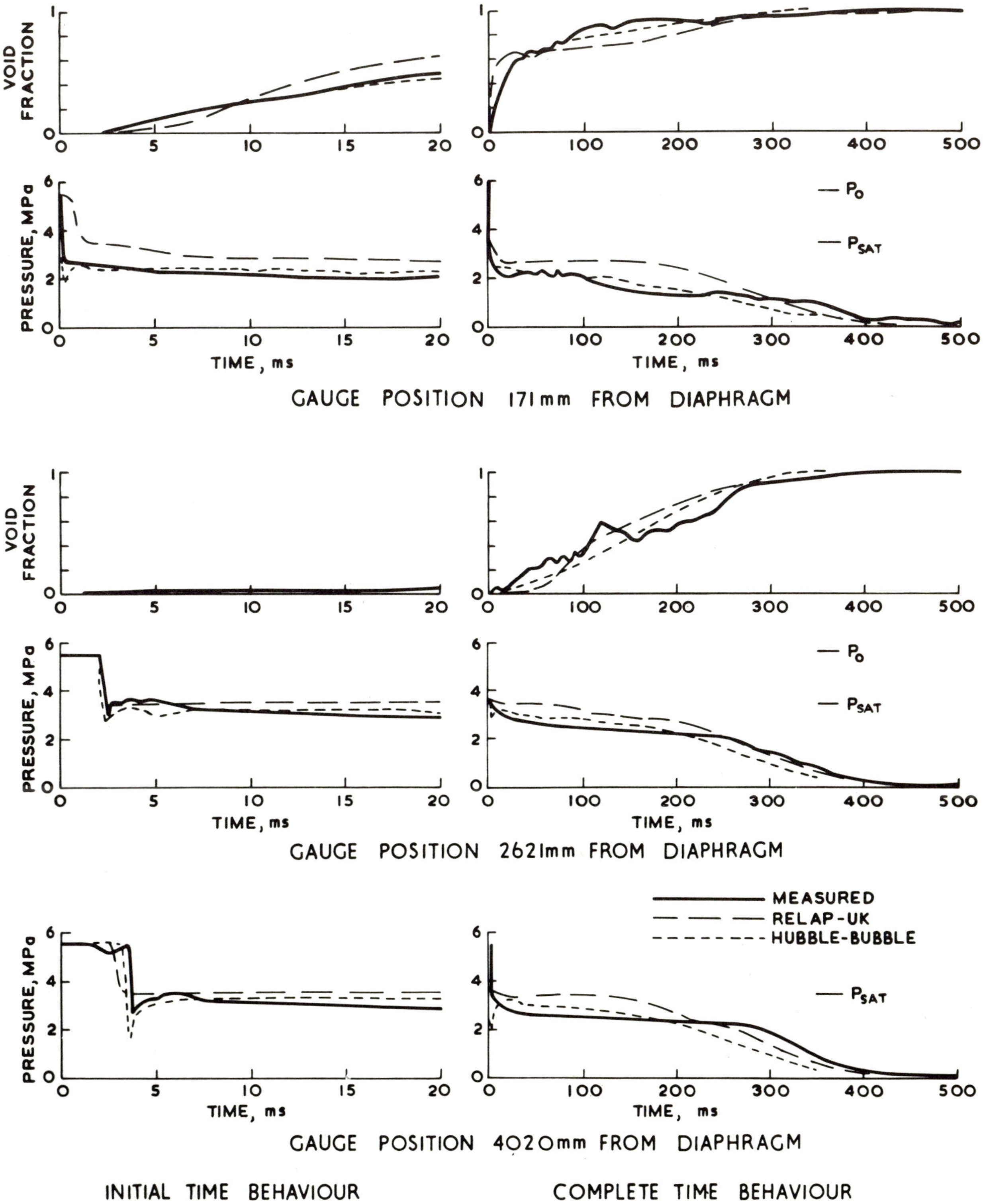


**Fig. 2: Depressurization studies—pressure and void behaviour constant diameter pipe 206 mm i.d., 4096 l**

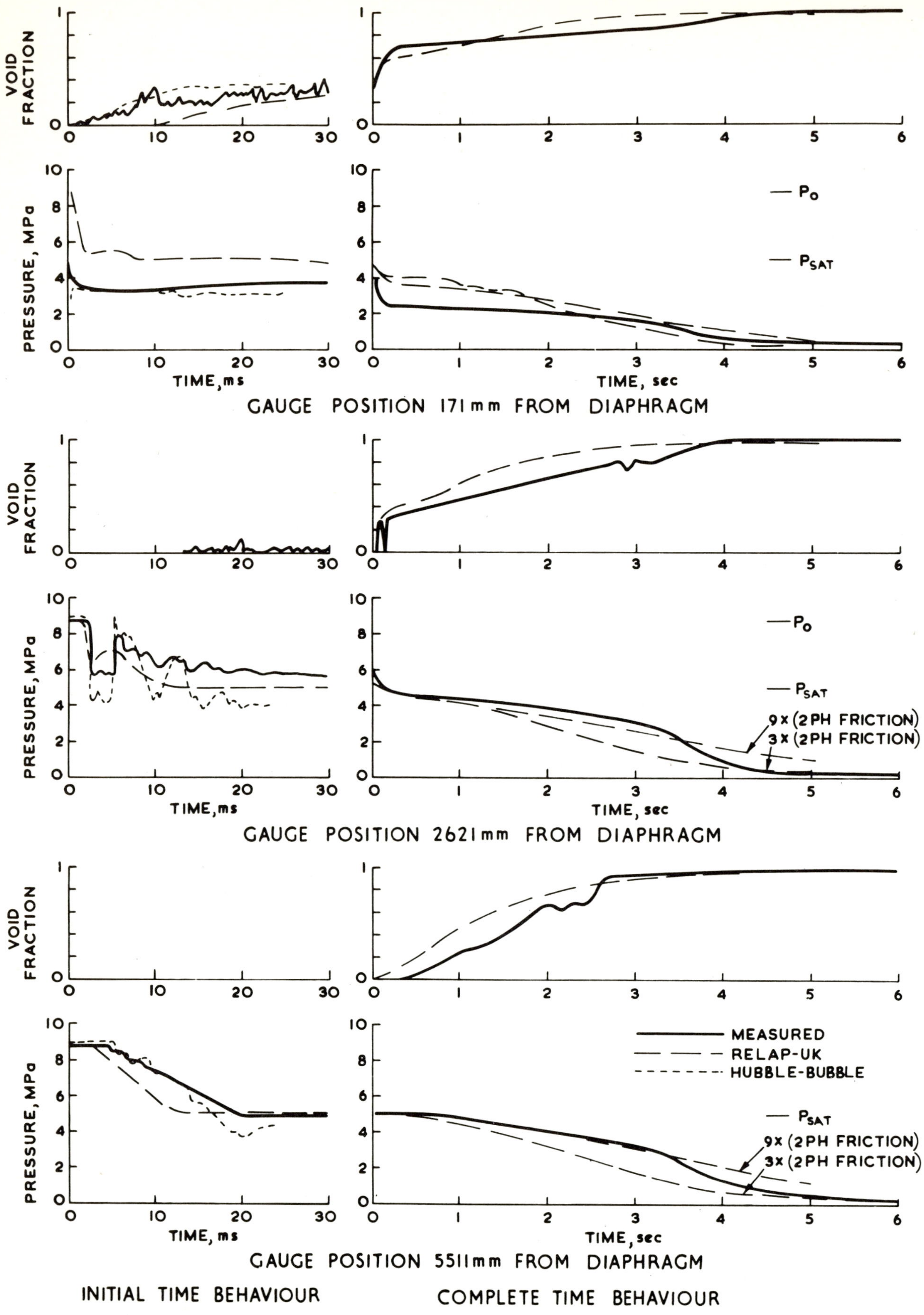


**Fig. 3:** Depressurization studies—pressure and void behaviour two diameter system 32 mm i.d., 4100 mm l., plus 132 mm i.d., 2280 mm l

C196/77

# THE INFLUENCE OF FUEL CHARACTERISTICS ON GAP CONDUCTANCE AND HENCE CLAD TEMPERATURE IN A LOCA

L. F. A. RAVEN, CEng and C. C. BAYLIS, BSc(Eng), MS, PhD
British Nuclear Fuels Ltd, Springfields, Nr. Preston

SYNOPSIS Gap conductance is one of the primary parameters required for the calculation of fuel and cladding temperatures during a LOCA as it determines the amount of stored heat in the fuel at the start of the transient calculation. This paper shows that by using controlled porosity fuel (CONPOR) instead of 'standard' fuel the minimum gap conductance during irradiation of the fuel is higher and therefore the maximum amount of stored heat at the start of the LOCA is reduced. This reduces the maximum predicted cladding temperature during the LOCA. The improved gap conductance is due to CONPOR fuel being more dimensionally stable than 'standard' fuel.

## INTRODUCTION

1. The maximum temperature attained by the cladding during a LOCA (Loss of Coolant Accident) is determined by the relative heat balance between the heat sources and heat sinks. The stored heat within the fuel pellet is one of the major heat sources and, for a given linear rating, the temperature distribution within the pellet is dependent on the thermal resistance between the pellet surface and inside bore of the cladding. (Usually the term "gap conductance" is used in preference to thermal resistance and is the reciprocal of the latter).

2. This paper describes the way in which the characteristics of the fuel pellet play an important part in determining the gap conductance, during the irradiation life of the fuel, and hence the stored heat.

## GAP CONDUCTANCE

3. The gap conductance is calculated, in general, by dividing the conductivity of the gas mixture in the pellet/clad gap (see figure 1) by the width of the gap (giving units of $W/m^2K$). If contact between pellet and cladding occurs during irradiation a second term primarily involving the contact pressure is included. However, if the fuel designer is concerned mainly with maximum clad temperatures the contact situation is of less interest.

4. Initially fuel pins are helium filled and the release of fission gases during irradiation degrades the thermal conductivity of the gas mixture thus lowering the gap conductance. Fission gas release is a function of the fuel temperature, the higher the temperature the greater the release. Hence any method of improving the gap conductance reduces the fission gas release which in turn helps to maintain the existing gap conductance.

5. The pellet/clad gap is a function of the dimensional changes in the pellet and cladding during irradiation. A predominant dimensional change is caused by densification of the pellet during the early stages of irradiation. Densification is caused by the unstable small pores ($<1$-$2\mu m$) sintering during irradiation thus reducing the volume of the pellet and increasing the gap.

6. It is known that if a LOCA should occur at the time of complete densification of the pellet, and before fission gas bubble swelling within the pellet becomes predominant, the highest clad temperature that could occur in a LOCA would occur then. Hence any means of reducing in-reactor densification and thus maintain the existing pellet/clad gap would maintain the existing gap conductance and reduce the expected maximum cladding LOCA temperature.

7. Maximum stored heat occurs in the pellet very early in the irradiation life of the fuel. The amount of fission gas present in the gap is of the order of 3% of the total gas volume and changes in the amount of fission gas present, at this time, have a secondary effect on the conductivity of the gas mixture and hence on the gap conductance. The main parameter which influences the gap conductance is the size of the gap between the pellet and cladding. CONPOR (Controlled Porosity) fuel has been designed to reduce in-reactor densification and limits the maximum gap size during the early stages of irradiation.

## CONPOR FUEL

8. It is a necessary design feature of oxide fuels to provide some porosity to accommodate induced fission product swelling which occurs during irradiation. Typical oxide fuels provide most of their porosity as small ($<5\mu m$ dia.) pores. In contrast, CONPOR fuel, developed by BNFL (British Nuclear Fuels Ltd.) contains most of its porosity as large spherical pores ($>10\mu m$, typically 20-40$\mu m$) randomly distributed in a high density matrix of $UO_2$, (see fig. 2a, 2b).



 

9. Under irradiation small pores are reduced by sintering at a greater rate than large spherical pores. Densification, therefore, is more rapid in fuel containing small pores than fuel with large spherical pores as used in CONPOR fuel, (ref. 1).

10. CONPOR fuel is manufactured by producing a high density matrix of $UO_2$ (up to 99% theoretical density is obtainable under commercially viable sintering conditions) which contains a controlled porosity additive capable of introducing voids of the required size and amount. The family of additives developed by BNFL are known by the generic name of CONPOR.

## IRRADIATION PERFORMANCE OF CONPOR FUEL

11. In-pile experiments
The effectiveness of CONPOR fuel has been tested by means of a series of irradiation experiments described in (ref. 2). The density changes in fuel irradiated in the SGHWR are shown in table 1 for "standard" and CONPOR fuel. The table clearly shows that CONPOR fuel densifies very much less than standard fuel.

12. Over 689 fuel pins have been irradiated in controlled porosity fuel experiments covering a wide range of burn-ups (up to 42 500 MWD/MTM) and in all cases the percentage change in density is very small. Even at low density CONPOR fuel is more stable than the standard fuel.

## COMPUTER CODE PREDICTIONS OF FUEL PERFORMANCE

13. The peak cladding temperature during a postulated LOCA is predicted by means of computer codes which contain models of the physical processes existing in and around the fuel. Before a LOCA transient prediction can be performed an estimation of the stored heat is needed and this is obtained by applying a heat balance between fuel and coolant using the linear rating at the time of the LOCA in combination with the gap-conductance which is determined by using BNFL's fuel performance code HOTROD (ref. 3).

## HOTROD CODE PREDICTION/OBSERVATION COMPARISON

14. Figure 3 shows the comparison of observed and predicted porosity in CONPOR fuel after 18 500 MWD/MTM at 36.7 kW/m (mean). The agreement is good.

15. Observed and predicted temperature distributions for a LWR (Light Water Reactor) fuel are shown in figure 4. Figure 4(a) is based on an EPRI (Electrical Power Research Institute - ref. 4) study of fuel temperatures during two ramps (1.1 W/cm per min) experienced by an experimental fuel element in the Halden Reactor. Figure 4(b) shows the variation of fuel centre temperature of another experimental fuel pin irradiated in the same reactor. The predicted temperatures agree well with the measured temperatures.

16. Measured end of life cold diametral gaps (also taken from the EPRI study) have been compared with predictions for three axial locations of an experimental fuel pin. The comparison is shown in table 2.

## COMPARISON OF PREDICTED FUEL CENTRE TEMPERATURES AND RADIAL GAPS FOR CONPOR AND STANDARD FUEL

17. Figure 5 shows a plot of predicted fuel centre temperatures and radial gaps for CONPOR and standard fuel. It clearly can be seen that use of CONPOR fuel reduces in pile radial gaps and fuel temperatures. The maximum gap due to densification is reduced by about $\frac{2}{3}$ rds from 4 x $10^{-2}$ mm to 1.5 x $10^{-2}$ mm. It has been shown (ref. 5) that it is possible to run CONPOR fuel at at least 15% higher rating without exceeding the temperature predicted for the 'standard' fuel.

## LOCA TEMPERATURES

18. Figure 6 is a plot of the predicted maximum cladding temperature as a function of time during a typical LOCA and shows that the peak cladding temperature is reduced by 114°C when CONPOR fuel pellets are used. (These curves correspond to the conditions of maximum stored heat indicated by the HOTROD runs used to construct figure 5).

## CONCLUSIONS

19. This paper has shown that the use of CONPOR oxide fuel with its built-in large stable pores reduces in-pile densification when compared with 'standard' oxide fuel. The reduction in densification reduces the amount of stored heat present at the start of a postulated LOCA, which in turn reduces the predicted maximum clad temperatures during the LOCA.

20. It is possible to run the CONPOR fuel at at least 15% higher rating without exceeding the temperature predicted for 'standard' fuel.

## ACKNOWLEDGEMENTS

The authors would like to thank NPC (Whetstone) Ltd for performing the LOCA clad temperature calculations.

## REFERENCES

1. ROGAN et al. Controlled porosity fuel. Proceedings of European Nuclear Conference, Paris. April, 1975.

2. RAVEN, L.F.A. The performance of BNFL controlled porosity nuclear fuel. Paper presented to the American Ceramic Society, 78th Annual Meeting, Cincinnati, Ohio, May, 1976 Paper 45-N-76.

3. RAVEN, L.F.A. A comparison of the HOTROD code predictions with P.I.E. data. A paper presented at American Nuclear Society meeting on water reactor fuel performance, Chicago, May, 1977.

4. ANDREWS et al. Light water reactor fuel rod modelling code evaluation - Phase II Topical report. CENPD - 218, April, 1976 (an EPRI study).

5. RAVEN, L.F.A. The impact of product reproducibility on the quality of nuclear fuel and its reliability in service. A paper presented to the American Ceramic Society 79th Annual Meeting April, 1977. (BNFL Memo. 831(S).

Table 1

Density changes in fuel irradiated in SGHWR

| Pin (Fuel) | Mean Rating (kw/m) | Time equivalent full power days | Burn-up MWD/MTM | Pre-irradiation density ($t/m^3$) | Post-irradiation density ($t/m^3$) | % change in density |
|---|---|---|---|---|---|---|
| 34 (Standard) | 10.5 | 685 | 4 760 | 10.49 | 10.55 | + 0.57 |
| | 23.0 | " | 12 920 | 10.49 | 10.52 | + 0.28 |
| | 34.4 | " | 17 000 | 10.57 | 10.39 | - 1.70 |
| | 36.7 | " | 18 500 | 10.57 | 10.38 | - 1.61 |
| 35 (CONPOR) | 10.5 | 685 | 4 760 | 10.23 | 10.25 | + 0.19 |
| | 23.0 | " | 12 920 | 10.26 | 10.27 | + 0.10 |
| | 34.4 | " | 17 000 | 10.23 | 10.21 | - 0.19 |
| | 36.7 | " | 18 500 | 10.23 | 10.18 | - 0.49 |

Table 2

Comparison of measured and observed cold gaps

| Axial location, mm | 304.8 | 508.0 | 711.2 |
|---|---|---|---|
| Measured cold diametral gap, mm | 0.089 | 0.071 | 0.071 |
| Predicted cold diametral gap, mm | 0.082 | 0.078 | 0.0861 |

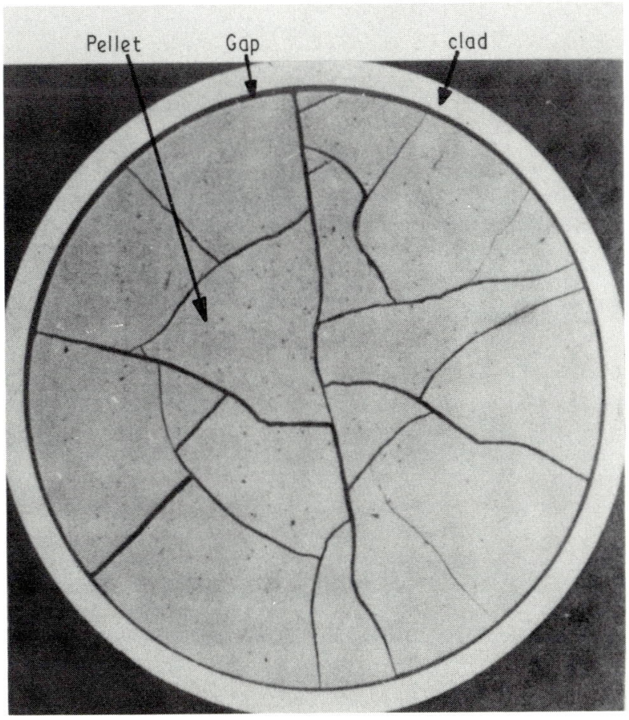


**Fig. 1: Cross section of fuel pin showing pellet clad gap and cracks ($\approx$ x 11, photographic reduction 56%)**

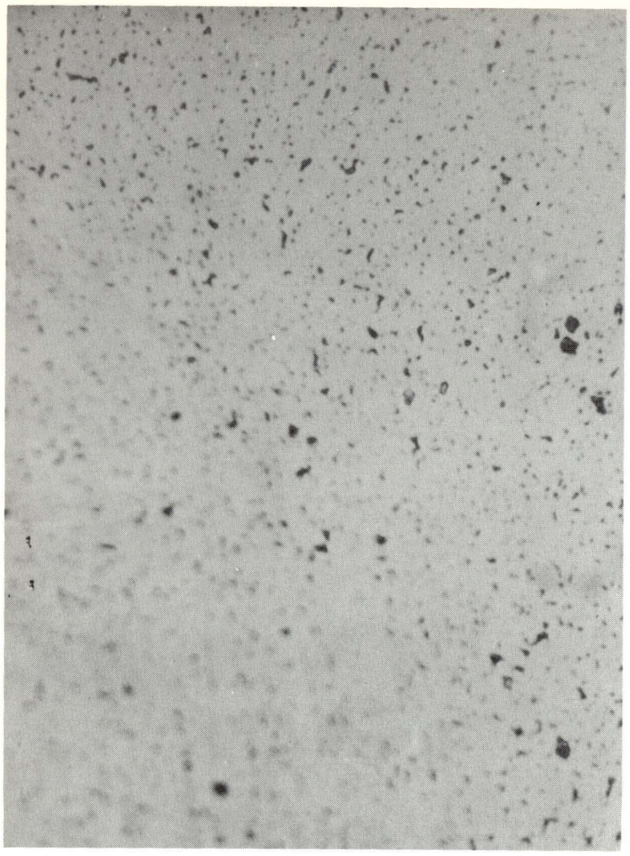

**(a) Standard fuel (≈ x 1 500, photographic reduction 56%)**

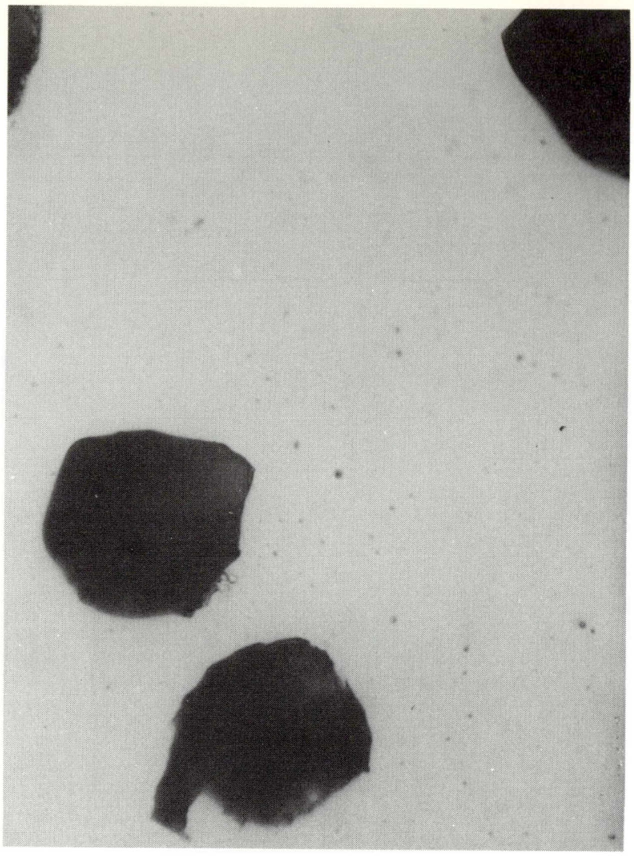

**(b) CONPOR fuel (≈ x 1 500, photographic reduction 56%)**

Fig. 2: Fuel porosity

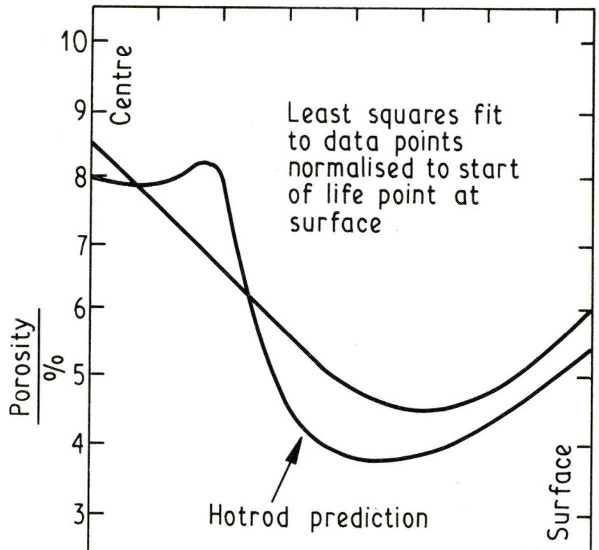


Fig. 3: Comparison of observed and predicted porosity in CONPOR fuel

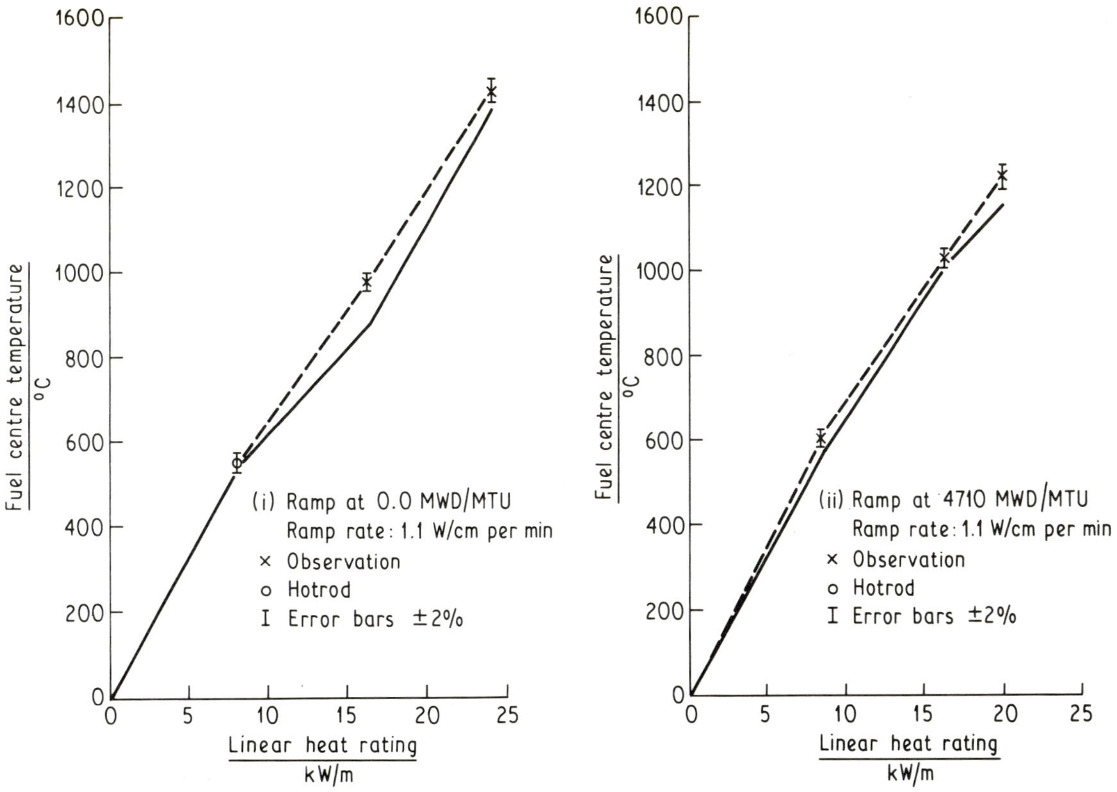


**Fig. 4(a): Predicted vs observed fuel temperatures (ramp conditions)**

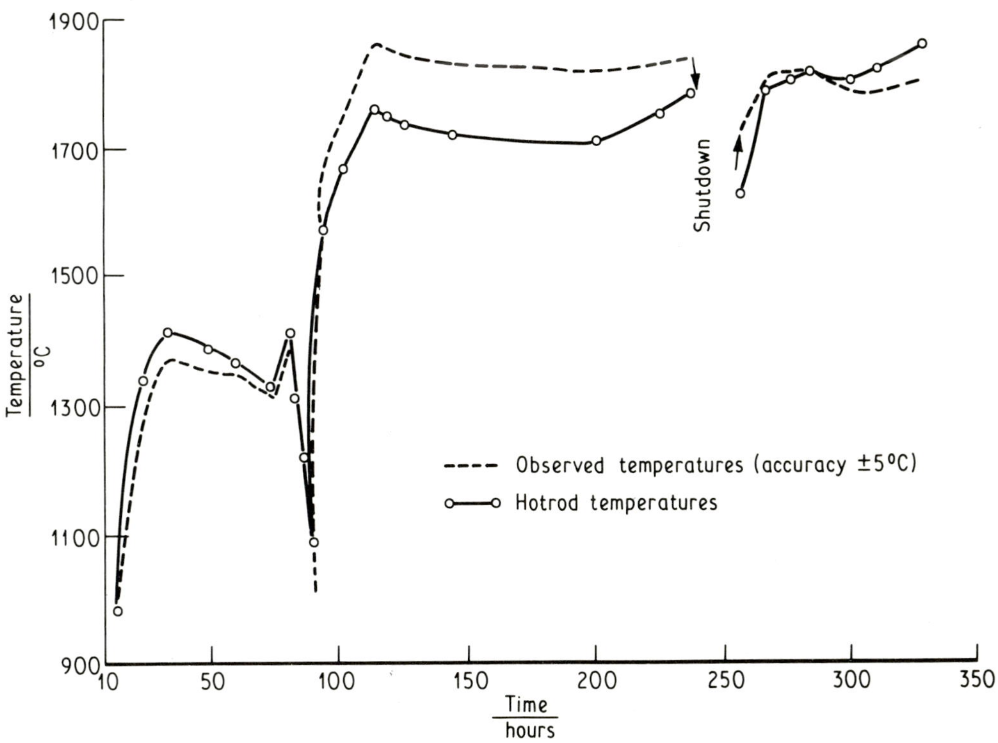


**Fig. 4(b): Time/temperature history for IFA 117 pins**

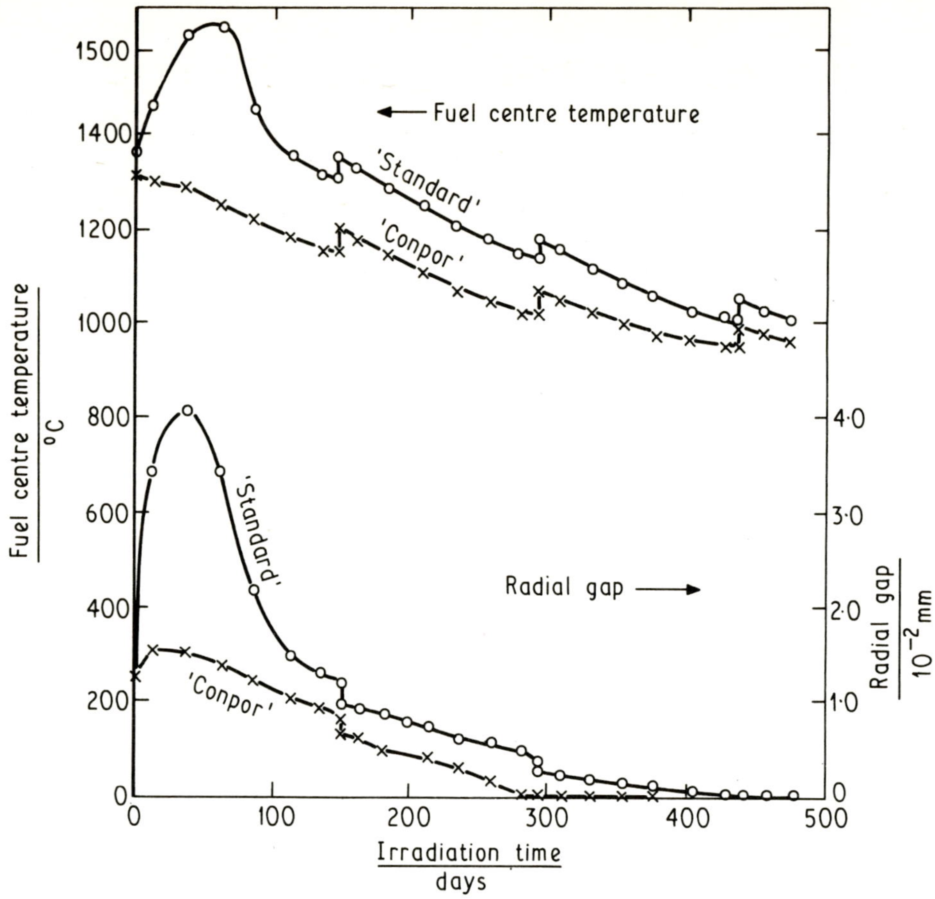


**Fig. 5: Comparison of fuel centre temperatures and radial gaps for standard and CONPOR fuels**

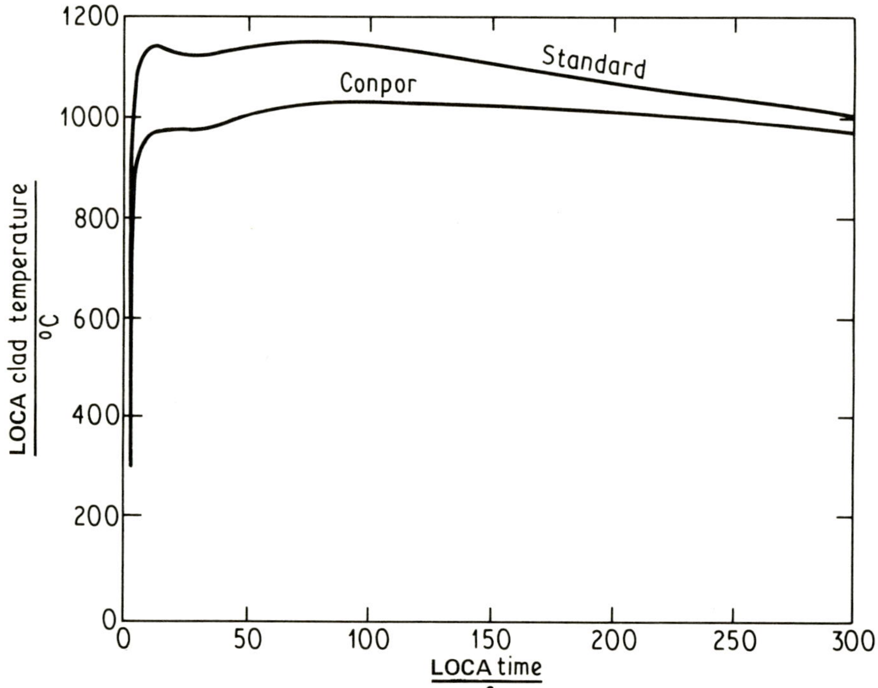


**Fig. 6: Comparison of LOCA clad temperatures for standard and CONPOR fuels**

C197/77

# ANALYTICAL APPROXIMATIONS TO THE REWETTING RATE OF OVERHEATED COMPOSITE FUEL PINS UNDER LOW PRESSURE, HIGH TEMPERATURE CONDITIONS

D. BLACKBURN, BSc, PhD, and K. G. PEARSON, BSc(Eng)
United Kingdom Atomic Energy Establishment, AEE Winfrith, Dorchester, Dorset

SYNOPSIS This paper presents analytic and finite difference results of the effect of stored heat in the fill material (eg $UO_2$) on the rewetting rate of overheated composite pins containing a gas-gap between the cladding and fill. Without a gas-gap a reduction of the rewetting rate by up to a factor of four compared with an empty tube is predicted in agreement with experimental results. The analytic method assumes that the conductivity of the filler is small compared to that of the clad, but plausible arguments are advanced for extrapolation to high conductivity filler, based on the observation that there is a broad minimum in the curve of re-wetting rate against the ratio of fill/clad thermal diffusivities.

NOTATION

| | |
|---|---|
| $B_i$ | Biot number, $= h\varepsilon/k$ |
| $c$ | specific heat |
| $\rho$ | density |
| $\alpha$ | thermal diffusivity |
| $h$ | heat transfer coefficient to liquid film |
| $h_g$ | gas-gap conductance |
| $a$ | pin radius |
| $\varepsilon$ | clad thickness |
| $T$ | temperature |
| $T_w$ | initial clad temperature |
| $T_0$ | sputtering temperature |
| $T_s$ | Coolant saturation temperature |
| $u_1$ | re-wetting speed of empty tube |
| $u_2$ | re-wetting speed of solid uniform rod |
| $x$ | axial co-ordinate |
| $r$ | radial co-ordinate in fill |
| $y$ | transverse co-ordinate in clad |
| $\beta$ | axial decay constant |
| $B_1^+$ | First coefficient in series expansion of axial temp distribution |
| $D_1$ | unknown coefficient |
| $q$ | $\left\{\beta\left(\frac{u}{\alpha_f}-\beta\right)\right\}^{1/2}$ |
| $\gamma$ | $\left\{\beta\left(\beta-\frac{u}{\alpha_c}\right)\right\}^{1/2}$ |
| $H$ | heat lost through water film with filler |
| $H_1$ | heat lost through water film with no filler |
| $\omega_1$ | inverse dimensionless re-wetting speed |
| $k$ | thermal conductivity |
| $\psi$ | dimensionless temperature |
| $\tilde{\alpha}$ | $\alpha_f/\alpha_c$ |
| $\tilde{\rho}$ | $(\rho c)_f/(\rho c)_c$ |
| $X$ | $u/u_1$ |
| $Z$ | $1-X$ |
| $\delta$ | gap parameter, |

SUBSCRIPTS

| | |
|---|---|
| $c$ | denotes clad |
| $f$ | denotes filler |
| $g$ | denotes gas-gap |
| $i$ | denotes interface |

INTRODUCTION

1. Most theoretical work on the rewetting of overheated fuel pins or electrical heater pins by a falling liquid film or by bottom reflooding has treated the pin as a single region eg as an annular region consisting of cladding alone (1), (2), (3), or as a solid rod (4), (5), this latter model being partially representative of high conductivity filler while the former approximates to very low conductivity filler.

2. Recent experiments with irradiated fuel and other composite heater pins (6), have shown a marked reduction in the re-wetting rate with even low conductivity fill material, at atmospheric pressure and with high surface temperatures, where re-wetting rates are of the order of mms/sec for empty tubes.

3. These results have been confirmed theoretically by Pearson (7) using a finite difference method which, however, is inappropriate for inclusion in safety codes such as RELAP, because of severe mesh restrictions. There is, therefore, an incentive to develop an analytical model (or a correlation) for the problem, which in its entirety would be a major task, so some approximation is desirable.

## PREVIOUS THEORIES

4. Pearson et al (8) made the first attempt at such a solution, using heat balance arguments involving the development of thermal boundary layers, but the results were not entirely satisfactory and moreover did not consider the effect of gas-gaps. An attempt was made to rectify the situation by Blackburn et al (9) using more realistic axial and radial temperature profiles to estimate the heat flux between filler and clad and taking into account a temperature jump across any gas-gap. Iterative solutions were required in general, but some simplified formulae were presented for 'high' and 'low' speed re-wetting, which compared favourably with the finite-difference results for low conductivity filler and typical gas-gaps.

5. More accurate finite-difference results, covering a wider range of fill/clad conductivities, have indicated shortcomings in the previous analytic approach which will be re-examined in the present paper and some alternative formulae presented.

## ANALYTIC MODELS

6. It will be assumed initially that perfect thermal contact exists between the fill and clad materials and that the common interface axial temperature distribution is given by:

$$T_w - T_i(x) = D_1(T_w - T_s)e^{-\beta x}, \quad x \geqslant 0 \tag{1}$$

where $T_w$, $T_s$ represent the initial (pre-quench) and final clad/fill temperatures. It may be noted that in the absence of any fill material (or for one with zero conductivity) $D_1 = B_1^+$ and $\beta = u_1/\alpha_c$, where $u_1$ is the re-wetting speed, $\alpha_c$ the clad thermal diffusivity and $B_1^+$ is a coefficient defined by Coney which can be simply calculated from the relation:

$$B_1^+ \simeq \omega_1^2 B_i/\psi^2 \tag{2}$$

using Coney's notation.

### Heat-balance approach

7. Let $r, x$ denote the radial and axial co-ordinates in the fill, $y, x$ in the clad, so that the clad outer and inner surfaces are at $y=0$ and $y=\varepsilon$ (or $r=a$) respectively, (where conventionally for thin cladding $\varepsilon/a \ll 1$, Cartesian co-ordinates have been used). With the usual assumption of no heat loss at $y=0$, the fill and clad temperature distributions appropriate to (1) are given by:

$$T_w - T(r,x) \simeq D_1(T_w - T_s)e^{-\beta x} I_0(qr)/I_0(qa), \; r \leqslant a \tag{3}$$

and

$$T_w - T(y,x) = D_1(T_w - T_s)e^{-\beta x}\frac{\cos\gamma y}{\cos\gamma\varepsilon}, \quad 0 \leqslant y \leqslant \varepsilon \tag{4}$$

where $q$, $\gamma$ are constants depending on $\beta$, $u$, $\alpha_c$ and $\alpha_f$ through the heat diffusion equation in the frame of reference moving with velocity $u$ of the quench front (assumed uniform) namely:

$$k_{c,f}\nabla^2 T = -(\rho c)_{c,f}\, u \frac{\partial T}{\partial x} \tag{5}$$

leading to

$$q^2 = \beta\left(\frac{u}{\alpha_f} - \beta\right), \quad \gamma^2 = \beta\left(\beta - \frac{u}{\alpha_c}\right) \tag{6}$$

8. If one now integrates equation (5) over the entire clad volume, for the 'filled' and 'empty' situations and subtracts the results, one obtains

$$(\rho c\varepsilon)_c(T_w - T_s)(u_1 - u) = \int_0^\infty k\frac{\partial T}{\partial r}\bigg|_{r=a} dx + \int_{-\infty}^0 k\frac{\partial T}{\partial r}\bigg|_{r=a} dx + H_1 - H \tag{7}$$

where $H, H_1$ denote the heat lost through the water-film in the 'filled' and 'empty' situations respectively. Since it would be extremely difficult to calculate even approximately the last three terms on the right hand side of (7), they will be neglected* on the plausible assumption that heat flowing into the clad from the fill on the wet-side, cannot easily diffuse 'upstream' to the dry-side. Thus (7) becomes approximately:

$$(\rho c\varepsilon)_c(T_w - T_s)(u_1 - u) \simeq \int_0^\infty k\frac{\partial T}{\partial r}\bigg|_{r=a} dx \tag{8}$$

9. The expression on the R.H.S. of (8) can be calculated using (3) and (4) to give

$$(\rho c\varepsilon)_c(u_1 - u) = k_f D_1 q\frac{I_1(qa)}{I_0(qa)}\int_0^\infty e^{-\beta x}dx \tag{9}$$

$$\simeq k_f D_1 q\frac{I_1(qa)}{\beta I_0(qa)} \tag{10}$$

and

$$(\rho c\varepsilon)_c(u_1 - u) = k_c D_1 \gamma\frac{\sin\gamma\varepsilon}{\cos\gamma\varepsilon}\int_0^\infty e^{-\beta x}dx \tag{11}$$

$$\simeq \varepsilon k_c B_1^+\left(\beta - \frac{u}{\alpha_c}\right) \tag{12}$$

---

* Finite difference estimates of the neglected terms indicate that they may contribute as much as 25-30 per cent of the first term in total.

using (6) and replacing $\tan\gamma\varepsilon/\gamma\varepsilon$ and $D_1$ by 1 and $B_1^+$ respectively, in the spirit of the perturbation approach. Solving (12) for $\beta$ we obtain

$$\beta\alpha_c = u_1 + \left(\frac{1}{B_1^+} - 1\right)(u_1 - u) \geqslant u_1 \qquad (13)$$

since $B_1^+ \leqslant 1$. We now see that the heat-balance approach implies that $\beta\alpha_c \simeq u_1$, since when $\omega_1 \leqslant 1$ (high speed), $B_1 \lesssim 2/3$ and $u_1 \sim u$; whilst when $\omega_1 >> 1$ (low speed), $B_1^+ \simeq 1$, $u << u_1$, both of which ensure that the product of the two factors on the R.H.S. of (13) is small compared with $u_1$. It was observed in Pearson's numerical work and used in his analytical model that $\beta \simeq u_1/\alpha_c$ for low conductivity filler, even though $u$ can be a factor of four or more lower than $u_1$. More recently, improvements in the finite difference solution have suggested that $\beta < u_1/\alpha_c$, at low speeds, which contradicts (13), implying errors in the heat-balance model.

10. The final equation for $u$, is now found from (10), ie

$$u = u_1 - \frac{k_f B_1^+}{(\rho c \varepsilon)_c}\left\{\frac{q I_1(qa)}{\beta I_0(qa)}\right\} \qquad (14)$$

with $\beta$ given by (13) or as a first approximation by $\beta = u_1/\alpha_c$. In general this requires an iterative solution, since $q$ and $\beta$ both depend on $u$, which is unknown, but closed form solutions of adequate accuracy may be obtained using the approximations:

$$\frac{I_1(qa)}{I_0(qa)} \simeq \frac{qa}{2.16 + 0.16\,q^2a^2}, \quad qa \leqslant 1 \qquad (15)$$

$$\simeq 1 - \frac{0.57}{qa}, \quad qa > 1 \qquad (16)$$

which reduce (14) to the solution of quadratic equations in $u/u_1$, when $\beta$ is replaced by $u_1/\alpha_c$

11. The range $qa > 1$ is appropriate to most values of $\omega_1$ of practical interest ($\lesssim 10$) with values of $\alpha_f/\alpha_c$, typical of $UO_2$ fill (or very low unrealistic values of $\alpha_f/\alpha_c$ and larger values of $\omega_1$); this follows from the definition of $q$ in (6) which implies that $qa$ is of order $\frac{a}{\varepsilon\omega_1}\left(\frac{\alpha_c}{\alpha_f}\right)^{\frac{1}{2}}$

Continuity condition method

12. An alternative to the approach outlined above, though perhaps less physically appealing, is to use the exact continuity of heat flux at $r = a$, namely:

$$k_f\, q\,\frac{I_1(qa)}{I_0(qa)} = k_c\,\gamma\,\frac{\sin\gamma\varepsilon}{\cos\gamma\varepsilon} \qquad (17)$$

in which $D_1$ does not appear; again replacing $\tan\gamma\varepsilon/\gamma\varepsilon$ by 1, (17) can be written in the form

$$u = \beta\alpha_c - \frac{k_f}{(\rho c\varepsilon)_c}\left\{\frac{q I_1(qa)}{\beta I_0(qa)}\right\} \qquad (18)$$

13. If we assume $\beta = u_1/\alpha_c$, as observed, (18) becomes

$$u \simeq u_1 - \frac{k_f}{(\rho c\varepsilon)_c}\left\{\frac{q I_1(qa)}{\beta I_0(qa)}\right\} \qquad (19)$$

which agrees with (14) when $B_1^+$ is replaced by unity, when (13) also becomes $\beta = u_1/\alpha_c$. Thus in view of the previous remarks about the correct value of $\beta$, there does not seem to be any particular merit in replacing $D_1$ by $B_1^+$, rather than by unity. Even so when $\alpha_f = \alpha_c$, (19) has the incorrect solution $u = u_1$, as before regardless of the value of $\beta$; the reason being that there is then no heat flux associated with the assumed axial profile and the further terms in the series must be considered.

Relationship with earlier models

14. Contact with Pearson's model can be made using the limiting case of $qa >> 1$ in (16) and using equation (19) of this work, when one obtains:

$$u \simeq u_1 - \frac{k_f q}{(\rho c\varepsilon)_c\,\beta} \qquad (20)$$

$$= u_1 - \frac{\alpha_c}{\varepsilon}\left\{\frac{\alpha_f}{\alpha_c}\left(\frac{u}{u_1} - \frac{\alpha_f}{\alpha_c}\right)\right\}^{\frac{1}{2}} \qquad (21)$$

which differs from Pearson's equation in that his factor $2/3\sqrt{\pi}$ is replaced by 1 and $(u/u_1)^{\frac{1}{2}}$ becomes $\left(\frac{u}{u_1} - \frac{\alpha_f}{\alpha_c}\right)^{\frac{1}{2}}$; arising respectively from the use of a linear rather than exponential heat flux profile and from the neglect of axial conduction.

15. If $u/u_1$ is denoted by $X$, (21) can be written in the form

$$X = 1 - \tilde{\beta}\omega_1\left\{\tilde{\alpha}(X - \tilde{\alpha})\right\}^{\frac{1}{2}} \qquad (22)$$

where $\tilde{\rho} = (\rho c)_f/(\rho c)_c$, $\tilde{\alpha} = \alpha_f/\alpha_c$ and $\omega_1 = \alpha_c/\varepsilon u_1$, which has the relevant solution

$$X = 1 + \tfrac{1}{2}\tilde{\alpha}\tilde{\rho}^2\omega_1^2 - \left\{(1+\tfrac{1}{2}\tilde{\alpha}\tilde{\rho}^2\omega_1^2)^2 - (1+\tilde{\alpha}^2\tilde{\rho}^2\omega_1^2)\right\}^{\frac{1}{2}} \qquad (23)$$

16. It is not difficult to show that $X$ has a minimum as a function of $\tilde{\alpha}$ when $X = 2\tilde{\alpha}$ and that $X_{min}$ is given by

$$X_{min} = (u/u_1)_{min} = \frac{1}{1+\frac{1}{2}\tilde{\rho}\omega_1} \qquad (24)$$

This contrasts with Pearson's equation in which $X$ decreases monotonically with increasing $\tilde{\alpha}$ as expected, though both forms exhibit an infinity in $\frac{dX}{d\tilde{\alpha}}$ at $\tilde{\alpha}=0$. The explanation of the minimum in the curve of $X$ versus $\tilde{\alpha}$ lies in the opposing effects of axial conduction in the fill and radial conduction to the cladding (at least in the model).

17. This was an unexpected result and prompted a series of finite-difference calculations to demonstrate whether the effect was a real one or not. The fill and cladding were assumed to have the same value of $\rho c$, so that when $\alpha_f = \alpha_c$, a result for a uniform solid cylinder was obtained as an additional check on purely geometrical effects on re-wetting rates. Fig 1 shows the finite-difference results and various analytic approximations for comparison.

18. It is interesting to note that $(u/u_1)_{min}$, given by (24) for this case, ( $\tilde{\rho}=1$, $\omega_1 = 3.2$ ) equals 0.385, very close to the value of 0.41 indicated in Fig 1. Since the analytic solution (23) breaks down as $\tilde{\alpha}\to 1$, it seems reasonable to use (23) for $\tilde{\alpha} \leq \frac{1}{2}X_{min}$, and given a suitable method of estimating $X_{\tilde{\alpha}=1}$, to fit a parabola between $\tilde{\alpha} = \frac{1}{2}X_{min}$ and $\tilde{\alpha}=1$, having a zero gradient at the minimum point. Such a method will be outlined in the next section.

## Suggested model for $\alpha_f \sim \alpha_c$

19. If $u_2$ denotes the re-wetting rate when fill and clad have identical properties then if the fill properties are changed slightly the new re-wetting rate $u$ is given by

$$u \simeq u_2 - \frac{k_f}{(\rho c\varepsilon)_c}\left\{\frac{q I_1(qa)}{\beta I_0(qa)}\right\} \qquad (25)$$

on similar arguments to the low $\alpha_f/\alpha_c$ approach and assuming $\beta \sim u_2/\alpha_c$, but since $qa$ is small in this case $I_1(qa)/I_0(qa) \sim qa/2$ by (15) and

$$\frac{u}{u_2} \simeq \frac{\tilde{\alpha}+2\delta^{-1}}{1+2\delta^{-1}} = \frac{\bar{k}}{\overline{\rho c}}\Big/\frac{k_c}{(\rho c)_c} \qquad (26)$$

using (6), where $\bar{k}$, $\overline{\rho c}$ are volume weighted averages and $\delta = (\rho c a)_f/(\rho c \varepsilon)_c$

20. Now the well known one-dimensional approximation for the re-wetting rate of a slab of thickness $\varepsilon$ is that $u \propto k/(\rho c\sqrt{\varepsilon k})$, whereas (26) contains no $K^{\frac{1}{2}}$ factor. It is conjectured that this is due to the use of the approximation $\beta \sim u_2/\alpha_c$ and that (26) should rather be

$$\frac{u}{u_2} \simeq \frac{(\rho c)_c}{\overline{\rho c}}\left(\frac{\bar{k}}{k_c}\right)^{\frac{1}{2}} \qquad (27)$$

if the one-dimensional approximation were strictly applied on a 'lumped-parameter' basis. At high Biot numbers and high surface temperatures of interest here, two-dimensional effects become important and (27) can be modified on the same 'lumped-parameter' basis to give

$$\frac{u}{u_2} \simeq \frac{(\rho c)_c}{\overline{\rho c}}\left\{\frac{\frac{\bar{k}}{k_c} + \frac{C}{\omega_{2,0}^2}}{1+\frac{C}{\omega_{2,0}^2}}\right\}^{\frac{1}{2}} \qquad (28)$$

where C is a Biot dependent correlating coefficient † fitted to Coney's tabulation given by

$$C = 0.2\left\{1 + \frac{2}{2+B_i} + \log_{10} B_i\right\} \qquad (29)$$

and $\omega_{2,0}$ is the 'one-dimensional' estimate of the inverse dimensionless re-wetting speed of the unperturbed rod based on an equivalent thickness of $\frac{1}{2}(a+\varepsilon)$.

21. Finally an expression for $u_2/u_1$ based on the same equivalent slab is

$$\frac{u_2}{u_1} = \left\{\frac{\frac{2\varepsilon}{a+\varepsilon} + \frac{C}{\omega_{1,0}^2}}{1+\frac{C}{\omega_{1,0}^2}}\right\}^{\frac{1}{2}} \qquad (30)$$

which for the case illustrated in Fig 1 is very close to 0.5, in good agreement with the finite-difference result of 0.49. ($\omega_{1,0}$ is defined in a similar manner to $\omega_{2,0}$ ).

## Effect of gas-gaps

22. If allowance is made for a temperature junp $\Delta T(x)$ across the gap, to connect the fill and clad temperature profiles and to preserve the heat flux as $h_g\Delta T(x)$, where $h_g$ is the gap-

† If $\omega = \omega_1$, represents the one-dimensional approximation given by $\omega_1^2 B_i = \psi(\psi-1)$, it has been shown that a good approximation for $\omega$ in general is given by: $\omega = \omega_1^2/(\omega_1^2+C)^{\frac{1}{2}}$

conductance, it can readily be shown that (14) is modified to:

$$u = u_1 - \frac{k_f B_1^+}{(\rho c \varepsilon)_c} \left\{ \frac{q I_1(qa)}{\beta I_0(qa)} \right\} \left[ 1 - \delta \left( 1 - \frac{u}{u_1} \right) \right] \quad (31)$$

where $\delta = k_c / \varepsilon h_g B_1^+ \omega_1^2$. For $\omega_1 \ll 1$, $\delta \gg 1$

and inspection of (31) shows that $\frac{u}{u_1} \sim 1 - \delta^{-1}$

23. Equation (31) requires iterative solution in general, but following Blackburn's previous approach and making further simplifications it can be shown that to a good approximation the solution of (31) can be expressed, when $\tilde{\alpha} \ll 1$ in the form:

$$Z \simeq \frac{1}{(\delta_{eff} + Z_0^{-1})} \quad (32)$$

where $Z$ denotes $(1 - \frac{u}{u_1})$, $Z_0$ the value of $Z$ when $\delta = 0$, $Z_1$ the value of $Z$ when $\delta_{eff} = \delta$ and $\delta_{eff}$ denotes $\delta(1 - Z_1)$. $Z_0$ may be evaluated as $(1 - X)$ from Equation (23), or conservatively as $(1 - X_{min})$ leading to $Z_1 \sim (\delta + 1 + 2\tilde{p}^{-1}\omega_1^{-1})^{-1}$. To be consistent with the earlier remarks $\delta$ should be evaluated with $B_1^+$ replaced by unity.

THE FINITE-DIFFERENCE METHOD

24. This was discussed in the earlier paper by Pearson et al so only the developments since then will be described. The boundary conditions at the extreme axial mesh points now take into account the exponential decay of the asymptotic axial temperature profiles, on both the wet and dry sides, by calculating, for an assumed re-wetting speed, these decay constants from the smallest roots of the transcendental equations resulting from the two-region radial temperature profiles and associated boundary conditions. These roots can be markedly different from those without filler material, which were used in the earlier finite difference solutions to determine the distance from the wet front where the temperature was considered asymptotic.

25. The new method has resulted in some downward revision of the re-wetting rates obtained previously particularly at low speeds. Good agreement has been found between the asymptotic analytic and numerical radial temperature profiles on the wet and dry side confirming the adequacy of the revised boundary condition. Fig 2 illustrates the geometry, co-ordinate system and equations used to derive the above boundary conditions expressed in non-dimensional form.

COMPARISON OF ANALYTIC AND FINITE-DIFFERENCE RESULTS

Cases with gas-gaps

26. The only results available to date, to serve as a check on the gas-gap treatment are those quoted in Table 2 of the paper by Pearson et al corresponding to Zircaloy clad/$UO_2$ filled pins. The effect of the new boundary conditions in the finite-difference method is not known at the present time for gaps but is expected to reduce the re-wetting rate somewhat. The columns headed A and A* refer respectively to the analytic values based on equation (32) using the predicted value of $Z_0$ and the value given by the original zero-gap finite difference result for consistency. The revised speeds are given in brackets, and the cases refer to a gap-conductivity of 0.0023 watts/cm°C.

Table 1

Finite-difference v analytic re-wetting rates (m/sec x $10^{-3}$)

| Gap width (mm) | Biot number | f-d | A | A* |
|---|---|---|---|---|
| ∞ | 100 | 11.75 | 11.75 | 11.75 |
| ∞ | 20 | 2.94 | 2.94 | 2.94 |
| 0.1 | 100 | 10.94(10.5) | 10.84 | 10.96 |
| 0.1 | 20 | 1.32(1.15) | 1.24 | 1.28 |
| 0.05 | 100 | 10.64(10.3) | 10.21 | 10.57 |
| 0.05 | 20 | 1.09(.95) | 1.00 | 1.05 |
| zero | 100 | 8.94(8.4) | 7.31 | 8.94 |
| zero | 20 | 0.85(.79) | 0.79 | 0.85 |

27. With the rather limited evidence available it is difficult to draw firm conclusions but one may say that the method appears promising, though further finite-difference or experimental results would be desirable.

Effect of clad temperature - no gas-gaps

28. Reliable finite-difference solutions are presented in the following Table again for $UO_2$/Zircaloy clad pins of slightly different dimensions, in which the physical properties have been averaged over appropriate temperature distributions though this is incidental to the comparison with the analytic methods.

29. Also included are values calculated from the iterative solution of the heat-balance method, equation (14) with the $B_1^+$ term included in $\beta$ and also values using an empirically adjusted value of $B_1^+$ i.e $B^+_{1,eff} = B_1^+/(1-\tilde{\alpha})^2$, to compensate for deficiencies in the heat-balance method, which tend to underestimate the total fill-to-clad heat flux.

Table 2

Finite-difference v analytic re-wetting rates (m/sec x $10^{-3}$)

| Clad Temperature °C | 1500 | 1000 | 700 | 500 |
|---|---|---|---|---|
| $u_1$ | 2.8 | 4.05 | 7.23 | 12.15 |
| $\omega_1$ | 5.5 | 3.15 | 1.84 | 1.17 |
| $\tilde{\alpha}$ | 0.112 | 0.176 | 0.187 | 0.202 |
| $u$ (equation 14) | 0.88 | 1.80 | 4.14 | 9.35 |
| " ( " $B^+ \to B^+_{1,eff}$) | 0.76 | 1.49 | 3.47 | 8.56 |
| " (equation 23) | 0.65 | 1.38 | 3.11 | 6.45 |
| " (equation 24) | 0.60 | 1.35 | 3.08 | 6.42 |
| " (f-d) | 0.74 | 1.44 | 3.30 | 8.12 |

30. The low-speed estimates of $u$ using equation (23) could easily be improved by taking the next term in the expansion of $I_1(qa)/I_0(qa)$ into account, since it merely contributes a constant additional term to the right-hand side of (21) which can still be reduced to the solution of a quadratic equation. The calculation of the minimum point does not lead, however, to simple formulae.

## CONCLUSIONS

31. Analytic methods of estimating the rewetting rate of composite rods, including gas-gaps, caused by stored heat in the filler material flowing into the cladding, assuming low conductivity filler have been presented. Results have been compared with more accurate calculations based on finite-difference methods and limitations of earlier formulations have been pointed out. Suggestions for extending the range of conductivity ratios of fill and clad have been made based on the unexpected result that there is a shallow minimum in the curve of rewetting rate with fill/clad conductivity.

32. A simple approximate formula for estimating this minimum re-wetting rate has been given which could be used as a conservative estimate for practical fuel rods. While the results are encouraging overall, further experimental and numerical work is desirable for both high conductivity fill and controlled gas-gaps.

## REFERENCES


1. YAMANOUCHI A. 'Effect of Core Spray-cooling in transient state after loss of coolant accident'. J.Nucl.Sci.Technol 5 (1968) 547.

2. DUFFEY R.B. and PORTHOUSE D.T.C. 'The Physics of Re-wetting in water reactor emergency core cooling'. Nuc.Eng.Des. 25 (1973) 379-394.

3. CONEY M.W.E. 'Calculations on the Re-wetting of Hot Surfaces'. Nucl.Eng.Des. (1974) 31(2) 234-245.

4. BLAIR J.M. 'An Analytical Solution to a Two-dimensional Model of the Re-wetting of a Hot Dry Rod'. Nucl.Eng.Des. 32 (1975).

5. YEH H.C. 'An analysis of rewetting of a nuclear fuel rod in water-reactor emergency core cooling'. Nucl.Eng.Des. 34 (1975) 317-372.

6. PIGGOT B.D.G and DUFFEY R.B. 'The Quenching of Irradiated Fuel Pins'. Nucl.Eng. Des. 32 (1975) 182-190.

7. PEARSON K.G. and LOVELESS J. 'Calculations on the Effect of Pellet Filling on the Re-wetting Rate of Overheated Fuel Pins'. AEEW - M1473 March 1977.

8. PEARSON K.G., PIGGOT B.D.G. and DUFFEY R.B. 'The Effect of Thermal Diffusion from Fuel Pellets on Re-wetting of Overheated Water Reactor Pin'. Nucl.Eng.Design (1977) 41.2

9. BLACKBURN D., PEARSON K.G. and SHIRES G.L. 'Theoretical Studies of the Influence of Filler Material Gas Gap and Cladding Material on Re-wetting Rate'. European Two-Phase Flow Group Meeting, 1976. Erlangen 31 May - 4 June 1976. AEEW - R 1100.

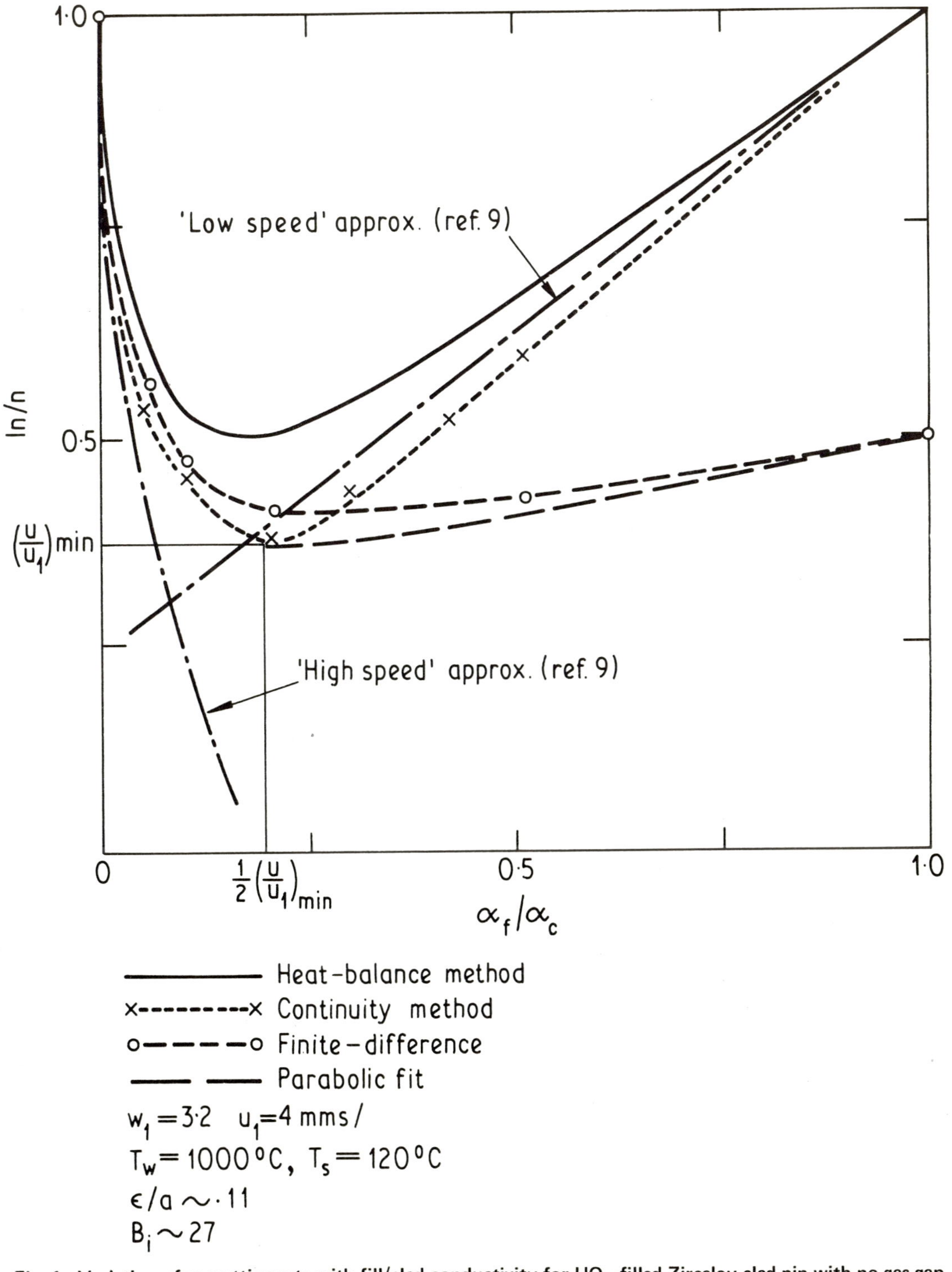


**Fig. 1: Variation of re-wetting rate with fill/clad conductivity for $UO_2$-filled Zircaloy clad pin with no gas-gap**

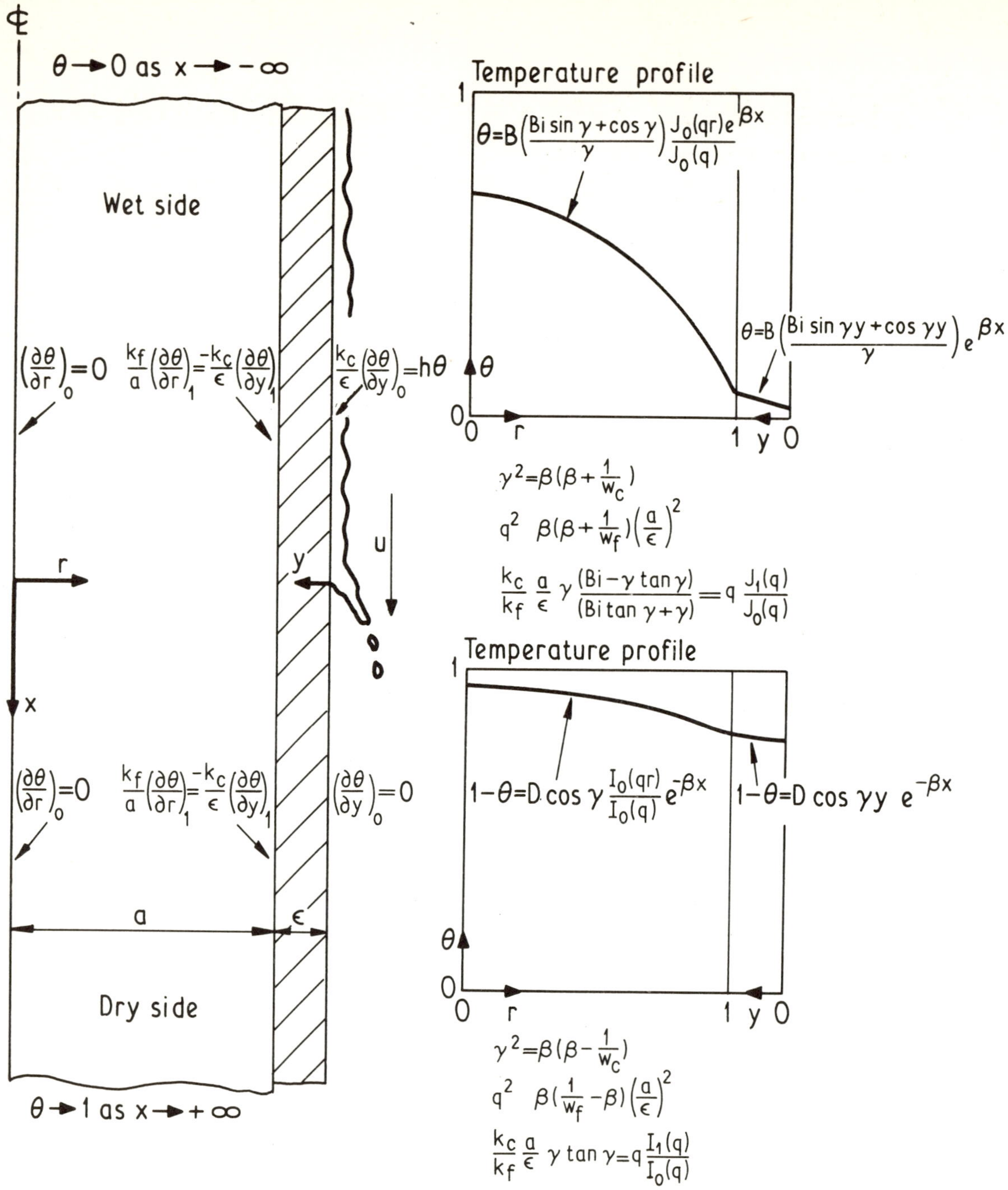


| | |
|---|---|
| Clad | $\dfrac{\partial^2\theta}{\partial x^2} + \dfrac{\partial^2\theta}{\partial y^2} + \dfrac{1}{w_c}\dfrac{\partial\theta}{\partial x} = 0$ |
| Fuel | $\dfrac{\partial^2\theta}{\partial x^2} + \left(\dfrac{\partial^2\theta}{\partial r^2} + \dfrac{1}{r}\dfrac{\partial\theta}{\partial r}\right)\left(\dfrac{\epsilon}{a}\right)^2 + \dfrac{1}{w_f}\dfrac{\partial\theta}{\partial x} = 0$ |
| Where | $w_c = \dfrac{k_c}{\rho_c c_c \epsilon u}$ $\quad w_f = \dfrac{k_f}{\rho_f c_f \epsilon u}$ $\quad B_i = \dfrac{h\epsilon}{k_c}$ |

**Fig. 2: Equations and temperature profiles**

C198/77

# BASIC STUDIES ON THE COOLING OF LWRs BY IN-CORE SPRAYS

S. A. FAIRBAIRN, BSc, PhD, P. C. HALL, BSc, B. D. G. PIGGOTT, BSc, MSc,
T. HEALEY, BSc, PhD, and R. B. DUFFEY, BSc, PhD
General Electricity Generating Board, Berkeley Nuclear Laboratories, Berkeley, Glos, GL 13 9PB

SYNOPSIS This paper reviews BNL studies of the modelling of the spray cooling of fuel pins during a LOCA. In particular, analyses of convective heat removal processes, due to droplet sprays, and of quench front propagation are discussed. The effect on these analyses of fuel pin bundle geometry and distortion is examined. The objective is to help provide the physical models and understanding needed to support the extrapolation of rig data to reactor conditions.

NOTATION

| | |
|---|---|
| $A_b$ | Blockage Area. |
| $A_s$ | Area of Undistorted Sub-Channels. |
| $A_{TOT}$ | Total Fuel Channel Area. |
| B | Numerical Constant in Equation (4) |
| Bi | Biot Number (Heat Transfer Coefficient $\times \delta/K_c$). |
| c | Specific Heat. |
| $d_e$ | Equivalent Diameter. |
| f | Friction Factor. |
| $f_B$ | Pressure Loss Coefficient for Blocked Region. |
| $f_s$ | Pressure Loss Coefficient for Undistorted Region. |
| $h_B/h_{B\infty}$ | Ratio of Blocked Sub-Channel Heat Transfer Coefficient in Confined Bundle to that for a very Large Core. |
| $h_u$ | Heat Transfer Coefficient in Undistorted Sub-Channels. |
| j | Counterflow Velocity. |
| K | Thermal Conductivity. |
| L | Latent Heat. |
| $\dot{m}$ | Mass Flux. |
| n | Creep Rate Sensitivity. |
| Q | Creep Activation Energy. |
| R | Gas Constant. |
| T | Temperature. |
| $T^*$ | $(T_o - T_s)/(T_w - T_o)$. |
| $u^*$ | Dimensionless Rewetting Rate = $\left(\rho \frac{c}{K}\right)_c \delta u$ |
| $\Gamma$ | Film Flow per Unit Perimeter. |
| $\gamma_t$ | Diametral Expansion at Time t, including Thermal Expansion. |
| $\delta$ | Clad Thickness. |
| $\mu$ | Viscosity. |
| $\rho$ | Density. |
| $\Psi$ | $(T^* + 1)/T^*$. |
| $\sigma_t$ | Instantaneous Hoop Stress referred to Undeformed Clad Dimensions. |

SUBSCRIPTS

| | |
|---|---|
| c | Clad. |
| CF | Counterflow. |
| f | Fuel. |
| l | Liquid. |
| s | Cooling Water. |
| st | Steam. |

INTRODUCTION

1. With numerous large scale experimental facilities in operation for studying Emergency Core Cooling (E.C.C.) performance, the most pressing question is how data from these rigs will relate to reactor conditions. There is, therefore, a strong case for developing physical models to enable extrapolations from experimental to reactor conditions to be made within a framework of basic physical understanding. The work reported in this paper describes the construction of such physical models for in-core spray systems from experience gained in small scale experiments at Berkeley Nuclear Laboratories (B.N.L.).

MECHANISMS OF PRE-QUENCH COOLING

2. One function of in-core sprays (SGHWR, BWR) is to provide sufficient pre-quench heat transfer to avoid unacceptable clad temperatures. Fundamental data on spray cooling of surfaces above the 'Leidenfrost temperature' are scarce (Refs. 1-5), so that reactor models are necessarily crude.

3. However, the work of Hall at B.N.L., has now identified an important pre-quench cooling mechanism (Ref. 6). From an expression for the steam pressure generated by evaporation beneath a sessile Leidenfrost drop (Ref.7), a single droplet equation of motion was obtained. This enabled Hall (Ref.6) to estimate the integrated heat removal by conduction across the steam gap between the droplet and the hot surface. The predicted evaporation was < 1% of the droplet in agreement with previous experiments (Refs. 1,2) on droplet cooling. However, to account for the observations in Ref. (6) of heat removal from a hot ($\sim 700^oC$) steel strip bombarded by aerosol droplets, droplet evaporation of 20 - 200% would be required. Hall postulated that convection to

the air entrained by the spray is the dominant cooling mechanism and produced a correlation based on this hypothesis.

4. The applicability of this theory can be demonstrated using the heat transfer data of Eriksson (Ref.8) obtained by spray water injection into an annulus, of area A, containing a heated pin. Eriksson gives no details of droplet velocity so an initial value, $V_0$, of 5 $ms^{-1}$ has been assumed. The velocity, V, of the entrained steam, at the heater surface of length, $\ell$, has been derived from a simple momentum balance, viz:-

$$\dot{m}_\ell(V_o - V) = \dot{m}_{st}V + 2A\rho_{st}V^2 f(\ell/d_e) \qquad (1)$$

The calculated convective heat transfer to steam at velocity V for various pressures (see Fig.1) reproduces the trend of Eriksson's data. This is significant because the trend is relatively insensitive to the assumed spray characteristics. Similar calculations done on unpublished SGHWR data have also had reasonable success.

## CLAD DEFORMATION AND ASSOCIATED THERMAL TRANSIENTS

5. The previous section has considered spray cooling in the context of undeformed fuel. However, the thermal and pressure transients accompanying a postulated LOCA may give rise to fuel tube ballooning if the conditions are sufficiently arduous. Experiments, both at BNL and elsewhere, have confirmed that both the diametral expansion and axial extent of the balloons are sensitively dependent on the detailed pressure and thermal history experienced by zircaloy tubes (Refs.9, 10, 11). In view of the variety of postulated LOCA transients, we have given attention to the development of analytical procedures for calculating both clad deformation (Refs.12,13,14) and the associated thermal behaviour of distored pins in fuel bundles (Ref.15).

6. Our latest model for predicting clad deformation, which has been tested against a wide range of experimental data (Ref.12), assumed that the principal contribution to ballooning arises from thermally activated creep. Using appropriate mechanical property data for α and β phase zircaloy (Refs.16,17), numerical solution of the equation

$$(2n - 1)^{-1}(1 - \{1 + \gamma_t\}^{1 - 2n}) =$$
$$= \sum B\, \sigma_t^n \exp(-Q/RT)\Delta t \qquad (2)$$

enables the occurrence of ductile rupture in zircaloy tubes to be calculated for any specified thermal and pressure history. An internally-pressurised zircaloy tube, subject to the notional external pressure and heating transients shown in Fig.2 is predicted to balloon and rupture after 100 and 200 seconds for internal pressures of 3.0 and 0.9 $MNm^{-2}$, respectively.

7. The subsequent temperature history of ballooned tubes contained in a fuel element assembly may be evaluated by interpolating between the respective thermal transients for undistorted and adiabatically heated fuel (Ref.15). The ratio of the respective heat transfer coefficients is derived from a flow by-pass model. For reactors using confined fuel assemblies (SGHWR, BWR and CANDU) a two stage calculation is required (Ref.15) in which local flow redistribution and then channel-to-channel by-pass are taken into account. This approach is best illustrated by considering the ratio of heat transfer in confined and infinite array assemblies. We have extended the analysis of Ref.15 to give the following result:-

$$\frac{h_B}{h_{B\infty}} = \left[ \left\{ \left[1 + \left(\frac{f_s}{f_B}\right)^{\frac{1}{2}} \frac{A_B}{A_s}\right] \frac{A_s}{A_{TOT}} \right\}_{local} \times \left\{ \frac{f_B}{f_s} \right\}^{\frac{1}{2}}_{core} \right]^{-m} \qquad (3)$$

The 'local' term allows for the restraining effect of an assembly boundary on flow by-pass close to the blockage. The 'core' term takes account of flow by-pass through parallel channels, but has little effect until the flow resistance of the blockage is comparable with the by-pass resistance. Details of recirculation and turbulence caused by the blockage are assumed the same in confined and infinite array assemblies, and do not, therefore, appear explicitly in (3). Effects of radiation to the pressure tube or channel box are not considered, since these are assumed to be relatively unimportant for fuel pins in the centre of a bundle. The present model, based on axial flow considerations, should be equally applicable to fuel bundles undergoing spray cooling, since entrainment by the spray and steam generation within the assembly will give rise to significant axial convection.

8. High temperatures can be expected downstream of small regions of gross distortion because they promote extensive by-pass of the coolant flow. Having estimated when ballooning can occur (at 100 or 200 seconds into the transients of Fig.2), we have calculated the resulting thermal transients in the blocked region for two assumed levels of heat transfer reduction (see Fig.2). The reductions of 25% and 75% correspond to 70% and 90% blockage in 5% of the fuel element sub-channels, conditions chosen to maximise the by-pass flow. Significantly increased temperatures can thus be predicted for not unrealistic estimates of local blockage due to clad distortion.

## DISTRIBUTION OF SPRAY WATER

9. The cooling rates experienced by individual fuel pins in a bundle depend on the spray water distribution amongst them. Distribution measurements are extremely difficult in large rigs and so our experiments have been performed with small clusters of heater pins to facilitate visual observations (Ref.18). These experiments have used a single jet normal to vertical pins in open air to provide simple tests of an SGHWR spray cooling configuration, and give information of general interest.

### Single Pin Experiments

10. Even this most basic of experiments is difficult to analyse but useful qualitative information was obtained. Typical results for an unheated pin are shown in Fig.3a-c. There is a critical flow rate above which water even at normal impact detaches from the pin and at which flow down the pin is proportional to the cosine of the impact angle, θ, (see Fig.3d,f). On oblique impact water left the pin at the rear as a flat horizontal water sheet or spray (see Fig. 3e, f for angular spread). At high velocities, jet disintegration at the point of impact generated

fine conical sprays.

Pin Bundle Experiments

11. Tests with a cold six-pin sector of a three-ringed 36 pin bundle showed that the distribution from a single jet was non-uniform and sensitive to both jet orientation and velocity. There was a trend to improve distribution as jet velocity was increased because a greater fraction of the jet flow detached from the pins hit directly, and was available to other pins in the bundle. Distribution was found to be unaffected by spacer grids.

12. The distribution in a hot (700$^{o}$C) cluster was inferred from the measured quench time of the individual pins. The flow rate was estimated from a calibration curve of quench rate against flow for a single pin. The sensitivity to jet orientation meant reproducibility of the cluster tests was about ± 25%, but within these limits the hot and cold distributions are similar. This is because once the pins rewetted, near the jet a 'cold' flow condition was set up in ∿5 secs. Patchy intermittent wetting was observed on some pins presumably due to intermittent water delivery as differential thermal contraction distorted the bundle. However, if wetting delays occurred (Ref.19) the jet 'bounced' off hot pins on a steam cushion and the distribution was markedly different from that after quenching was established. Spacer grids were observed not to affect the passage of established quench fronts. However, when aerosol droplets, like those produced in jet disintegration and quenching, were directed at a spacer grid new quench fronts progressed from it. Once established quenching proceeds at a rate largely determined by thermal conduction in the pins (Ref.20), so this is examined next.

THERMAL DIFFUSION IN FUEL PINS DURING QUENCHING

13. Early work at BNL, and elsewhere, has concentrated on essentially the same model for the quenching of metal plates and cylinders. The rewetting front has been assumed to be circumferentially symmetric, at a well defined temperature, $T_O$, and moving at a constant rate, u. Heat transfer to the coolant has been described by a suitably averaged heat transfer coefficient for the wetted regions (Refs.21,22) and taken as zero in the hot dry regions. Many and various computer (Refs.23,24,25) and analytic (Refs.20,22,26,27,28) solutions for the temperature field have now been obtained. To compare these analyses, the non-dimensionalised rewetting rates, u*, are plotted against the Biot number, Bi, as shown in Fig. 4 for different initial values of the pre-quench metal temperature, $T_W$, as incorporated into the temperature group T*. For small Bi, all the solutions agree rather well but for large Bi, where two-dimensional temperature gradients become important and analysis is more difficult, the agreement is not so good. Simple analytical solutions derived for the limits of high and low Bi (Refs.26,27) follow the trends of the computer solutions. More general analytical solutions have been derived, in a completely different context, by Horvay (Ref.29).

14. Although the work quoted above can be used to correlate rewetting data (e.g. Ref.21) the problem to be solved for reactor safety calculations must include the well identified (Ref.30) but often neglected effects of heat release from the fuel pellets. As an initial attempt, we derived correction factors for the rewetting rates obtained from the above analyses using approximate heat balances for fuel and clad within approximate thermal boundary layers (Ref. 31). Rewetting times for fuel and clad together have been predicted up to three times longer than for clad alone, in approximate agreement with experiment and finite difference calculations (Refs.30,31). As a further development a perturbation theory has been presented (Ref.32) utilising the first term of an expansion for the temperature profile.

15. An alternative and more exact approach is to specify a functional form for the heat flux to the cooling water (Ref.33). We have found that the parameters in the heat flux function must be chosen carefully to avoid unphysical temperature profiles. Despite criticism (Ref.34), this method provides an exact analytical result, including pre-quench cooling, and can be extended to handle the effects of fuel and a gas gap (Ref.27). The expressions so derived give the same results as Ref.33 in the limit where the 'fuel' properties are the same as those of the clad.

EFFECTS OF COUNTER FLOW ON QUENCHING

16. It is unrealistic to suppose that there will be no disturbance to a water film other than disruption by the boiling in a narrow region (∿ 1 mm) at the quench front. Steam generated in this way, or by reactor depressurisation in a LOCA, can generate waves on the film and cause 'flooding' (Ref.35). This is an important effect in downcomer penetration studies and BWR top spray injection (Ref.37,38).

Counterflow Experiments

17. The effects of air, steam and two-phase counterflows on falling film quenching were investigated by experiments (Ref.36) using a vertically mounted stainless steel clad heater (14.2 mm o.d.) set in a silica tube (19.5 mm bore) which was open at the top. The film was supplied by water jetted onto the heater at its mid point. At zero counterflow the heater was quenched only by a falling film at a rate reducing as counterflow velocity was increased. At higher velocities, waves formed at the quench front nearly always bridged the annulus to form water plugs which were accelerated up to feed a climbing film, and then expelled from the top of the annulus. Only a climbing film was observed at still higher counterflows, when plugging was superceded by an annular flow regime (see Fig.5). At the same counterflow momenta, the same phenomena were observed for air and steam, except for condensation effects on sub-cooled films. For high wall temperatures and certain counterflows a wet patch was formed which did not spread. When the counterflow was a steam/water mixture produced by water filling the bottom of the silica tube, both falling and climbing films were always formed.

18. The result of doubling the annular flow area was to remove the region of co-existence of climbing and falling films. When the heater was mounted eccentrically in the annulus, the counterflow velocity needed to stop the falling film was increased by a factor of three.

Interfacial Shear Analysis

19. If the film flow is treated as laminar, then a simple force balance on an element of water in the film gives

$$u^*_{CF} = u^* \left[1 - 0.25 \left\{ \frac{g^2 \rho_\ell \mu_\ell \Gamma_o}{9} \right\}^{-\frac{1}{3}} f \rho_{CF} j^2 \right] \quad (4)$$

where the film flow rate has been related to the quenching rate, for sub-cooled water, by the empirical relation (Ref.20)

$$u^* \propto \Gamma \quad (5)$$

and the wave dynamics appear only as a friction factor. If the steam counterflow is estimated as the sum of the input flow and the net steam generated in quenching, then the steam velocity needed to stop a falling film is seen to be close to that required when the heater is cold (Ref.36). Equation (4) provides a reasonable correlation of results from downcomer penetration studies (Ref.37). We find greater difficulty in treating pin bundle experiments both because of the uneveness of the water distribution and its sensitivity to the more complex geometry.

## CONCLUSIONS

20. Although spray cooling poses a complex three-dimensional problem, the present work illustrates the capacity of simplified analytical approaches to improve our understanding.

21. Our experiments have demonstrated the importance of convection to steam entrained by spray water droplets, and illustrated the sensitivity of water distribution to jet orientation and pin bundle geometry.

22. The higher temperatures predicted as a consequence of clad ballooning during a LOCA indicate the importance of flow diversion on heat transfer in the region of local blockages. To test these calculations for spray cooling, heat transfer experiments with a partially blocked pin bundle are planned for future work at BNL.

23. Having shown the significant effects of fuel and a gas gap, we are now incorporating them into the fuel pin quenching analysis. This will improve predictions of rewetting rates.

24. Counterflow experiments have shown how wave growth in thin films can alter the water distribution and quenching behaviour of single pins, especially at the high temperature of interest.

## ACKNOWLEDGEMENT

This paper is published by permission of the Central Electricity Generating Board.

## REFERENCES

1. Wachters, L.H.J. and Westerling, N.A.J. 'The Heat Transfer from a Hot Wall to Impinging Water Drops in the Spheroidal State', Chem. Eng. Sci., 1966, 21, 1047-1056.

2. Wachters, L.H.J., Smulders, L., Vermulen,J.R. and Kleiweg, H.C. 'The Heat Transfer from a Hot Wall to Impinging Mist Droplets in the Spheroidal State' Chem. Eng. Sci., 1966, 21, 1231-1238.

3. Pedersen, C.O. 'An Experimental Study of the Dynamic Behaviour and Heat Transfer Characteristics of Water Droplets Impinging upon a Heated Surface', Int. J. Heat Mass Transfer, 1970, 13, 369-381.

4. Toda, S. 'A Study of Mist Cooling - Thermal Behaviours of Liquid Films Formed from Mist Drops on a Heated Surface at High Temperatures and High Heat Fluxes', Tech. Reports of Tohoku Univ., 1971, 36, No. 2, 299-350.

5. Gaugler, R.E. 'Experimental Investigation of Spray Cooling of High Temperature Surces', Ph.D. Thesis, Carnegie Institute of Technology, Pittsburg, 1966.

6. Hall, P.C. 'The Cooling of Hot Surfaces by Water Sprays', CEGB Report, RD/B/N3361, 1975.

7. Gottfried, B.S. and Bell, K.J. 'Film Boiling of Spheroidal Droplets', Ind. and Eng. Chem. Fundamentals, 1966, 5, 561-568.

8. Eriksson, S.O. 'DSF - P1: An Emergency Core Cooling Experiment for BWRs', Trans. European Nuclear Conf., Paris, 1975, 494-496.

9. Emmerich, K.M., Juenke, E.F. and White, J.F. 'Failure of Pressurised Zircaloy Tubes during Excursions in Steam and in Inert Atmospheres', Applications Related Phenomena in Zirconium and its Alloys. American Society for Testing and Materials, Philadelphia, 1969, STP 458, 252-268.

10. Hobson, D.O., Osborne, M.F. and Parker, G.W. 'Comparison of Rupture Data from Irradiated Fuel Rods and Irradiated Cladding', Nucl. Technol., 1971, 11, 479-490.

11. Hardy, D.G. 'High Temperature Expansion and Rupture Behaviour of Zircaloy Tubing',ANS Topical Meeting on Water Reactor Safety, Salt Lake City, Utah, 1973, 254-272.

12. Healey, T., Evans, H.E. and Duffey, R.B. 'Deformation and Rupture of Pressurised and Transiently Heated Zircaloy Tube', J. Br. Nucl. Energy Soc., 1976, 15, 247-251.

13. Clay, B.D., Healey, T. and Redding, G. 'The Deformation and Rupture of Zircaloy-2 Tubes during Transient Heating', presented at CSNI Specialist Meeting, Spåtind, Norway, 13-16 Sept. 1976 and CEGB Report RD/B/N3512, 1976.

14. Jones, P.M., Gittus, J.H. and Hindle, E.D. 'CANSWEL: A Computer Model of Clad Behaviour during a Loss-of-Coolant Accident', presented at CSNI Specialists Meeting, Spåtind, Norway, 13-16 Sept. 1976.

15. Hall, P.C. and Duffey, R.B. 'A Method of Calculating the Effect of Clad Ballooning on Loss-of-Coolant Accident Temperature Transients', Nucl. Sci. Eng., 1975, 58, 1-20.

16. Clay, B.D. and Redding, G. 'Creep Rupture Properties of Alpha Phase Zircaloy-2 Cladding Relevant to the Loss-of-Coolant Accident', J. Br. Nucl. Energy Soc., 1976, 15, 252-256.

17. Clendenning, W.R. 'Primary and Secondary Creep Properties of Zircaloy Cladding at Elevated Temperatures of Interest in Accident Analyses', Trans. of 3rd. Int. Conf. on Structural Mechanics in Reactor Technology, London, 1975. Commission

of the European Communities, DG XIII, Brussels, 1975, Division C, paper 2/6.

18. Piggott, B.D.G. 'An Experimental Study of Water Distribution from a Jet to a Single Pin and Pin Bundles', CEGB Report RD/B/N3432, 1976.

19. Piggott, B.D.G., White, E.P. and Duffey, R.B. 'Wetting Delay due to Film and Transition Boiling on Hot Surfaces', Nucl. Eng. Des., 1976, 36, 169-181.

20. Duffey, R.B. and Porthouse, D.T.C. 'The Physics of Rewetting in Water Reactor Emergency Cooling', Nucl. Eng. Des., 1973, 25, 379-394.

21. Piggott, B.D.G. and Porthouse, D.T.C. 'A Correlation of Rewetting Data', Nucl. Eng. Des., 1975, 32, 171-181.

22. Yamanouchi, A. 'Effect of Core Spray Cooling in Transient State after Loss-of-Coolant Accident', J. Nucl. Sci., Tech., 5, 547-558.

23. Coney, M.W.E. 'Calculations on the Rewetting of Hot Surfaces', Nucl. Eng. Des., 1974, 31, 246-259.

24. Andersen, J.G.M. and Hansen, P. 'Two-dimensional Heat Conduction in Rewetting Phenomena', Report No. NORHAV-D-6 Danish Atomic Energy Commission Research Establishment, Risø, Denmark, 1974.

25. Yeh, H.C. 'An Analysis of Rewetting of a Nuclear Fuel Rod in Water Reactor Emergency Core Cooling', Nucl. Eng., Des., 1975, 34, 317-322.

26. Tien, C.L. and Yao, L.S. 'Analysis of Conduction Controlled Rewetting of a Vertical Surface', J. Heat Transfer Trans. ASME, 1975, 97, 161-165.

27. Fairbairn, S.A. 'The Conduction Equation in Quenching Analyses', to be published.

28. Blair, J.M. 'An Analytical Solution to a Two-dimensional Model of the Rewetting of a Hot Dry Rod', Nucl. Eng., Des., 1975, 32, 159-170.

29. Horvay, G. 'The Effect of Discontinuous Biot Number on the Temperature Distribution in Moving Slabs', Proc. Int. Heat Transfer Conf., Boulder, Colo., 1961, 7-17.

30. Piggott, B.D.G. and Duffey, R.B. 'The Quenching of Irradiated Fuel Pins', Nucl. Eng. Des., 1975, 32, 182-190.

31. Pearson, K.G., Piggott, B.D.G. and Duffey, R.B. 'The Effect of Thermal Diffusion from Fuel Pellets on Rewetting of Overheated Water Reactor Pins', Nucl. Eng. Des., to be published.

32. Blackburn, D., Pearson, K.G. and Shires, G.L. 'Theroetical Studies of the Influence of Filler Material Gas Gap and Cladding Material on Rewetting Rate', European Two-Phase Flow Group Meeting, Erlangen, 31 May-4 June, 1976, Paper A4.

33. Edwards, A.R. and Mather, D.J. 'Some UK Studies Related to the Loss-of-Coolant Accident', Topical Meeting on Water Reactor Safety, USAEC Conf. -730304, Mar. 26-28, 1973, 720-739.

34. Sun, K.H., Dix, G.E. and Tien, C.L. 'Effect of Precursory Cooling on Falling-Film Rewetting', J. Heat Transfer, Trans. ASME, 1975, 97, 360-365.

35. Hewitt, G.F. and Hall-Taylor, N.S. Annular Two Phase Flow, 1st edition 1970 (Pergamon Press, Oxford).

36. Duffey, R.B., Ackerman, M.C., Piggott, B.D.G. and Fairbairn, S.A. 'The Effects of Countercurrent Single and Two Phase Flows on the Quenching Rate of Hot Surfaces', Int. J. Multiphase Flow, 1977, to be published.

37. Block, J.A. "CREARE Downcomer Penetration Program', Presentation to 3rd Water Reactor Safety Information Meeting, Washington, 1975.

38. Griebe, R.W. and McConnell, J.W. 'BWR-FLECHT Tests Project. Final Report on Atmospheric Pressure Stainless Steel Experiments', Aerojet Nuclear Co. Report, ANCR - 1115, 1973.

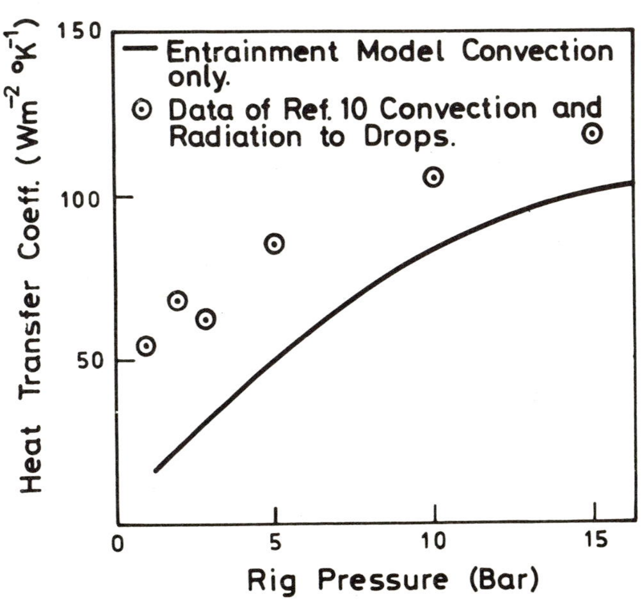


Fig. 1: Comparison of entrainment model with data of Ref 10

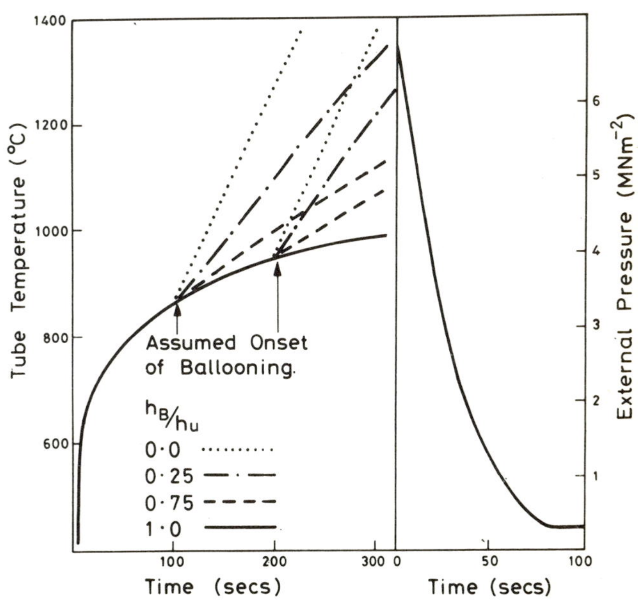


Fig. 2: Notional reactor transients and the effect of ballooning

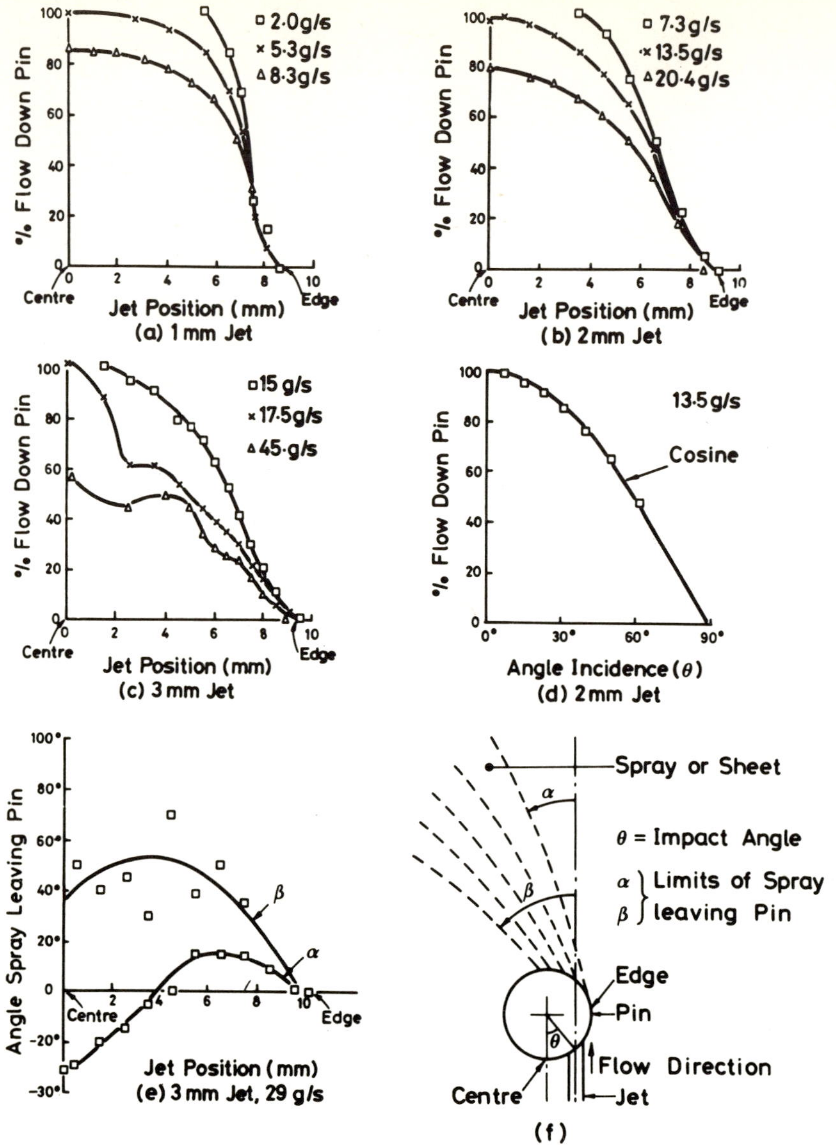


Fig. 3: Water distribution from a horizontal jet to a cold 15.9 mm O.D. pin

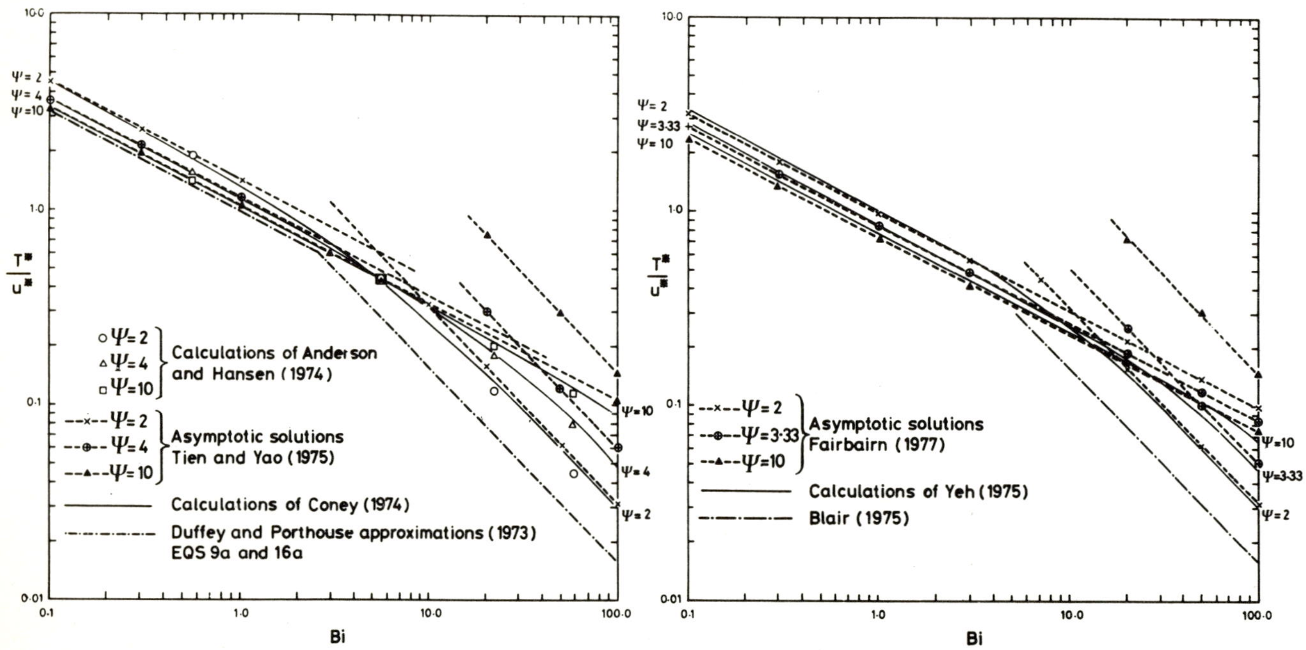


(a) Flat plate

(b) Solid cylinder

Fig. 4: Comparison of quenching analyses

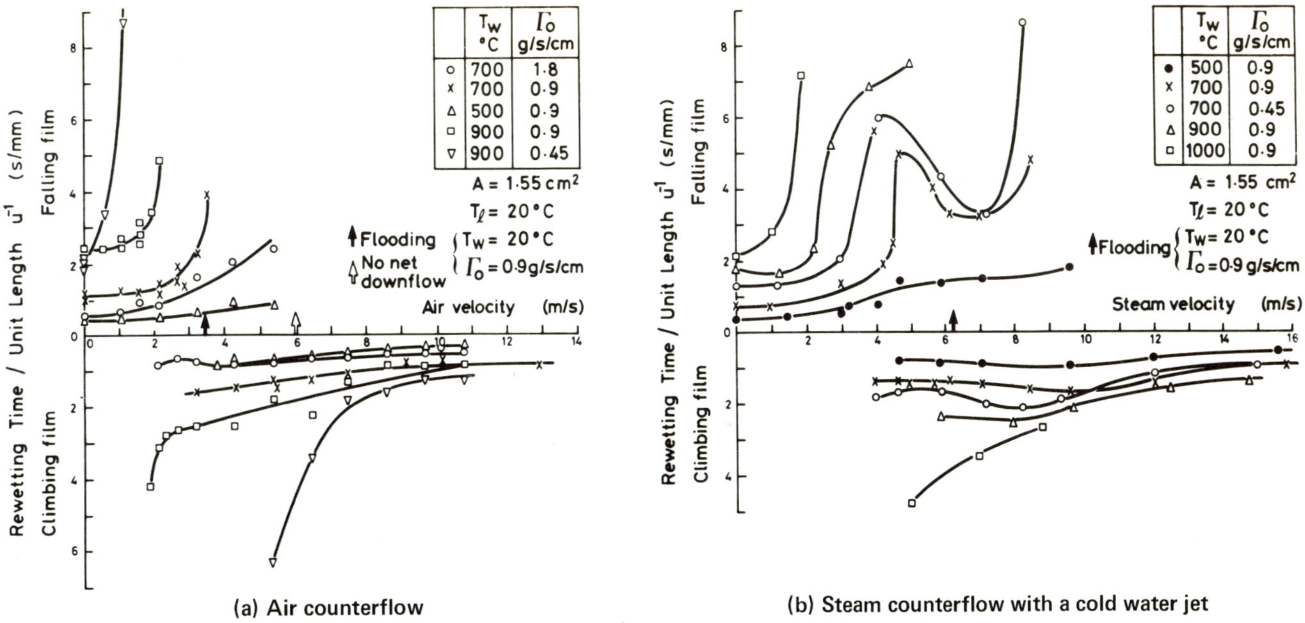


Fig. 5: The effect of counterflow on rewetting velocity

C199/77

# TRANSIENT BOILING HEAT TRANSFER AND BURNOUT CAUSED BY RAPID DEPRESSURIZATION

A. SAKURAI, DSc, DEng, M. SHIOTSU, BEng, and K. HATA, BEng
Institute of Atomic Energy, Kyoto University, Kyoto, Japan

SYNOPSIS The transient boiling heat transfer and the burnout phenomena caused by a rapid pressure release for a 1.2 mm dia. platinum wire heater which was initially in steady nucleate boiling in a pool of water under pressurized condition were studied. The transient experiments were performed for the saturated and subcooled initial conditions at the pressures of 0.29, 0.59, 1.08 and 1.57 MPa. The pressure reduction before the initiation of bulk boiling in water can be expressed as an exponential form and the pressure reduction period was varied from 5 to 850 ms. The transient boiling heat transfer process follows the steady nucleate boiling curves and their extensions for each pressure during the pressure reduction, whether bulk boiling in water occurs or not during the period, and then it greatly departs from the curve. The departing point and the heat flux at the point were defined as the transient DNB point and the transient DNB heat flux respectively. Transient DNB heat fluxes under the condition of no bulk boiling in water coincided with the corresponding steady critical heat fluxes for the pressure reduction period longer than 50 ms, but those for the period shorter than 50 ms were distributed from that higher than to that lower than the steady critical heat flux, depending on the initial heat flux levels. An empirical equation to express this relation was presented. When pressure is reduced below the saturation pressure, water is superheated and bulk boiling occurs. In this case, the transient process reached the DNB point at lower heat flux level than the corresponding steady critical heat flux. The lower heat flux limit for the transient process to reach the DNB point in the presence of superheated water was 70 to 87 % of the corresponding steady critical heat flux, for the pressure reduction period ranging from 5 to 850 ms.

## INTRODUCTION

1. The investigation of transient boiling heat transfer and burnout caused by a rapid pressure release is important for the fundamental understanding of the transient boiling phenomenon on the fuel surface of a water-cooled nuclear reactor due to a loss of coolant accident and has been investigated by some workers (Ref. 1, 2, 3). However, existing data were found to be insufficient for accurate prediction of the transient boiling heat transfer and burnout under pressure transients.

2. The purpose of this study is to make clear on what conditions the transient heat transfer and the transient DNB heat flux differ from the corresponding steady ones by widely varying the parameters such as the initial pressure, initial water temperature, pressure reduction rate and initial heat flux.

## NOTATION

| | |
|---|---|
| $e$ | Exponential function |
| Min. $q_{DNB}$ | Minimum DNB heat flux |
| $P$ | System pressure in absolute |
| $P^*$ | System pressure in gauge |
| $P_{atm}$ | Atmospheric pressure |
| $P_{in}$ | Initial pressure in absolute |
| $P^*_{in}$ | Initial pressure in gauge |
| $P_{sat}$ | Saturation pressure in absolute for initial water temperature |
| $q$ | Heat flux |
| $q_{DNB}$ | Transient DNB heat flux |
| $q_{ic}$ | Steady critical heat flux at initial pressure |
| $q_{in}$ | Initial heat flux |
| $q_{st}$ | Steady critical heat flux |
| $T_L$ | Water temperature |
| $T_{Lin}$ | Initial water temperature |
| $T_s$ | Surface temperature of test heater |
| $T_{sat}$ | Saturation temperature |
| $t$ | Time |
| $\Delta T_{sat}$ | Excess surface temperature of test heater beyond saturation temperature (= $T_s - T_{sat}$) |
| $\alpha$ | Boiling heat transfer coefficient (= $q/\Delta T_{sat}$) |
| $\tau_1$ | First stage pressure reduction period |
| $\tau_2$ | Second stage pressure reduction period |

## APPARATUS

3. The experimental facility used is shown schematically in Fig. 1. It consists of a pressure vessel, a pressure releasing device, a

vapor condenser, a test heater, a heating device and the instrumentation system.

### Pressure vessel

4. A cylindrical stainless steel pressure vessel of 20 cm in inner diameter and 60 cm in height with three 4 cm sight ports was constructed. Auxiliary heating of the deionized water in the vessel is available with a stainless steel sheathed immersion heater at the bottom of the vessel. The vessel has two pressure release lines of 5 cm in inner diameter from the vapor section of the vessel; one is from the upper lid of the vessel to the atmosphere through a rupture disc device and the other is from the side of the vessel to the vapor condenser through another rupture disc device.

### Pressure releasing device

5. A conventional pressure releasing device using one rupture disc and a gaff for artificially breaking the disc was used on the line from the upper lid of the vessel. Another type of a conventional device using two rupture discs was used on the line from the side of the vessel to the vapor condenser. For the transient experiments whose initial water temperatures were below 373 K, the system pressure was directly released to the atmosphere and the pressure reduction rate was varied by varying the water level. For the case of initial water temperature beyond 373 K, a large amount of steam generated by bulk boiling was led to the vapor condenser where it was rapidly condensed by the spray cooling. Pressure reduction rate in this case was varied by varying the orifice diameter from 6 to 50 mm.

### Vapor condenser

6. The vapor condenser is a horizontally mounted cylindrical tank of 45 cm in diameter and 170 cm in length. It has 12 spray nozzles and a water feed loop.

### Test heater

7. A platinum wire heater of 1.2 mm in diameter and 64 mm in length was supported horizontally in the vessel. In order to prevent the error in the measurement of the average temperature due to the terminal effect, two fine platinum wires were spot welded as potential taps at 7 mm apart from each terminal.

### Heating device

8. Heating current to the test heater was supplied by a power amplifier whose input signal was controlled by a high-speed analogue computer so that the heat generation rate of the heater accorded with the desired one (Ref. 4, 5). So it was possible with this device to give a constant heat generation rate throughout an experimental run in spite of the rapid variation in the heater resistance due to the variation in the heater temperature. The instantaneous average temperature of the test heater was also calculated by the computer from signals of the bridge circuit including the heater. As soon as it rose up to the preset value, the input signal of the power amplifier was shut off to prevent the heater from melting.

### Instrumentation

9. A fast response electromagnetic oscillograph was used to record the system pressures at the vapor and the liquid sections, bulk water temperature from a 0.3 mm diameter sheathed thermocouple, the current to the test heater, the voltage difference between the potential taps on the test heater and the output voltage from the double bridge circuit including the heater. The frequency response of the oscillograph was flat from 0 to 1 kHz, -3 dB at 2 kHz. This was sufficient for the pressure reduction rates tested. The instantaneous electrical resistance of the heater was determined from the current to the heater and the output of the bridge which had been balanced for the initial water temperature before the experimental run. The average temperature of the heater was calculated from the measured value of the resistance with the aid of the previously calibrated resistance-temperature relation. The heat generation rate of the heater was determined from the current to the heater and the voltage difference between potential taps on the heater. The heat flux from the heater is the difference between the heat generation rate per unit surface area and the rate of change of energy storage in the heater obtained from the faired average temperature-time curve. It was assumed that the surface temperature of the heater was uniform. The surface temperature of the heater was calculated from the non-steady thermal conduction equation in the heater having a time dependent heat generation rate and the heat flux as a boundary condition at the surface.

## EXPERIMENTAL

### Transient

10. Pressure transient experiments were performed for the following three cases.
Case 1: Initial water temperature was below 373 K and no bulk boiling in water occurred during the pressure reduction.
Case 2: Initial water temperature was saturated and the bulk boiling occurred soon after the initiation of pressure reduction.
Case 3: Initial water temperature was subcooled but higher than 373 K and, as a result, the bulk boiling occurred on the way of the pressure reduction.

11. The initial pressures in each case ranged from 0.29 to 1.57 MPa. In case 1, the pressure reduction can be expressed throughout an experimental run by $P^* = P^*_{in} e^{-t/\tau_1}$, where $P^*$ and $P^*_{in}$ are the system pressure and the initial system pressure in gauge. In Case 2 and 3, the pressure reduction before the initiation of bulk boiling is similar to that in Case 1. $\tau_1$ was varied from 5 to 850 ms. After the initiation of bulk boiling, the pressure recovers to a certain value lower than the initial pressure and then reduces with an approximate exponential function of the period $\tau_2$. The variation in $\tau_2$ with the above mentioned variation in $\tau_1$ was from 0.9 to 34 s in Case 2 transients of $P_{in}$ = 1.08 MPa.

### Steady state

12. The steady boiling curve and its critical heat flux to be compared with those of Case 1 transients were obtained for the initial water temperature of the corresponding transient under pressures ranging from the atmospheric pressure $P_{atm}$ to the initial pressure of the transient experiment, $P_{in}$. The steady boiling curve and critical heat flux to be compared with those of Case 2 transients were obtained for the saturated water under pressures ranging from $P_{atm}$ to $P_{in}$. The steady boiling curve and critical heat flux to be compared with those of Case 3 transients were obtained for the saturated water under pressures ranging from $P_{atm}$ to the saturation pressure $P_{sat}$ at the initial water temperature, and for the initial water temperature under pressures ranging from $P_{sat}$ to $P_{in}$.

## RESULTS AND DISCUSSION

### Typical experimental data for three cases

13. Typical experimental data for transients caused by a sudden breaking of rupture discs are shown for Cases 1, 2 and 3 respectively in Fig. 2(a), (b), (c). Initially, the test heater is in steady state nucleate boiling with a constant heat flux $q_{in}$ and the water of temperature $T_{Lin}$. In Fig. 2(a), the pressure reduction period is 5.3 ms. The heat flux first increases and then decreases. On the other hand, the heater temperature first decreases and then increases to burnout. In Fig. 2(b), the system pressure first reduces with $\tau_1$ = 23 ms, but when it reaches to a certain value, bulk boiling begins to occur and the pressure recovers to a certain value lower than the initial pressure and again decreases with $\tau_2$ = 1.7 s to the atmospheric pressure. It is clearly shown in the figure that the water temperature $T_L$ is about 10 K higher than the saturation temperature at the system pressure. This means that the water is in a superheated condition for a certain time duration after the start of the pressure reduction. The heat flux first increases and then decreases nearly to the initial value and holds the value for some time and then rapidly decreases. The heater temperature rapidly reduces at first, increases gradually and then rapidly increases to burnout. It should be noted that burnout occurs in the time duration where superheated water with vigorous bulk boiling exists. The experimental data in Fig. 2(c) are similar in the pressure variational form to those in Fig. 2(b). The first pressure reducing range is extended because of the difference between the initial pressure and the saturation pressure at the initial water temperature. The transient process in Case 3 transient reaches the DNB point before or after the initiation of bulk boiling depending on the initial heat flux. The former case corresponds to Case 1 transient and the latter to Case 2 transient. The data in Fig. 2(c) belong to the former.

### Transient boiling heat transfer

14. Transient boiling heat transfer data were compared with the steady ones under pressures ranging from $P_{in}$ down to $P_{atm}$. This comparison was made for the results of Cases 1, 2 and 3. Typical compared results for the three cases are shown on the $\alpha/q^{2/3}$ vs. P plane in Fig. 3(a), (b), (c) respectively, where α, q and P are the boiling heat transfer coefficient, the heat flux and the system pressure respectively. α is defined as $\alpha = q/\Delta T_{sat}$, where $\Delta T_{sat}$ is the excess heater surface temperature beyond saturation.

15. The steady value in Fig. 3(a) was obtained from the measured steady nucleate boiling curve for the water temperature of 343 K under each pressure. Steady value in Fig. 3(b) was from the steady nucleate boiling curve for the saturated condition under each pressure. Steady value in Fig. 3(c) was from the steady nucleate boiling curve for the water temperature of 456.4 K under each pressure ranging from 1.08 to 1.57 MPa, and for the saturated water under each pressure below 1.08 MPa. Dotted lines show the upper and lower limits of the steady values when 1 K increase or decrease from the measured heater temperature was supposed.

16. Transient heat transfer data for $P_{in}$ = 1.08 MPa, $T_L$ = 343 K, $\tau_1$ = 5.3 ms and for ratios of initial heat flux to the steady critical heat flux $q_{in}/q_{ic}$ equal to 0.86, 0.72, 0.42 and 0.10 are shown in Fig. 3(a) as typical results of Case 1 transients. As shown in the figure, the transient process very closely follows the steady value until it reaches a certain point, where it departs from the steady value. This departing point and the heat flux at the point are defined as the transient DNB point and the transient DNB heat flux respectively. When the initial heat flux is high, the transient process until the DNB point includes a certain pressure range where the transient heat flux is higher than the steady critical heat flux. Even in such a region, the value of $\alpha/q^{2/3}$ is consistent with the steady one. This means that the transient heat flux in such a region is on the linear extension of the steady nucleate boiling curve on log q vs. log $\Delta T_{sat}$ graph. This extended line was defined by the authors as the quasi-steady nucleate boiling curve (Ref. 4, 5). All experimental runs for the pressure reduction period ranging from 5 ms to 850 ms showed the same trend.

17. Transient boiling heat transfer data for $P_{in}$ = 1.08 MPa, saturated initial water temperature, $\tau_1$ = 23 ms and for $q_{in}/q_{ic}$ = 0.66 and 0.62 are shown in Fig. 3(b) as typical illustrations of Case 2 transients. It is shown in this figure that, in spite of the vigorous bulk boiling, the transient process closely follows the steady values until the transient process departs from the steady value at the transient DNB point. The heat flux at this point is also defined as the transient DNB heat flux. Experimental results for $\tau_1$ ranging from 5 to 850 ms with $\tau_2$ ranging from 0.9 to 34 s showed the same trend.

18. Transient heat transfer data for $P_{in}$ = 1.57 MPa, $T_{Lin}$ = 456.4 K (saturation temperature at 1.08 MPa), $\tau_1$ = 6 ms and for $q_{in}/q_{ic}$ = 0.85 and 0.49 are shown in Fig. 3(c) as typical results of Case 3 transients. Transient boiling heat transfer process in this case also follows the steady values as is expected from the results of Case 1 and Case 2 transients.

### Transient DNB heat flux

19. Transient DNB heat flux is determined as the heat flux at the departing point of the transient process from the steady heat transfer values on $\alpha/q^{2/3}$ vs. P plane as shown in Fig. 3(a), (b), (c). In this section, the transient DNB heat fluxes $q_{DNB}(P)$ for the transients of Cases 1, 2

and 3 are obtained under various conditions and are compared with the corresponding steady critical heat fluxes.

20. Transient DNB heat flux for $P_{in}$ = 1.08 MPa, $T_L$ = 353 K, $\tau_1$ = 5.6, 11 and 22 ms and for $q_{in}/q_{ic}$ ranging from 0.09 to 0.93 are shown in Fig. 4(a) as typical illustrations of Case 1 transients. Steady critical heat flux for the same water temperature is also shown in the figure. As is seen in the figure, the transient DNB heat fluxes for the same pressure reduction period are continuously distributed from that higher than to that lower than the corresponding steady critical heat flux, depending on the initial heat flux. This trend is clearer for the shorter pressure reduction period. For the period longer than 50 ms, transient DNB heat fluxes almost coincide with the corresponding steady critical heat fluxes. The relation between the heat flux and pressure for typical two runs, Run 1 and Run2, are shown in the figure. The heat flux becomes higher than the initial value owing to the release of stored thermal energy from the heater accompanied with its temperature decrease during the pressure reduction. In the case of Run 1, the heat flux becomes higher than the steady critical heat flux. As mentioned in the preceding section, the transient heat transfer process successively passes on the steady nucleate boiling curve and its extension for each pressure during pressure reduction. The same trend was observed for $P_{in}$ = 1.08 MPa and $T_L$ = 343, 353, 363 and 373 K, $P_{in}$ = 0.59 MPa, $T_L$ = 353 K.

21. Transient DNB heat fluxes for $P_{in}$ = 0.59 MPa, $T_{Lin}$ = 431 K (saturated), $\tau_1$ = 260 ms and for $q_{in}/q_{ic}$ ranging from 0.67 to 0.91 are shown in Fig. 4(b) as typical examples of Case 2 transients. The relation between q and P during the transient is also shown in the figure. The steady critical heat flux in the figure is for saturated water under each pressure. In the transients of Case 2, the pressure first decreases exponentially according to $P^* = P^*_{in}e^{-t/\tau_1}$ and then recovers to a certain value lower than the initial value and again decreases as shown in Fig. 2(b). Water is in superheated condition from $P_{in}$ down to the certain value; vigorous bulk boiling takes place in this pressure range. The initiation pressure of bulk boiling and the range of existence of superheated liquid are shown in the figure. It should be noted that the transient DNB point occurs nearly at the same pressure, about 0.55 MPa, in the range of superheated water when the initial heat flux is reduced from $q_{in}/q_{ic} \fallingdotseq 1$ down to a certain value. Further decrease in the initial heat flux makes the transient process pass through the range and causes the transient DNB heat flux to occur at lower pressure, the value of the DNB heat flux is almost consistent with the steady critical heat flux. In summary, the transient process for the initial heat flux higher than a certain value induces the transient DNB point in the time duration of superheated water where vigorous bulk boiling occurs. This lower limit of the transient DNB heat flux is called in this work the minimum DNB heat flux, Min. $q_{DNB}$, which is lower than the steady critical heat flux for the corresponding pressure. In the same experiment, except that an obstacle made of glass was used to prevent the bulk boiling void from directly striking the test heater, the lower DNB heat flux in the time duration of superheated water was not observed and the transient process passed through the duration and reached the DNB point at a lower pressure, the heat flux at which was almost consistent with the corresponding steady critical heat flux. It was assumed from this that major cause of the transient DNB heat flux being lower than the steady critical heat flux in the duration would not be the direct effect of the superheated water but would be the abrupt change in the two phase water structure by the occurrence of the vigorous bulk boiling in the superheated water. Similar Min. $q_{DNB}$ was observed in the experiments for $P_{in}$ = 1.08 MPa and for $\tau_1$ ranging from 18 to 800 ms.

22. Transient DNB heat flux for $P_{in}$ = 1.57 MPa, $T_{Lin}$ = 456.4 K (saturation temperature at 1.08 MPa), $\tau_1$ = 6 and 26 ms and for $q_{in}/q_{ic}$ ranging from 0.38 to 0.83 are shown in Fig. 4(c) as typical illustrations of Case 3 transients. The q vs. P curves during the pressure reduction are shown for Run 1 and Run 7 in the figure. Steady critical heat fluxes are obtained for $T_L$ = 456.4 K under pressures ranging from 1.08 to 1.57 MPa and for saturated water under pressures ranging from $P_{atm}$ to 1.08 MPa. These are shown in the figure for comparison. The transient DNB points for Runs 1 to 3 and for Runs 10 and 11 occur before the initiation of bulk boiling. Then these runs are similar to those of Case 1. The transient DNB points for Runs 4 to 9 and Runs 12 to 15 are after the initiation of bulk boiling. Transient DNB heat fluxes for these runs are generally lower than the steady critical heat fluxes. These runs are similar to those of Case 2. The value of Min. $q_{DNB}$ was obtained for these runs as well as in Case 2 transients. In summary, Case 3 transients are similar to Case 1 transients for high initial heat flux and are similar to Case 2 transients for low initial heat flux. The same trend was observed in the transients for $P_{in}$ = 1.57 MPa and for $T_{Lin}$ = 431 K (saturation temperature at 0.59 MPa).

23. The transient DNB heat fluxes of Case 1 transients and those of Case 3 transients under the condition of no bulk boiling for the tested conditions of initial pressure, initial liquid temperature and of pressure reduction period are related to the steady critical heat flux by the following empirical formula as shown in Fig. 5:

$$q_{DNB}(P^*)/q_{st}(P^*) = (1 + 0.01\, P^{*\,1.8}_{in}\tau_1^{-0.8}) \times (P^*/P^*_{in})^{4.1\times10^{-3}\tau_1^{-1}(1.1+\log P^*_{in})} \qquad (1)$$

where $P^*$ and $P^*_{in}$ are in MPa and $\tau_1$ is in second.

24. The value of $q_{DNB}/q_{st}$ shown in Fig. 5 are continuously distributed depending on the initial heat flux from values above to below unity. The thermal energy in the test heater is released accompanied by a decrease in the heater temperature during the course of pressure reduction. It is supposed that the trend shown in Fig. 5 might be the result of superposition of a trend to increased $q_{DNB}$ due to the rapid increase in the heat flux and of a trend to decreased $q_{DNB}$ due to the heater temperature decrease from the initial high value to the lower. The former effect would be similar to the effect on $q_{DNB}$ as that observed by the authors for the case of rapid increase in the heat generation rate (Ref. 4, 5).

25. The ratio of the minimum DNB heat flux to the steady critical heat flux, Min. $q_{DNB}/q_{st}$, obtained in Case 2 and Case 3 transients is shown

versus $\tau_1$ in Fig. 6. As shown in the figure, the value of Min. $q_{DNB}/q_{st}$ slowly increases with the increase in the period.

26. In the PWR loss of coolant accident, Case 3 transient will occur. The first stage pressure reduction period may become shorter than 50 ms in the hot-leg rupture accident. In such a case, the fuel burnout due to the stored energy release may occur at a time soon after the initiation of pressure reduction when the fuel rods are still in enough water. In the BWR loss of coolant accident, the water will be superheated and vigorous bulk boiling to compensate the superheated state will occur soon after the initiation of pressure reduction as observed in the Case 2 transients in this study. DNB heat flux may occur in the time duration of superheated water since Min. $q_{DNB}$ in the duration was shown to be lower than the steady critical heat flux. Superheat of water will need to be considered for the safety evaluation of water cooled reactors.

## CONCLUSIONS

27. The transient boiling heat transfer process caused by a rapid depressurization for the test heater initially in the steady state nucleate boiling condition in a pool of water successively passes on the steady boiling curve for each pressure including its extension, whether the bulk boiling occurs or not during the pressure release, and departs from the curve at transient DNB heat flux $q_{DNB}$.

28. The ratio of $q_{DNB}/q_{st}$ for the transient DNB point under condition of no bulk boiling for $\tau_1$ shorter than 50 ms is distributed from values above to below unity depending on the initial heat flux level. For $\tau_1$ longer than 50 ms, transient DNB heat flux almost coincides with the corresponding steady critical heat flux. An empirical equation to express the transient DNB heat flux was presented.

29. When the system pressure is reduced to that lower than the saturation pressure at the initial water temperature, the water is superheated in a certain time duration at which bulk boiling in water commences. In this case, the whole transients for the initial heat flux higher than a certain value reach the transient DNB point in the time duration of superheated water. The lower limit of the transient DNB heat flux in the time duration of superheated water is 70 to 87 % of the corresponding steady critical heat flux, for $\tau_1$ ranging from 5 to 850 ms.

## ACKNOWLEDGMENTS

The authors gratefully acknowledge the financial support of this study by Toray Science Foundation. They also wish to express their thanks to Mr. T. Yano for his assistance in the experimental work.

## REFERENCES

(1) Howell, J. R., and Bell, K. J., "An experimental Investigation of the Effect of Pressure Transients on Pool Boiling Burnout," Chem. Eng. Progr. Symposium Ser., Vol. 59, No. 41, 1963, pp. 88-95.

(2) Cermak, J. O., Farman, R. F., Tong, L. S., Casterlin, J. E., Kokolis, S., and Matzner, B., "The Departure from Nucleate Boiling in Rod Bundles during Blowdown," Journal of Heat Transfer, Trans. ASME, Series C, Vol. 92, November 1970, pp. 621-627.

(3) Aoki, S., Inoue, A., and Kozawa, Y., "Transient Boiling Crisis during Rapid Depressurization," Heat Transfer 1974, Vol. IV, B6.3, Scripta Book Co., Washington, D.C., pp. 250-254.

(4) Sakurai, A., Mizukami, K., and Shiotsu, M., "Experimental Studies on Transient Boiling Heat Transfer and Burnout," Heat Transfer 1970, Vol. V, B3.4, Elsevier Pub. Co., Amsterdam, 1970.

(5) Sakurai, A., and Shiotsu, M., " Transient Pool-Boiling Heat Transfer," ASME Paper 74-WA/HT-41, ASME Winter Annual Meeting, New York, N. Y., November 1974.

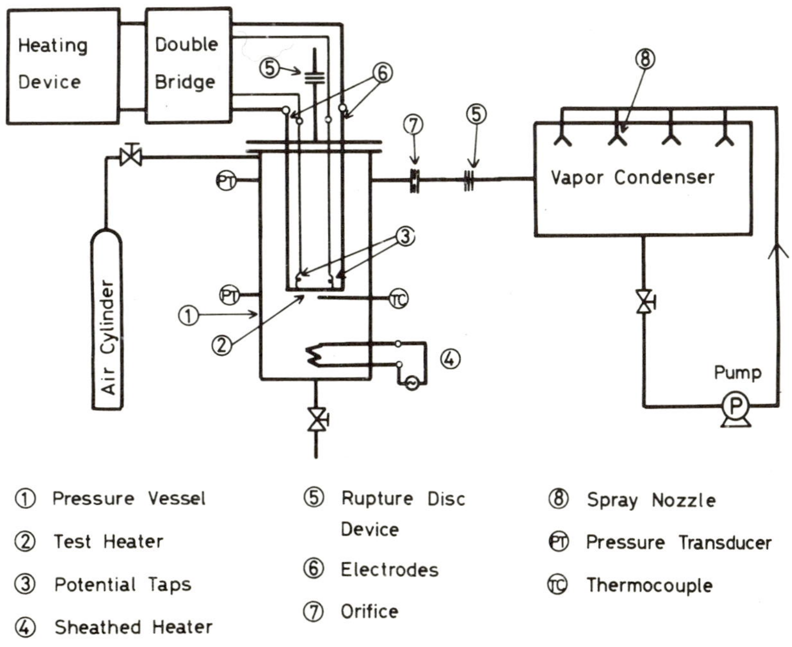


Fig. 1: Schematic diagram of experimental facility

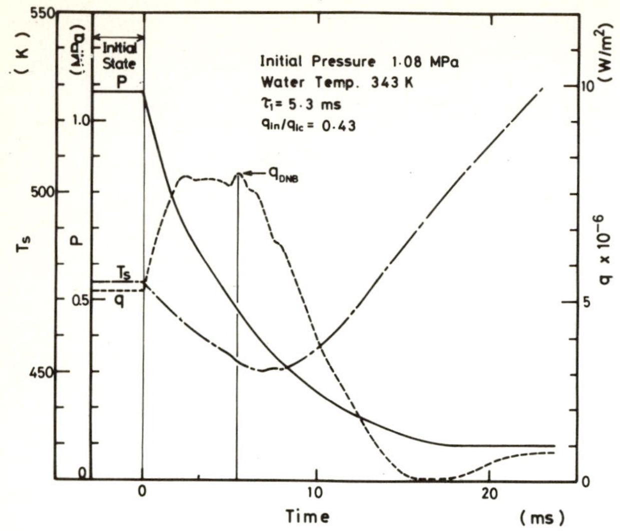


(a)

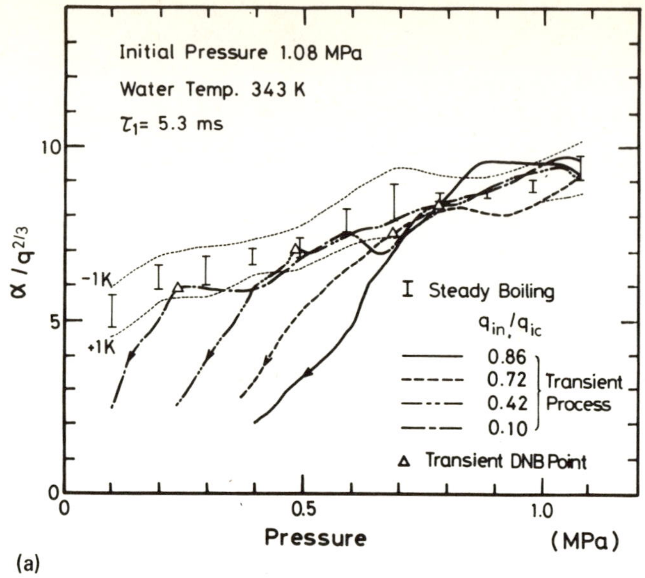


(a)

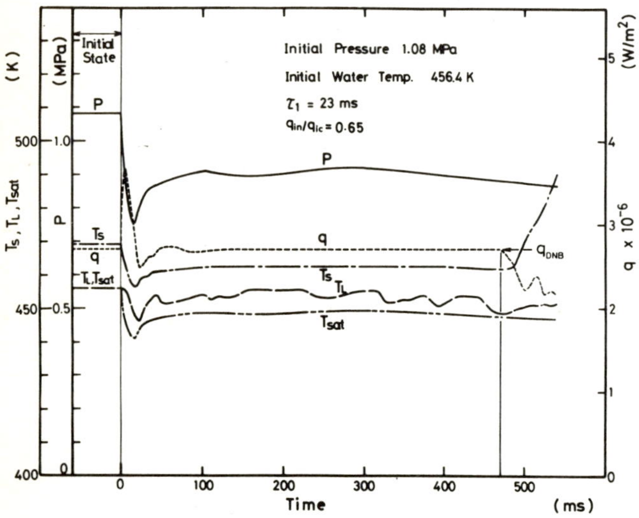


(b)

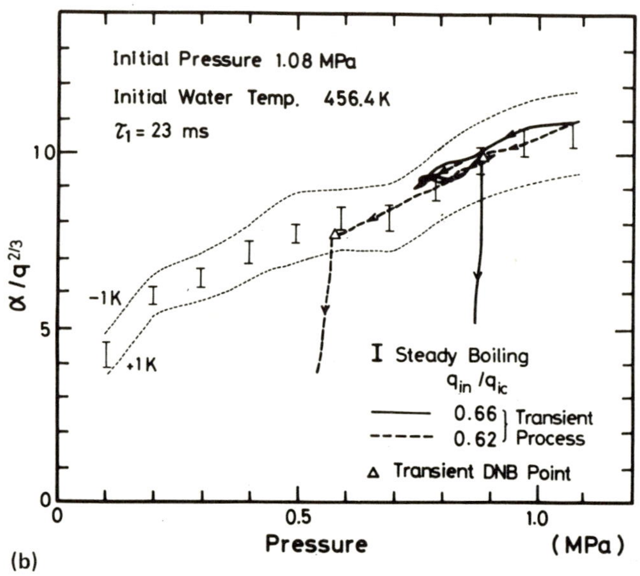


(b)

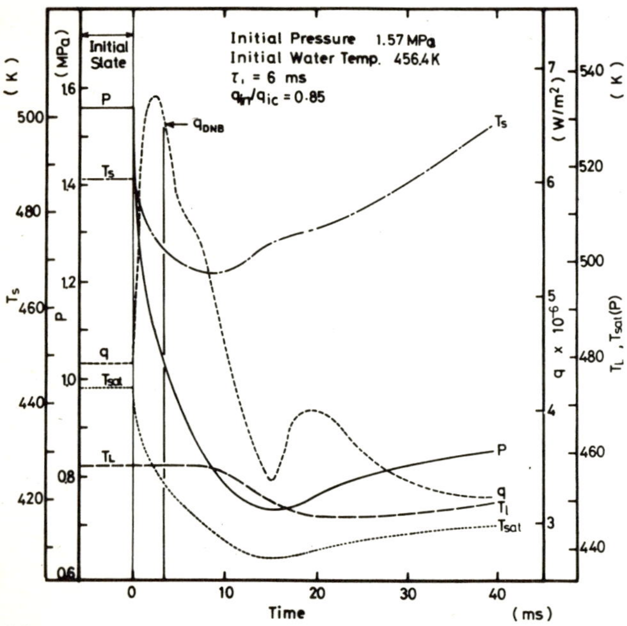


(c)

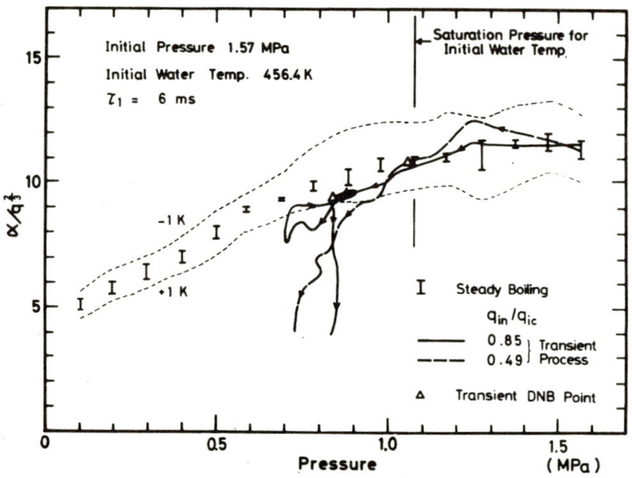


(c)

Fig. 2: Typical changes due to a pressure release in the system pressure P, heater surface temperature $T_S$, heat flux q, water temperature $T_L$ and saturation temperature of water $T_{sat}$

Fig. 3: Typical results of transient boiling heat transfer compared with the corresponding steady boiling heat transfer data

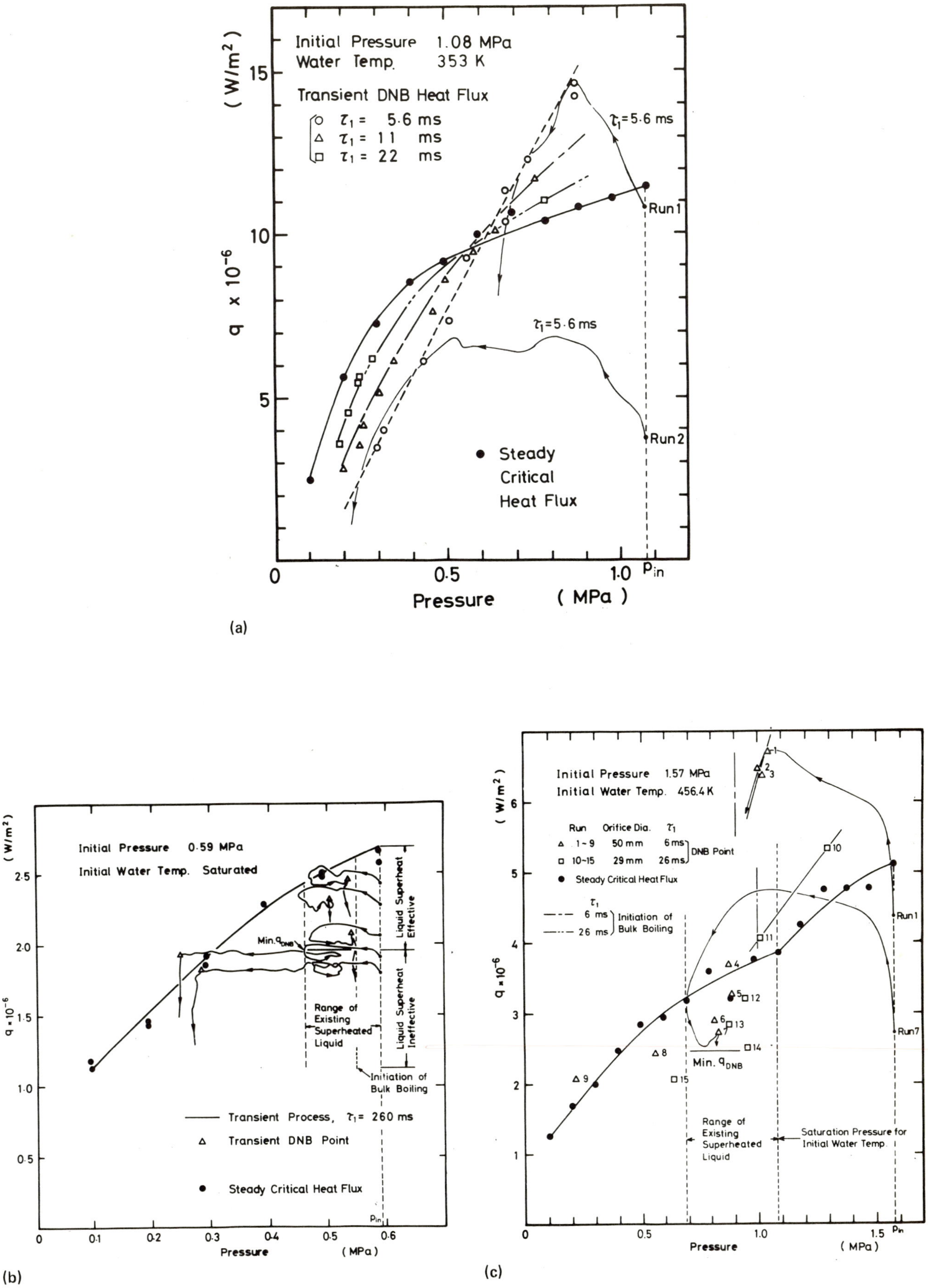


Fig. 4: Typical results of transient DNB heat flux compared with corresponding steady critical heat fluxes

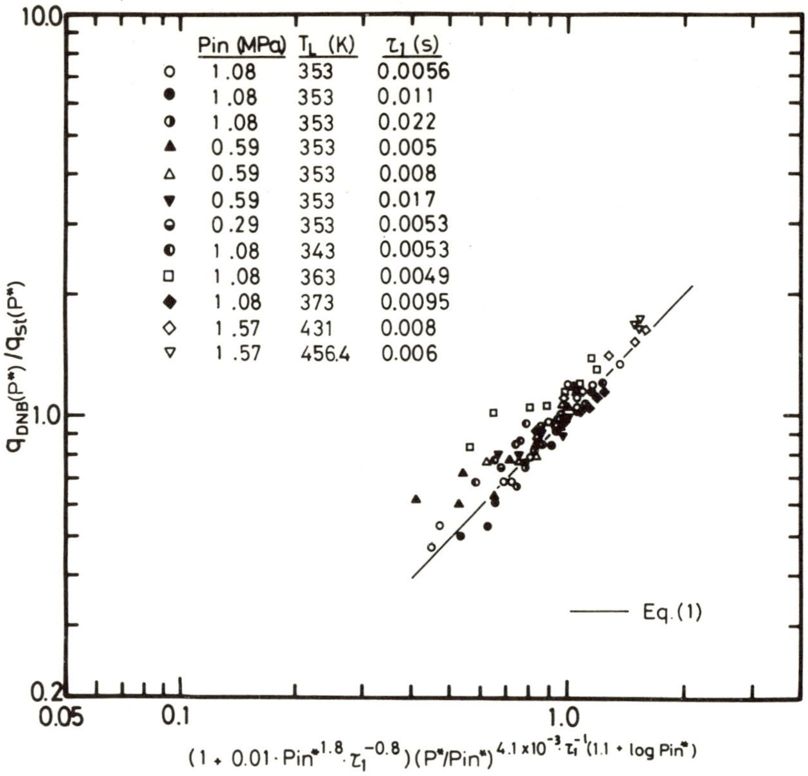


Fig. 5: Empirical equation, Eq. (1), compared with the experimental data

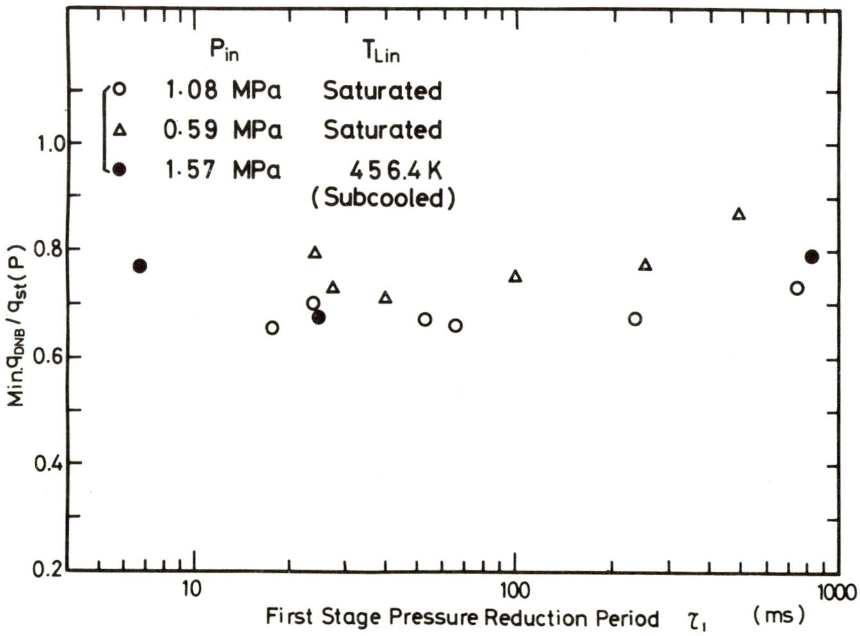


Fig. 6: Min. $q_{DNB}/q_{st}$ versus $\tau_1$ in the time duration of existing superheated water

C200/77

# DETERMINATION OF MASS FLOW RATE AND QUALITY USING A TURBINE METER AND A VENTURI

R. FRANK, J. MAZARS, R. RICQUE
Centre D'études Nucléaires, 38041 Grenoble Cedex, France

SYNOPSIS It is necessary, in blowdown experiments, to measure accurately the mass flow rate, the quality and the void fraction in two phase flow, at some points of the installation. A method of determining the mass flow rate and the quality is proposed using a venturi and a turbine meter. These two instruments are studied and calibrated in steady state two phase flow. A model of interpretation of each of these devices is proposed and the procedure used to determine the mass flow rate and quality is explained and illustrated in some cases.

NOTATION

$f$ rotationnal frequency of the turbine meter
$\lambda$ flow coefficient of the turbine meter
$M$ mass flow rate
$Q$ volumetric flow rate
$\rho$ density of the fluid
$x$ quality of the flow
$\alpha$ void fraction, measured with the fort acting valves.
$\alpha_T$ void fraction calculated at the turbine
$\alpha_V$ void fraction calculated at the venturi
$\Delta_P$ pressure drop in the venturi
$K$ flow coefficient of the venturi in single phase flow.
$K_G$ flow coefficient of the venturi in for gas
$K_L$ flow coefficient of the venturi for liquid

INTRODUCTION

The necessity of better understanding the behavior of nuclear reactors during a loss of coolant accident has led to the construction of a large number of out of pile blowdown experiments all around the world. In this type of experiments, with rapidly changing transient two phase flow one has to measure accurately mass flow rate and quality on the break pipe.

In the Heat Transfer Laboratory in Grenoble the laws of transient heat transfer between the cooling fluid and the fuel rods are studied on a steam water loop OMEGA. In this experiment we depressurize a heated test section placed between two capacities on which breaks are opened. This test section is either a tube or a bundle of 36 stainless steel rods which simulate the fuel rods.

The local flow conditions necessary to establish the heat transfer laws are calculated by numerical codes. These codes are adjusted with experimental measurements of mass flow rate, quality and void fraction performed at both ends of the test section. The accurate measurement of these quantities is not trivial and needs :

- a good knowledge of the physical laws relating the signals observed to the characteristics of the flow.
- an accurate calibration of the measurement device in flow conditions as similar as possible to those realised in a blowdown experiment.

A loop called DEDIF, working with an argon water mixture has been constructed in Grenoble to study and calibrate two phase flow measurement devices which are to be used on the Omega installation.

Two devices are necessary to measure simultaneously the mass flow rate and the quality, one of them can to a certain extent, be a void fraction measurement. However in the design of OMEGA we had chosen to use, in a first step the association of a venturi and of a full flow turbine meter to determine mass flow rate and quality, the void fraction measurement (by a $\gamma$ ray attenuation technique) being considered as a separate measurement. Therefore we have decided to study a venturi and a full flow turbine meter on DEDIF.

The procedure used is the following :
Each device is calibrated in two phase flow to obtain :

- First a correlation between quality, void fraction, mass flow rate on one hand and the measured quantity, the frequency of the turbine meter or the pressure drop in the venturi, on the other hand.
- Second a relation between the void fraction and the quality which, at a given pressure, is a characteristic of the device.

The set of parameters studied is the following :

- Pressure from 18 to 83 bar which in a steam water flow corresponds to a range from 40 to 140 bar.
- Quality from .03 to .75
- Void fraction from .3 to .99
- Mass flow rate from .1 to .8 kg/s

## I - DESCRIPTION OF THE EXPERIMENTAL DEVICE

1. General arrangement of the loop
We have chosen to calibrate our measurement devices on a gas-water loop, working at room temperature, to avoid the technological difficulties of a steam water loop working at 155 bar and 350°C.

The similarity condition between the gas water and the steam water mixtures is that the density ratio of the two phases of the mixture has the same value.

Chosing a mixture of water and argon leads to an equivalent pressure range from 1 to 92 bar to simulate steam water mixtures up to 155 bar.

 

On this loop we can measure separately :
- The liquid mass flow rate $M_L$ on the water loop
- The gas mass flow rate $M_G$ on the argon loop
- The void fraction $\alpha$ by a hold up technique on the two phase flow test section.

The test section has been placed vertically as in Omega, to avoid stratification in steady state flow.

2. Description of the test section
The test section is a stainless steel tube of 26 mm of inner diameter. It is schematically represented in figure 1.

Following the fluid one finds :
- an orifice plate which is used to minimize the effect of gas and water injection on the measurement devices, and make the flow more homogeneous.
- a venturi
- a fluid bearing turbine meter
- a system of quick closing valves which gives a measurement of the void fraction, at the top of the test section by a hold up method. The closing time of these mechanically coupled valves is 50 milliseconds.

## II - STUDY AND CALIBRATION OF THE TURBINE METER

The turbine meter which has been studied here is a fluid bearing turbine meter which has been developped specially for blowdown experiments in our laboratory. This turbine meter is specially well adapted to measure the large and rapid changes of the mass flow rates which occur during a blowdown experiment.

In single phase flow the rotational frequency of the turbine meter is proportional to the volumetric flow rate for a given viscosity of the fluid :

$$f = \lambda Q = \lambda \frac{M}{\rho}$$

where Q is the volumetric flow rate
M the mass flow rate
$\rho$ the density of the fluid
and $\lambda$ the flow coefficient of the turbine meter.

We shall try to find a formulation of the density of the two phase fluid which allows us to write f, the rotational frequency of the turbine meter, in a similar way as for a single phase fluid. In all this paper we shall make the assumption of a separate phase flow.

a) Theoretical model
Rouhani (1) has developed, for a turbine, a model taking into account different velocities for the gas and the liquid. If one writes the equilibrium of momentum between gas, liquid and turbine one shows that the rotational frequency of the turbine can be expressed as :

$$f = \lambda_{2\emptyset} M \left[ \frac{x^2}{\alpha \rho_G} + \frac{(1-x)^2}{(1-\alpha)\rho_L} \right] = \lambda_{2\emptyset} \frac{M}{\rho_{2\emptyset}}$$

where $\rho_{2\emptyset}$ is the so called "momentum density"
and $\lambda_{2\emptyset}$ is the flow coefficient of the turbine in two phase flow.

To show the validity of this model one must show that, at least at a given pressure, the flow coeffient $\lambda_{2\emptyset}$ is a constant. Therefore we have plotted for each pressure the frequency of the turbine versus the two phase volumetric flow rate

$$Q = M \left[ \frac{x^2}{\alpha \rho_G} + \frac{(1-x)^2}{(1-\alpha)\rho_L} \right]$$

These curves are represented in figure 2 for two pressures : P = 28 bar and P = 83 bar.

We can see that the experimental points lie about a straight line the slope of which is the flow coefficient $\lambda_{2\emptyset}$.

The value of $\lambda_{2\emptyset}$ is 2,8. $10^5$ $m^{-3}$ at P = 83 bar and 3.$10^5$ $m^{-3}$ at P = 28 bar. The calibration value $\lambda$ found in a water flow was 2,8 $10^5$ $m^{-3}$. This shows that one can apply Rouhani's model with accuracy better than 5 %, the flow coefficient $\lambda_{2\emptyset}$ being in a first approximation :
- constant at each pressure
- almost independant of the pressure and equal to the single phase flow coefficient.

We are now justified to use Rouhani's model with a constant flow coefficient $\lambda_{2\emptyset}$ = 2.8 $10^5$ $m^{-3}$ and we can recalculate the void fraction at the turbine knowing M, X and P which will give the calibration curve of the turbine in two phase flow. This calibration curve $\alpha_T(X,P)$ is shown in figure 3 for the above mentioned pressures P = 83 bar and P = 28 bar. It appears that the void fraction in the turbine is independant of M in the whole range which was studied.

Comments
1) A more accurate examination of the experimental results shows that $\lambda_{2\emptyset}$ is not quite constant specially at low pressures. However an error of 20 % on $\lambda_{2\emptyset}$ gives a variation of about 5 % in the determination of the quality.
2) In the previous set of 2 equations three quantities : $\alpha_M$, X and M are unknown. This means that if one quantity, the void fraction for instance, is measured with another device the mass flow rate and the quality can be deduced from the turbine signal.

## III - CALIBRATION OF THE VENTURI

As for the turbine the behavior of a venturi in single phase flow is well known : the pressure drop along the converging part writes :

$\Delta P = \frac{K\ M^2}{\rho}$ where K only depends on the geometry and on the Reynolds number for an incompressible fluid. We have used a venturi with a diameter ratio of .6. As for the turbine we have tried to find a model in two phase flow, taking into account the different velocities of the two phases.

Such a model was established by Chisholm (2). Taking into account a drag force between the gas and the liquid, in dispersed or separated flow distribution he finds the following correlation :

$$\frac{\Delta P_{2\emptyset}}{\Delta P_L} = 1 + \left[ \gamma \sqrt{\frac{\rho_G}{\rho_L}} + \frac{1}{\gamma} \sqrt{\frac{\rho_L}{\rho_G}} \right] \sqrt{\left(\frac{\Delta P_G}{\Delta P_L}\right)} + \frac{\Delta P_G}{\Delta P_L}$$

where $\gamma$ is the velocity ratio of the two phases
$\Delta P_{2\emptyset}$ is the pressure drop in the venturi in two phase flow corrected of the elevation between the pressure taps.
$\Delta P_L$ the pressure drop which would be due to the liquid alone
$\Delta P_G$ the pressure drop which would be due to the gas alone
$\rho_G$ the density of the gas
$\rho_L$ the density of the liquid

If the Reynolds number of the gas and the liquid, calculated at the throat of the venturi, are both in the range from $10^4$ to 5.$10^5$ then the flow coefficients for the two phases can be considered as equal and one can write.

$$\Delta P_L = K \frac{(1-X)^2 M^2}{\rho_L}$$

$$\Delta P_G = \frac{K\, X^2\, M^2}{\rho_G}$$

and the above correlation writes :

$$\Delta P_{2\phi} = K\, M^2 \left[ \frac{x^2}{\alpha_V \rho_G} + \frac{(1-x)^2}{(1-\alpha_V)\rho_L} \right]$$

where $\alpha_V$ is given by the well known relation

$$\gamma = \frac{x}{1-x} \; \frac{1-\alpha_V}{\alpha_V} \; \frac{\rho_L}{\rho_G}$$

and K is the single phase flow coefficient of the venturi.

The pressure drop in the venturi is written in the same way that in a single phase flow :

$$\Delta P_{2\phi} = K \frac{M^2}{\rho_{2\phi}}$$

and $\rho_{2\phi}$ is, as for the turbine the momentum density.

$$\rho_{2\phi} = \left[ \frac{x^2}{\alpha_V \rho_G} + \frac{(1-x)^2}{(1-\alpha_V)\rho_L} \right]^{-1}$$

where $\alpha_V$ is the mean void fraction in the venturi

As for the turbine one can now calculate the void fraction at the venturi which will certainly be different from the void fraction measured with the quick closing valves system.

In this way we get the calibration curves $\alpha_V$ (X, P) of the venturi in two phase flow. This calibration curve is show in figure 4 for the pressures P = 83 bar and P = 28 bar.

comments :

1) The calculated values of the average void fraction in the venturi is not far from the value $\alpha$ measured with the quick closing valves system

$$\frac{(\alpha-\alpha_V)}{\alpha_V} \leqslant 12\ \%$$

This justifies the use of Chisholm's model.

2) The Reynolds number for the liquid in some cases is lower then $10^4$ ($M_L$ < 0,150 kg/s). In that case one must utilize Chisholm's relation in its original form.

## IV - CALCULATION OF THE MASS FLOW RATE IN OMEGA

A certain number of blowdown experiments have been performed on OMEGA with a tubular test section. Due to a technological deficiency of the turbine, measurements, at each end of the test section, were made only with a venturi and a $\gamma$ ray absorption system.

We have calculated the mass flow rate using the venturi signal, the calibration curve of the venturi and the void fraction $\alpha$ measured by the $\gamma$ ray system.

The procedure is the following :
One supposes that $\alpha = \alpha_V$ (which gives an error which can be of the order of 10 to 15 %).

From the calibration curve of the venturi one deduces the quantity.

From the equation :

$$\Delta P_{2\phi} = K\, M^2 \left[ \frac{X^2}{\alpha \rho_G} + \frac{(1-X)^2}{(1-\alpha)\rho_L} \right]$$

one computes the value of the mass flow rate.

Two different cases were examined :

- In the first one the break was made at the upper part of the test section. The calculation of the mass flow rate confirmed the physical assumption that a large amount of liquid remained in the system at the end of the blowdown.
- In the second case the break was made at the lower part. The calculated mass flow rate, integrated on the total time of the experiment is equal to the mass of liquid present in the depressurized volume, before the blowdown, with an accuracy better than 10 per cent. The calculated mass flow rate was, in a typical case 4,9 kg and the initial mass present was 5 kg.

## V - CALCULATION OF MASS FLOW RATE AND QUALITY

The following equations and relations have been established :

- For the turbine :

$$f = \lambda\, M \left[ \frac{X^2}{\alpha_M \rho_G} + \frac{(1-X)^2}{(1-\alpha_M)\rho_L} \right]$$

$$\alpha_M = \alpha_M\ (X,\ P)$$

- For the venturi :

$$P_{2\phi} = K\, M^2 \left[ \frac{X^2}{\alpha_V \rho_G} + \frac{(1-X)^2}{(1-\alpha_V)\rho_L} \right]$$

$$\alpha_V = \alpha_V\ (X,P)$$

We have four unknown quantities : M, X, $\alpha_V$ and $\alpha_M$, four relations and two measured quantities : the frequency of the turbine and the presure drop in the venturi.

To determine the values of M and X one uses an iterative procedure from an initial solution which is obtained by assuming that the slip ratio is equal to 1.

For this assumption the equations are reduced to :

$$f = Mo \left[ \frac{Xo}{\rho_G} + \frac{1-Xo}{\rho_L} \right]$$

$$\Delta P = K\, Mo^2 \left[ \frac{Xo}{\rho_G} + \frac{1-Xo}{\rho_L} \right]$$

which gives Mo and Xo. Using Xo in the calibration curves one gets $\alpha_{VO}$ and $\alpha_{MO}$. Using these values in the complete equations gives a new couple $X_1$, $M_1$. One iterates this procedure which generally converges with less than 20 iterations.

## VI - CONCLUSION

1/ We have established very simple functionnal laws for the venturi and the turbine in two phase flow.

2/ The calibration curves obtained in steady state for the venturi have been applied to transient conditions in Omega. The accuracy obtained in the mass flow rate determination was of the order of 10 per cent which is quite satesfactorily.

3/ We have shown that, in steady state flow, one can calculate the mass flow rate and the quality with a turbine and a venturi. The accuracy obtained has been shown to be of the order of 20 per cent for both quantities.

4/ However extensive studies and calibrations have to be made in transient conditions to verify if the results obtained in steady state are applicable in transient flow and with what accuracy.
This study will take place in the second half of 1977.

## REFERENCES

1. ROUHANI Application of the turbine type flowmeters in the measurement of steam quality and void. Symposium on in core instrumentation Oslo 15-19th June 1964.
2. CHISHOLM D. Journal Mechanical Engineering Science - Vol 9 n° 1.

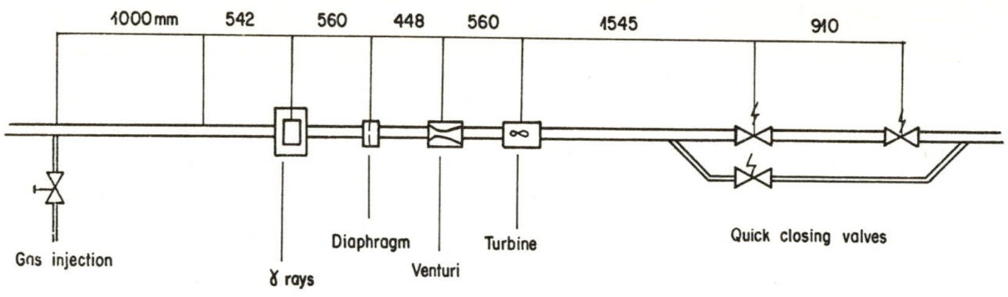


Fig. 1: Test section

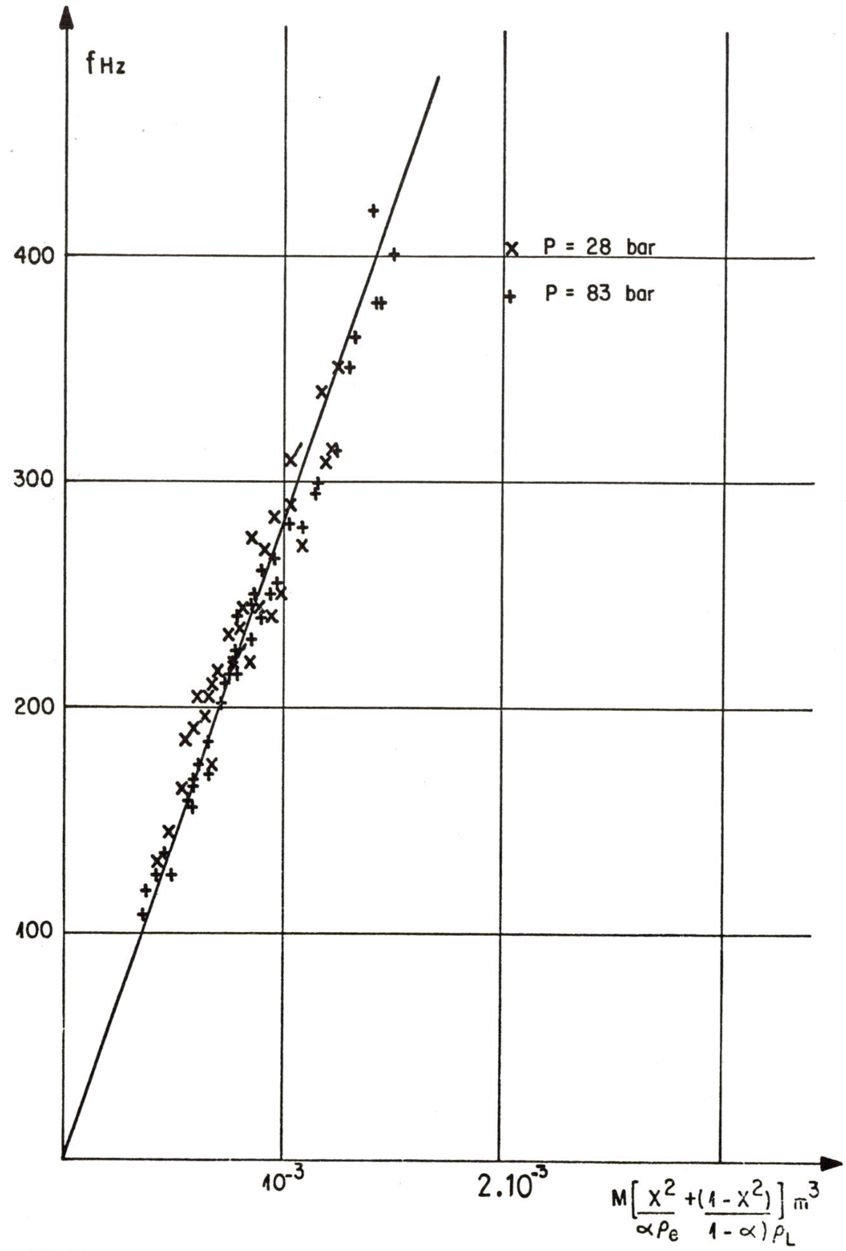


Fig. 2:

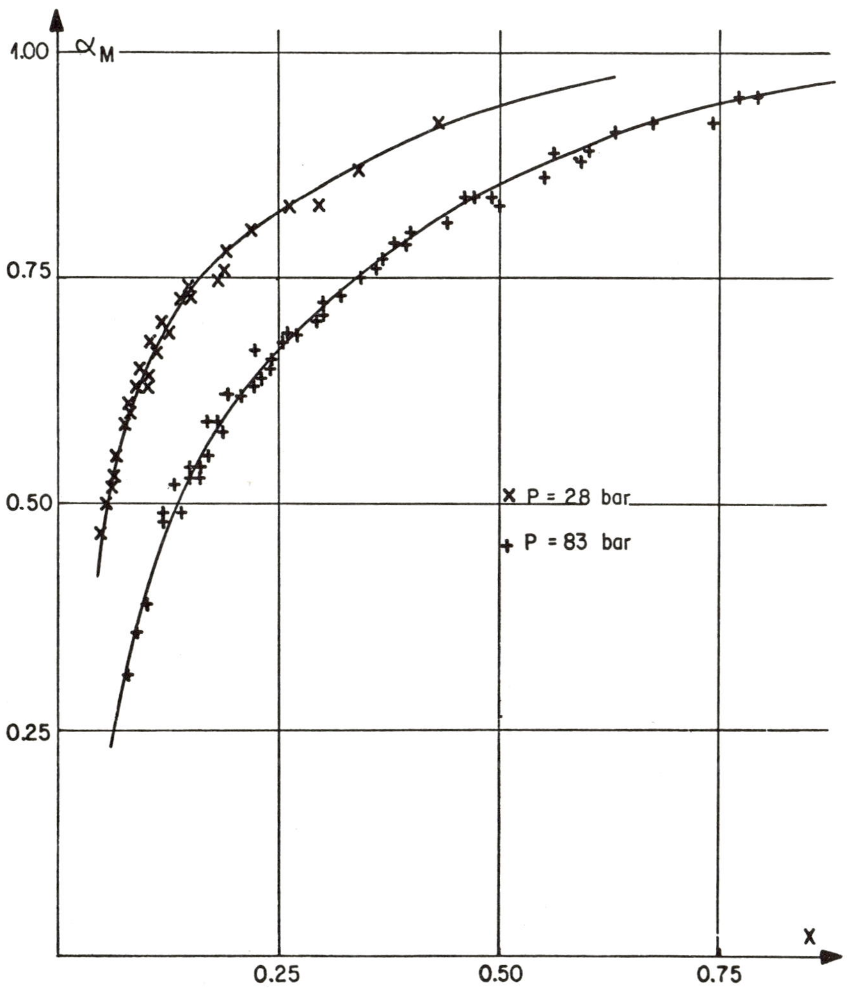


Fig. 3:

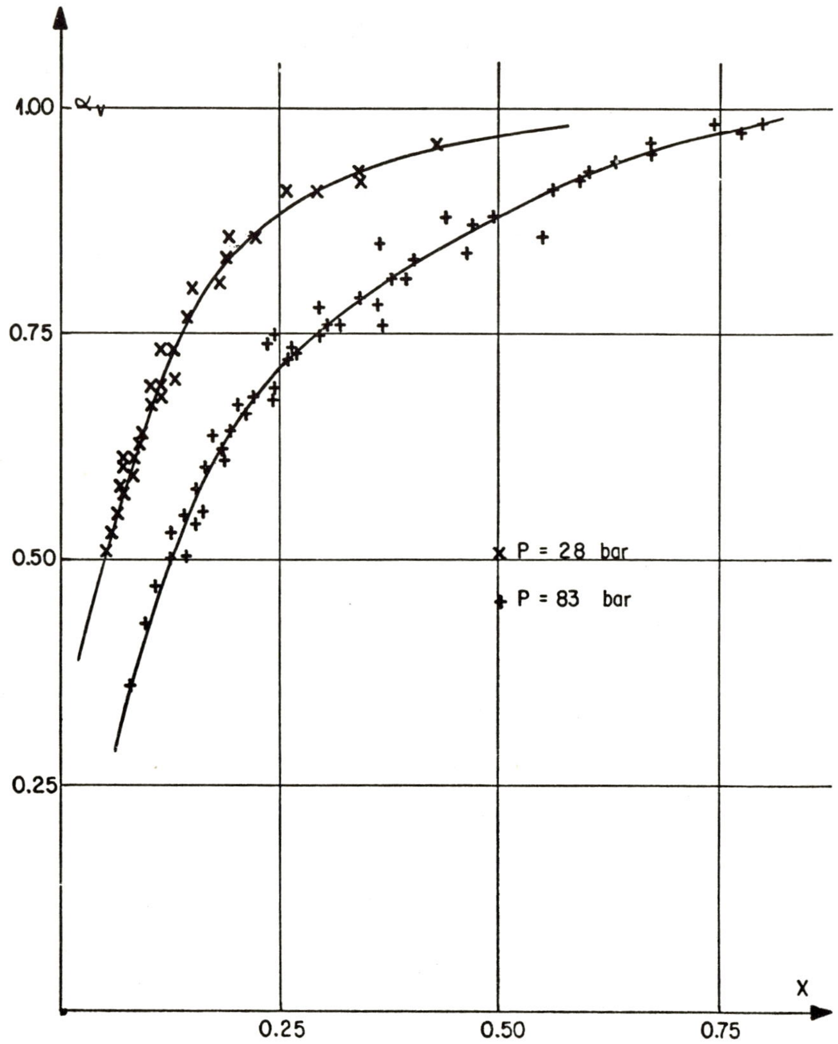


Fig. 4:

C201/77

# COMPUTER CODES IN WATER REACTOR SAFETY: PROBLEMS IN MODELING OF LOSS-OF-COOLANT ACCIDENT

S. FABIC, MIMechE, MS, PhD
United States Nuclear Regulatory Commission,
Division of Reactor Safety Research, Washington, DC, USA

SYNOPSIS Problems in the Best Estimate modeling of the loss-of-coolant accident for Pressurized Water Reactors are reviewed, for both the existing codes and advanced system's codes now under development.

## INTRODUCTION

1. The purpose of this paper is to describe the problem areas already encountered or envisaged, in developing computer codes in water reactor safety. In order to make this tractable the author has elected to restrict this survey to the discussion of codes for Best Estimate analyses of the postulated loss-of-coolant accident, (LOCA) and, further, applicable to the primary coolant system of a pressurized water reactor (PWR). Some of the same observations are also applicable to the codes describing the boiling water reactor LOCA, and also to the Containment System codes.

2. LOCA analyses for licensing purposes contain bounding, conservative effects which do not require the details and complexities which the Best Estimate analyses must address. The Best Estimate (BE) analyses attempt to describe the physical processes in the most realistic manner possible. Their purpose is to quantify the degree of conservatism contained in the licensing calculations.

3. The reader interested in the present status of modeling of transient two-phase flow, applications, and characterization of the existing reactor safety codes, on a world-wide basis, is referred to a chapter in Reference 1, written by the same author.

## MATHEMATICAL MODELS OF TWO-PHASE FLOW

### Background

4. Initial efforts in the mathematical modeling of two-phase flow drew heavily from the experience gained in modeling single phase flows, resulting in a widespread use of the homogeneous equilibrium model (HEM). As the name implies the two phases were considered to be intimately mixed, in thermal and mechanical equilibrium with each other. The governing equations were those of the single-phase fluid and the equation of state for this pseudo-fluid was based on the well known "steam tables".

5. HEM is still in widespread use for the analysis of complex systems and, especially, for licensing calculations in which various conservative assumptions are introduced, that minimize the requirements for the detailed tracking of the dynamic evolution of each phase. Various experiments conducted within the last few years in test facilities whose geometries attempt to simulate the thermo-hydraulic behavior of the light water reactors, during LOCA, illustrated the limitations of the HEM simplifications when applied to the Best Estimate predictions of certain aspects of LOCA behavior, such as the ECC injection phase, refill, reflood, and small break analysis. In addition, the one-dimensional and or lumped parameter flow representations, commonly used in LWR LOCA predictions, were also found to be restrictive for BE analyses. In paragraphs 27 through 48 these restrictions and their consequences are discussed for various regimes of LOCA and for various PWR system components.

6. A typical example of the application of the HEM model to LWR systems LOCA is the first version of RELAP-4 code (ref. 2). This code has, since its release in 1974, undergone several modifications intended to extend its applicability to BE analyses. Some of the changes made had to do with the attempt to account for the flow inhomogeneity, especially counter-current flow and phase separation; the latter is very important for analysis of small breaks. Others were made to force better agreement with test data in the area of the break flow, pump degradation, and certain heat transfer processes. A great deal of effort was spent in developing calculational methods to eliminate non-physical, calculational anomalies such as "water packing" and "water stacking." The former causes the code to calculate unrealistically high fluid pressures when the control volume fills completely with liquid. The "stacking" problem results when the bubble rise model is applied to separate phases within a string of vertically oriented control volumes. With this model each such control volume contains a vapor/liquid interface - which is unrealistic unless the "junctions" represent some significant flow obstructions.

7. To help cure this problem the RELAP-4/MOD-5 code (ref. 3) employs a separate convective flux enthalpy, for each phase, in the mixture energy equation; this requires a constitutive equation for the difference between the vapor and the liquid velocities, $V_r$. The constitutive equation for $V_r$ is expressed in terms of the void fraction, $\alpha$, calculated for the junction between the two adjacent control volumes and based on the quasi-steady flow model in the

bubbly and the churn-turbulent flow regimes. For the flow regimes leading to "flooding" i.e., when the upflowing steam prevents further downflow of liquid, the relationship between $V_r$ and $\alpha$ is obtained empirically from the separate effects tests on ECC bypass phenomena. Such tests clearly show a strong dependence of "flooding" on the liquid temperature (subcooling). This particular effect is not modeled with the RELAP code on the local basis. Similar extensions were also applied to other existing codes, such as SATAN and RELAP-UK.

8. A fairly extensive table listing the existing codes and those under development, as of Spring 1976, can be found in Reference 1. The table also indicates the code applicability, the hydraulic model and the numerical solution technique. The same reference also discusses various mathematical models of two-phase flow, with the emphasis on the general homogeneous equilibrium, the drift flux, and the two-fluid models. Two auxiliary tables were also given, showing a relationship between the author's shorthand description of various models, the number and type of the implied restrictions, the number and type of the needed constitutive equations and, finally, the listing of the computed variables which could be compared with test data.

9. For a more complete treatment of the mathematical modeling of two-phase flow the reader should consult Ishii's book (ref. 4) or J. M. Delhaye's Ph.D. thesis (ref. 5) and several of his and J. Bouré's subsequent papers (ref. 6, 7). An in-depth review of the averaging operations required to make the two-phase flow field equations computationally tractable was recently given by Delhaye and Achard (ref. 8), while J. Boure presented a sobering review of problems in the mathematical modeling of two-phase flow (ref. 9). Bouré pointed out how "patching" of the existing models (as described at the beginning of this section) can lead to inconsistencies and, at best, to codes which are, one way or another, "adjusted" to agree with experimental data obtained with a particular test facility. On the other hand, Bouré also finds that our ability to properly model transient two-phase flow is also lacking. Researchers are, for example, not in agreement as to the representation of the induced mass effects in the field equations and the modeling of the constitutive laws contains a great deal of guesswork, largely because of the inability to measure the details which the more complex models need to describe. Thus, we see that existing codes have significant limitations and expert opinion is divided on how the future codes ought to be formulated, let alone solved. What then is the way out of the dilemma?

10. An expanded effort to resolve these questions has been undertaken by numerous researchers all over the world, resulting in a considerable proliferation of computer codes. For obvious reasons it is impossible to present here the various paths this world-wide effort is taking. Consequently the discussion in this paper will be limited to the path(s) sponsored by the United States Nuclear Regulatory Commission, Division of the Reactor Safety Research.

## CURRENT EFFORTS IN DEVELOPMENT OF BEST ESTIMATE CODES

### Improvements in Existing Codes

11. A statement was made above to the effect that patching up of the existing codes has questionable value regarding their truly predictive capability. Opponents of this approach have stated that gradual improvements in the RELAP code, based on comparisons with the existing test data on few smaller scale integral facilities, are nothing more than "tuning". It could be argued, however, that this "tuning" process could yield several benefits: (a) The code modifications are not accepted unless there is some physical justification involved, as well as better agreement with test data. This is a process of learning; whether (and which) models for two-phase friction, minor losses, heat transfer coefficients, critical flow, etc., previously formulated on the basis of the steady-state models or experiments, have applicability under transient conditions; (b) Comparisons of test data obtained from integral systems which differ in volume by a factor of 30 so far indicate very similar behavior. Hence, the learning or the code adjustments process based on comparisons with the small scale integral test data has considerable value; (c) A series of the code sensitivity analyses can be undertaken with an adjusted code to see the effects of variations in the models and/or parameters which were adjusted and/or which are still uncertain. This can provide valuable information for planning of future separate effects tests; to concentrate on phenomena and components which significantly affect the peak clad temperature; (d) A systems code should not be more complex than is necessary. For those regimes of transients and for those system components for which one-dimensional simple models do a satisfactory job, employment of multi-dimensional, two-fluid models would be unduly wasteful. One might argue that this process ought to be performed top-down; i.e., by performing the most complex analyses first and then gradually relaxing (simplifying) the models and recording the effects. However, while the advanced codes, more acceptable from the scientific standpoint, are not yet fully operational, variants of the RELAP code exist today that can be used for that purpose; (e) We have not yet exhausted the potential of the RELAP type code. Significant improvements, based on physics, are still possible for analyzing various regimes of a LOCA (e.g., PWR and BWR reflood) and are being currently implemented in RELAP-4/MOD 6 and RELAP-4/MOD 7, respectively.

### Advanced Systems Codes

12. Our most advanced Best Estimate systems code is TRAC, being developed at the Los Alamos Scientific Laboratory (LASL) and scheduled for initial public release at the end of 1977. In this code an attempt is being made to eliminate the known limitations of RELAP. TRAC employs a 3-dimensional representation of the whole reactor vessel and a one-dimensional representation of the remaining components of the PWR primary coolant loop. It employs the full two-fluid (2V2T) model for the interior of the reactor vessel and a full non-equilibrium drift flux (1VD2T) model for the remaining loop components.

The code is fully modularized to allow special treatment for system components that require it, in the description of the constitutive equations. TRAC will also allow consideration of non-condensible gas as it issues from the exhausted ECC accumulators. The companion paper by Reed and Kirchner presented at this conference describes the conservation and the constitutive equations considered in TRAC and their solution method.

13. We are also working towards the future generation of the licensing codes, based on more advanced thermo-hydraulic models although retaining a one-dimensional and the lumped parameter geometric representation for the sake of computational speed. There are two contenders for this purpose: THOR, being developed at Brookhaven National Laboratory (ref. 10), and RELAP-5 being developed at the Idaho National Engineering Laboratory (INEL). THOR is based on the classical Drift Flux model which accounts for the drift of the vapor velocity with respect to the velocity of the mixture's center of mass, and for thermal non-equilibrium of the dispersed phase only. RELAP-5, on the other hand, will compute the individual phase velocities and will, like THOR, consider partial non-equilibrium.

Advanced Component Codes

14. These have two basic purposes: one family, represented by LASL's K-FIX and K-TIF codes, serves as a test bed for innovation in the detailed modeling of multi-dimensional two-phase flow processes. Both of these codes are based on the two-fluid model, being the off spring of LASL's KACHINA code (hence the prefix K). K-FIX employs the Fully Implicit Exchange numerics while K-TIF deals with Two Incompressible Fields. Numerical simulation of choked flow is presently being studied with K-FIX while K-TIF had, so far, been used exclusively for the studies of ECC penetration and bypass in PWR downcomers. Both K-FIX and K-TIF are also being used for developmental work in the field of modeling of interfacial transport terms where new ideas for their representation (and the solution techniques) are being tested and compared against sophisticated model development experiments. Reference 11 deals with the data sources for Code Verification and describes the differences between the Integral, Separate Effects, and the Model Development experiments.

15. Other Component Codes are represented by LASL's SOLA-FLX, SOLA-DF, and by Pacific Northwest Laboratory's COBRA-4, which have been described during the 4th WRSR Yearly Information Meeting held in September 1976. The most recent experimental version of the COBRA code is named COBRA-DF. It employs the drift flux, full non-equilibrium model and its field equations are solved in the axial and the lateral direction. During 1977 COBRA-DF is being used exclusively for examination of Upper Head Injection of water during a LOCA in a PWR. Figure 1 illustrates USNRC's plans for the development of the advanced codes for analyses of postulated LOCA, ATWS, and RIA accidents in LWR systems, excluding the hydro-elastic codes and the containment systems codes.

## GENERAL PROBLEMS IN MODELING OF TWO-PHASE TRANSIENTS

Basic Problems with the Existing Codes

16. Concerning thermo-hydraulic models, the following is observed: HEM has been found to be fairly satisfactory for representing the blowdown phase, prior to ECC injection, for large breaks. Introduction of slip for the purpose of extending the code usefulness to small breaks, ECC injection, and reflood periods has been helpful but not sufficient; the thermal equilibrium assumption proves to be unrealistic for the analysis of refill and reflood periods while the slip or drift simulation is not well suited for modeling of counter-current flows.

17. Representation of the reactor vessel interior geometry is inadequate. The lumped parameter codes (such as RELAP) address satisfactorily the scalars involved in the mass and energy balances. The vector quantities involved in the momentum equation cannot be properly represented in lumped parameter codes, except for purely one-directional flows. Calculation of the momentum flux term is omitted when more than two junctions join one control volume. Hence, although the control volumes (or lumps) can be interconnected in arbitrary ways which may lead the observer to conclude that multi-dimensional flows can be handled - the fact is that the multidimensional transport of momenta cannot be calculated. In multi-dimensional flow situations the fluid inertia is calculated, in RELAP, in a way which gives freedom to the code user to change its contribution at will. The piping network representation of multi-dimensional geometries is shown to be accurate only for single-phase flows and under certain restrictions placed on the network arrangement (ref. 12). For other situations satisfactory solution can be obtained through experimentation with different types of networks, i.e., comparisons with test data. Purely predictive capability for untried geometries cannot be relied upon. The basic difference between a piping network and a true multidimensional representation is the omission of the cross terms in the spatial acceleration, which correlate the momentum fluxes between flow directions. System volume, lengths, and flow areas cannot be preserved simultaneously. Problems also arise in connecting single flow channels to the network representing some multidimensional region, especially if the former has a larger flow area than the sum of the network pipes joining with it.

18. Scores of technical articles on two-phase flow deal with the subject of frictional pressure drop under steady flow conditions. In most cases, the viscous forces are separated from the buoyancy and from spatial acceleration defined by the HEM model, in interpreting the experimental results. Two-phase multipliers are then defined, empirically or semi-empirically, to be used in one-dimensional flow modeling. The codes which rely on HEM use these correlations in a consistent manner. One should bear in mind, however, that there are very few instances in actual LWR plants where straight, long pipes are employed. Majority of the flow passages in an LWR are significantly different from the geometries for which the correlations have been developed. Application of steady-state correlations to flow transients is wide spread. It can be justified,

on scientific grounds, only when the temporal acceleration term becomes insignificant.

19. In the past, codes were written in a very general manner, with no particular attention being given to the peculiarities of various system components. The recent experience with the BE analyses has shown the necessity for modular codes which differentiate among the system components. In most cases these differences arise in the constitutive equations which are dependent on the boundary conditions and/or on the shape of the component. In some cases, however, even the type and number of field equations can vary among components. Patching the existing codes to account for these differences has proven to be a very difficult process.

Basic Problems with Two-Fluid Modeling

20. The largest share of the two-fluid model difficulties is associated with description of the interfacial transport terms for mass, momentum, and energy. In Reference 7 Bouré has dwelled upon the difficulties in their mathematical formulations. We shall examine these problems from an engineering point of view which identifies a basic common problem concerning the definition of the flow topology and the extent of the time and space averaged interfacial area between vapor and liquid.

21. Topology of the steady two-phase flow in vertical and horizontal pipes has been studied by several researchers (see well known texts by Tong, Wallis, and Hewitt for references). Variety of flow regimes were identified and "maped" in terms of the superficial velocities of each phase and/or other fluid properties. It is not expected that the existing correlations which describe the boundaries between the adjacent flow regimes, will also be valid during transients. Proximity of bends, Tees, and sudden flow area changes - a prevalent situation in PWR main coolant pipes - will also tend to invalidate the existing correlations.

22. Flow regime maps obtained for pipes are not applicable to the multi-dimensional flow regions. In the multi-dimensional modeling of the reactor vessel interior some computational cells may carry relatively clearly defined fluid; e.g., bubbly (or pure) liquid, or the dispersed droplets flow (or pure vapor). Other cells, however, may carry irregular, wavy interfaces. For the dispersed liquid flow regime one may be able to define a representative droplet size from the critical Weber number while, in bubbly flow, the representative bubble sizes in the bulk liquid may be related to the turbulence level and pressure. Existence of the bubbly or the dispersed droplets flow regimes can be associated with the low and the high local void fractions, respectively. Intermediate values indicate the presence of irregular interfaces.

23. Interfacial transport of mass, momentum, and energy must be related in some way to the local flow regime, hence void fraction, and to the "scale" of the dispersed phase (or to the extent of the irregular interface). When the computational cells also contain solids such as fuel rods, additional flow regime criteria must be employed. For example, the fuel rod surface temperature during reflood may be too high to allow liquid to wet the rods, resulting in the "inverted annular" flow regime. Such flow regimes play an important part in defining the solid/fluid heat and momentum transfer. It is clear that, in such cases, the knowledge of local void fraction and of phase velocities is insufficient to determine the flow topology; it needs to be supplemented by information concerning the solid surface temperature, fluid temperature, the geometric configuration of the solids contained within the computational cell, the wetting criteria, the liquid entrainment criteria (for wavy liquid surfaces), and maybe others.

24. Some type of solids such as fuel rods, control rods, guide tubes, support columns, instrument thimbles, vessel walls, core support plates, pipe walls, core support barrel walls, and various baffles and grids, are very likely to be present in every computational cell. Since the geometric configuration of the solids plays an important part in determining the flow regime it can be concluded that it will be impractical to attempt formulation of generalized criteria for the local flow regime, hence for the local interfacial transport terms, and for the solid/fluid momentum and energy exchange laws. These will have to be specialized to the specific geometries of various primary coolant system and reactor vessel components which, in turn, are to be described by the specific computer code modules. Hence the need for modularized codes.

25. Many experiments are conducted with test geometries which have the right shape; yet, the sizes may be smaller than full scale or some scale distortions may be present. When the models or correlations are normalized to obtain agreement between the calculated and the measured properties, care must be taken to ensure that the models or correlations also account for scale effects, to justify extrapolation to full scale.

26. Multi-dimensional formulation should be very helpful in accounting for the effects of gross scales such as the overall length, circumference, areas, and volumes of fluid flow passages. The scale information alluded to above concerns the effects within any given computational cell which, for practical reasons, cannot be further resolved.

27. It is envisaged that for a BE analysis of PWR LOCA about 1000 computational cells will be employed to represent the geometry of the primary coolant system flow passages. It is hoped that the important flow effects within the vessel downcomer will be resolvable by accounting only for axial and circumferential flow directions. In that case effects of the downcomer gap size will have to be accounted for in the constitutive equations for the transport terms. Their models will have to be validated through comparisons with test data obtained with test facilities featuring different sizes (scales). This is just one of many examples concerning various system components which are presently being studies in NRC sponsored experimental and analytical research efforts. Very detailed Component Codes are being utilized to establish which important effects must be incorporated in the Systems Codes.

## SPECIFIC PROBLEM AREAS RELATED TO PWR LOCA ANALYSIS

### Break Flow

28. RELAP: A sequence of the quasi-steady break flow models is being used, dependent on the "upstream stagnation quality." All other effects (of nozzle length, size, shape, etc.) are lumped into a "discharge coefficient." Break nodalization also plays a significant role.

29. TRAC: The same dynamic flow model is used at the break as in the interior flow channels, discharging against the containment pressure. One-dimensional calculation is employed with a fine axial mesh. For shaped nozzles and orifices it appears that, again, a "discharge coefficient" may have to be employed. However, the multi-dimensional Component Codes such as K-FIX could be used to "calibrate" the TRAC's break flow model, to give it predictive capabilities. Some researchers are not satisfied with the manner in which the break flow is solved in TRAC, pointing to the criteria presented in Reocreux's thesis (ref. 13).

### ECC Injection

30. During the accumulator discharge phase the flow pattern in the main coolant pipe in the vicinity of the injection Tee is extremely complex, involving a high degree of thermal non-equilibrium; especially since the discharged liquid contains significant acounts of dissolved gas which comes out of solution. It is intended to study this process in detail with three-dimensional component codes and learn how to produce a one-dimensional simplification for use in TRAC, where the ability to treat thermal non-equilibrium and unequal phase velocities will be very helpful.

31. In the RELAP code the inflows into the injection Tee control volume are lumped and the thermal equilibrium assumption drives the local pressure too low. That also leads to the previously mentioned water packing problem.

### ECC Bypass and Penetration in PWR Downcomer

32. RELAP-4: The lumped parameter model cannot describe this process from first principles. For any given nodalization scheme the subsidiary correlations must be attuned to test data. Hence, no predictive capability can be guaranteed for geometries for which test data do not exist. Even when the liquid penetration is adjusted through empirical correlations, the liquid reaching the lower plenum is not allowed to be subcooled. The assumption of thermal equilibrium also causes faster local depressurization and provides a source of artificial flow oscillations.

33. TRAC: Downcomer is modelled multi-dimensionally; hence, the fluid momenta are properly accounted for, in all important directions. The 2V2T model is essential here; nevertheless, the problems in defining the proper constitutive equations remain to be solved.

### Lower Plenum

34. During the reverse core flow some liquid fraction will tend to separate due to centrifugal action while the flow is turning around the lower edge of the core support barrel. As ECC downcomer penetration starts to fill the lower plenum the reverse flowing core steam will entrain some of the lower plenum liquid. The ECC liquid in the lower plenum will slosh, leading to periodic entrainment. A considerable amount of time may pass between the first occurence of the downcomer penetration and the time when the lower plenum liquid entrainment ceases and allows the ECC fluid to reach the reactor core inlet.

35. RELAP-4: One or more vertically stacked control volumes are utilized and the momentum flux ignored at junctions leading to the downcomer and the core. Phase separation is accounted for, with mixed success, through the use of a bubble rise model in which the void gradient and the bubble rise velocities are User specified. In some cases the vertical vapor drift model is employed. A highly simplified lower plenum entrainment correlation is used, based on steady-state test with quiescent water levels. Phase separation due to centrifugal action cannot be handled.

36. TRAC: The lower plenum is modeled in three dimensions, with the two-fluid model. The lower plenum entrainment has not yet been considered.

### Core Flow

37. The overall requirement is to have a predictive capability for the core-wide description of flow, enthalpy, and void distribution during all regimes of LOCA. These distributions directly affect the core-wide distributions of the local heat transfer coefficients, hence, core cooling. Flow redistribution must also be dealt with in the case of significant fuel clad ballooning which causes local flow blockage.

38. RELAP: Parallel stacks of the vertically connected control volumes are employed to represent the hot, average, and the "cold" regions of the core. Lateral interconnections were attempted on occasions; they oftentimes lead to numerical oscillations. Axial definition is too coarse for pinpointing hot spots. Hence, separate "hot channel" calculations are made, with a finer axial grid, to solve for the peak clad temperature. The lateral boundary conditions imposed by the adjacent channels cannot be considered. Flow blockage feedback cannot be handled directly. In the past, core quenching and liquid carry-over, during reflood, were computed on global basis utilizing empirical (FLECHT) correlations. Present trends are to consider local phenomena. A comprehensive set of heat transfer coefficients has been programmed in RELAP-4/MOD 6 for local application, based on simplified criteria for local flow regimes. An axially moving computational mesh is employed for heat transfer calculations, with different mesh sizes to allow characterization of steep axial temperature profiles and quench front propagation. Fine axial mesh, coincident with the heat conduction mesh, is used for calculation of fluid properties, in a quasi-steady manner, to account for steam superheat. This update is made after each time step of the system dynamics calculation. It introduces inconsistencies and could provide a source of significant numerical oscillations, unless damped through artificial means. The code cannot, presently perform calculations from blowdown through reflood, in a continuous manner.

39. TRAC: The core is viewed three-dimensionally with a two-fluid model. Mesh renodalization may have to be employed at the beginning of reflood to allow a more detailed tracking of the important flow regime boundaries. Reflood test data show that the presence of flow obstructions such as the upper core support plate and even the fuel spacer grids, can lead to earlier local quenching. This is attributed to the enhanced lateral droplet flow components and to "fin cooling" near those locations. It is not yet clear how these local phenomena will be accounted for; certainly not from first principles. Efficient computational schemes are still being researched for tracking of quench propagation. Criteria for defining the extent of the inverted annular flow regime have not yet been formulated, especially for the expected oscillatory type of the core inlet flow. Upper Head Injection in PWR's and BWR reflood presents a further difficulty since liquid films can propagate downwards against upflowing steam containing entrained droplets. Does this situation require a three-field model or could the falling films be tracked explicitelly with a side calculation?

## Upper Core Support Plate

40. Whether the liquid is being accumulated on top of that plate because of de-entrainment within the upper plenum (during reflood) or due to introduction of sprays (as in BWRs), is irrelevant. In both cases the question arises whether that liquid could be flowing down against the upflowing steam. Geometry of the flow passages through this plate, including the ends of the fuel assemblies, is extremely complicated. It is certain that the counter-current flow limitation offered by these passages cannot be described from first principles. Empirical correlations will have to be obtained from tests featuring the full scale hardware and for various reactor manufacturer's designs.

## Upper Plenum

41. The role of the upper plenum becomes particularly important during reflood when steam and the entrained liquid droplets enter the plenum region from below. The change in flow direction, towards the outlet nozzles, will be responsible for some de-entrainment of droplets. Much more important, however, is the effect of a veritable forest of the support columns and the control rod guide tubes among which the steam and droplets have to pass. These internals are expected to act as fairly efficient steam/water separators. The impinged droplets will form films which could either flow down or be re-entrained. Multi-dimensional calculation of the steam flow field will allow definition of the regions in which re-entrainment is likely to occur and the regions in which the falling films could propagate and form a liquid pool on top of the upper core support plate. Whether that pool will grow in height or not, will be established by the balance between (a) entrainment of the pool liquid by the vapor jetting upwards through the upper core support plate orifices, (b) addition of new liquid droplets coming from below, (c) removal of the pool liquid due to fallback into those core regions which allow counter-current flow, and (d) supply of liquid by the falling films within the upper plenum.

42. The need for considering the upper plenum de-entrainment process comes from the fact that the moisture carryover into outlet nozzles contributes to the additional steam generation inside the primary side of steam generators which, in turn, is responsible for "steam binding" i.e., the increased pressure on top of the core. The higher that pressure, the lower becomes the reflood rate which determines how fast the core is quenched. In licensing calculations the upper plenum de-entrainment process is ignored, resulting in the slowest reflood rate prediction. Since the role of the Best Estimate calculations is to quantify the degree of conservatism contained in the licensing calculations, these extremely complex phenomena need to be addressed in some way. It appears that it will be necessary to obtain test data with full scale upper plena and with the real internals hardware, depicting various manufactures's designs.

43. RELAP-4/MOD 6: Simple, global correlations are to be included for liquid de-entrainment, obtained from full scale test data. In the first version of this code a somewhat arbitrary de-entrainment "law" has been programmed with input specified magnitudes of its normalizing coefficients, pending the execution of the large scale tests. The liquid pool height is calculated through the bubble rise model. Pool liquid entrainment is also handled in a highly simplified manner. The fallback into the core is handled by the drift model, supplemented by the Wallis' flooding criterion.

44. TRAC: The flow processes in the upper plenum are viewed as three dimensional. Full scale test data will be used to normalize the constitutive equations for the momentum exchange between the fluid and the internals, as functions of the flow direction and the entrained liquid mass fraction. Local de-entrainment and re-entrainment of liquid on the support columns and guide tubes will first be studied, systematically, in separate effects tests - possibly on smaller scale - with the purpose of establishing sink and/or source terms for liquid fraction within the computational cells. The first version of TRAC will have very simplified forms of the required constitutive laws. Important large scale test data are not expected before 1979. More detailed views of the upper plenum de-entrainment process will also be attempted with the Component Codes to help resolve which important effects need to be introduced into future versions of TRAC. These studies will be complemented with sensitivity studies. The latter must be performed with the systems code since we are primarily concerned with the sensitivity regarding the peak clad temperature.

## Steam Generators (S.G.)

45. For a large break LOCA, the steam generator modeling details become important only during the reflood regime. Some of the liquid carryover which reaches the S.G. inlet nozzle will again be de-entrained as the fluid makes a turn and impinges on the inlet plenum walls and the tube sheet. The de-entrained liquid could be stored in the horizontal section of the hot leg as it drains from the S.G. inlet plenum - providing the steam velocities are not excessively high. In addition, thermal non-equilibrium could cause only partial vaporization of liquid droplets as they flow through the S.G. tubes.

46. None of these effects are being planned for inclusion in the RELAP code. Importance of these effects, for inclusion in TRAC, will first be studied with Component Codes. The first version of TRAC will not deal with the S.G. inlet plenum de-entrainment. Thermal non-equilibrium and vapor drift effects will, however, be included.

Main Coolant Pumps

47. Transient two-phase flow through centrifugal pumps is far too complex to be handled from first principles. At present, the full scale test data, obtained with the single-phase fluid (liquid) under steady flow condition, are utilized, together with the similitude laws, to form the pump performance curves. Two-phase degradation multipliers for the pump head and torque have been obtained empirically for a small scale pump and their use extrapolated to represent the two-phase behavior of large pumps. New research is underway in which both the steady and transient two-phase flow through pumps of various sizes (1/20,1/5,1/3 scale) is being studied experimentally and analytically. Semi-empirical correlations are being sought to correct for differences between the single-phase and the two-phase head and torque losses.

## CONCLUSIONS

48. LOCA analyses for licensing purposes use bounding, conservative models which do not require the details and complexities which a Best Estimate code must address. Because of significant limitations of the simplified models in existing codes, it is not possible to adequately quantify the degree of conservatism in the licensing calculations.

49. The advanced codes now under development avoid many important limitations characteristic of the existing codes. Nevertheless, many important obstacles remain to be overcome, particularly in defining interactions between vapor and liquid. Simple models employed in existing codes do not face that problem. Why then the urge to develop more complex codes? The basic intent for the advanced codes is to depart from the use of global correlations that are either obtained, from or normalized by, systems effects tests. Such global correlations are too dependent on the geometry (as well as the boundary conditions) of the test hardware which is frequently smaller than full scale of exhibits some scale distortions. It is hoped that the gross (overall) geometrical effects will be successfully resolved through the use of multi-dimensional models and that the needed correlations will be of "local" rather than the global nature. These "local" correlations will have to account for geometric effects which cannot be resolved by a reasonable discretization of the system geometry.

50. NRC is sponsoring a variety of the basic model development tests for examination of details of the interfacial exchange processes. Progress in this area is expected to be slow. Initial formulations, in TRAC, will be simple; they will be refined in a stepwise manner as knowledge increases. Our most important success, thus far, is in developing numerical methods which made the multi-dimensional solutions tractable, even with very complex thermo-hydraulic models. We have also been successful in developing an integral research plan with interdependent and complementary activities, involving very capable researchers in all of our National Laboratories.

## REFERENCES


1. FABIC, S. Review of the Existing Codes for Loss-of-Coolant Accidents. Advances in Nuclear Science and Technology, Vol. 10, Plenum Publishing Co., 1977.

2. MOORE, K. V. and RETTIG, W. H. RELAP 4: A Computer Program for Transient Thermal Hydraulic Analysis, ANCR-1127, 1973.

3. Aerojet Nuclear Co. Report RELAP-4/MOD 5: A Computer Program for Transient Thermal-Hydrualic Analysis of Nuclear Reactors and Related Systems; Users Manual (Vols. I, II, III). ANCR-NUREG-1335, 1976.

4. ISHII, M. Thermo-Fluid Dynamics Therory of Two-Phase Flow. Eyrolles, 1975.

5. DELHAYE, J. M. Équations Générales des Écoulements Diphasiques. C.E.A. -R-3429, 1968.

6. VERNIER, P. and DELHAYE, J. M. General Two-Phase flow Equations Applied to the Thermo-hydrodynamics of Boiling Nuclear Reactors, Acta Tecnica Belgica, EPE IV, 1-2, 1968.

7. BOURÉ, J. A. Mathematical Modeling of the Two-Phase Constitutive Equations. European Two-Phase Flow Group Meeting, Haifa, Israel, June 1-5, 1975.

8. DELHAYE, J. M. and ACHARD, J. L. On the Averaging Operators Introduced in Two-Phase Flow Modeling. Proceedings, OECD/NEA Specialists' Meeting on Transient Two-Phase flow, Toronto, Canada, 1976.

9. BOURÉ, J.A. Mathematical Modeling of Two-Phase Flows. Its Bases, Problems: A Review. Proceedings, OECD/NEA Specialists' Meeting on Transient Two-Phase Flow, Toronto, Canada, 1976.

10. WULFF, W. et. al. Development of a Computer Code for Thermal Hydraulics of Reactors (THOR). Brookhaven National Laboratory Quarterly Progress Reports, BNL-19978, 50455, 50458, 1975-76.

11. FABIC, S. Data Sources for LOCA Code Verification. Nuclear Safety Journal, Vol. 17, No. 6, p. 671, 1976.

12. FABIC, S. Two-and Three-Dimensional Fluid Transients. ANS Transactions, Vol. 13, No. 1, p. 386, 1970.

13. RÉOCREUX, M. Contribution à l'Etude des Débits Critiques en Écoulement Diphasique Eau-vapeur. Thèse Université Scientifique et Médicale de Grenoble, 1974.

Figure 1

USNRC Strategy For Development of Advanced Systems Codes

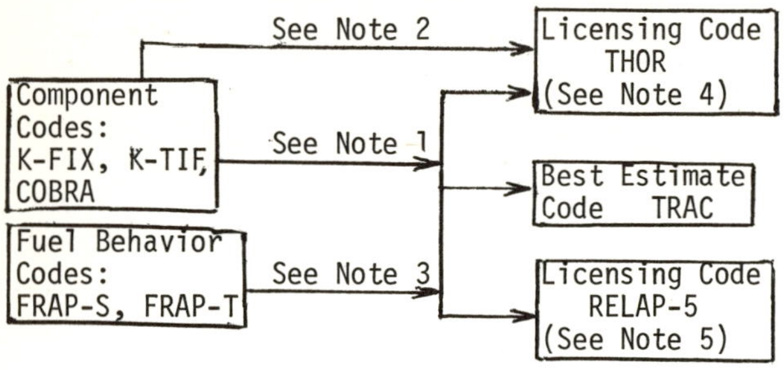


NOTES: 1. Verified models for interfacial transport processes and for wall/fluid interactions.

2. Spatial profiles of the dependent variables in multi-dimensional components for the range of the expected transient boundary conditions.

3. Sensitivity studies on the detailed fuel behavior codes to determine which important fuel behavior phenomena need to be modeled in systems codes.

4. Profile functions will be generated, for each system component, integrated in space for definition of the space-averaged dependent variables, and used to define the "junction" properties.

5. Pros-and-cons of the differences in physical modeling and the numerical solution techniques between THOR and RELAP-5 will be evaluated. This evaluation will determine which, if either, of the two codes will be developed further as a fast running code suitable for licensing calculations.

C202/77

# BLOWDOWN OF A RECIRCULATING LOOP WITH HEAT ADDITION

N. L. ARRISON, PhD, W. T. HANCOX, PhD, M.T. SULATISKY, MSc
Atomic Energy of Canada Ltd., Pinawa, Manitoba, Canada
S. BANERJEE, PhD
McMaster University, Hamilton, Ontario, Canada

SYNOPSIS Experimental data are presented for blowdown of a recirculating water loop containing two pumps, two heated sections and two heat exchangers arranged in a "figure-of-eight" geometry. The break size was varied and complete sets of data were obtained with pumps operating and bypassed. Predictions by the RODFLOW code are shown to be in good agreement with the data.

## INTRODUCTION

1. LOCA analysis codes have two distinct components:

- a mathematical structure which comprises the fundamental conservation laws, constitutive equations and boundary conditions, and the numerical solution scheme, and
- empirical content which includes transfer coefficients and component models (pumps etc).

2. To check systematically the mathematical structure and empirical content, fundamentally oriented experiments of gradually increasing complexity are needed. Edwards and O'Brien (ref. 1) have done experiments on simple tubes to investigate blowdown of initially stagnant high enthalpy water. Blowdown experiments, with added complications due to heat addition and initial flow, have been presented by Premoli and Hancox (ref. 2). This paper presents data obtained in blowdown experiments on a recirculating pressurized water loop, RD-4, which add further complicating effects.

3. The RD-4 loop is a small scale facility containing two horizontal heated sections, two single stage centrifugal pumps and two heat exchangers. The loop geometry and arrangement are of particular relevance to the verification of analytical procedures for pressure tube power reactors. However, the loop was sufficiently well instrumented and geometrically simple so that the data are of fundamental interest.

4. All the blowdown experiments were done starting with the same initial conditions. The break size was varied and a complete set of data obtained with the pumps operating, and another complete set with the pumps bypassed. Pressures, temperatures, void fractions and mass fluxes were measured at several locations around the loop. These data are compared with predictions made with the RODFLOW (ref. 3) code which is currently used in Canada for LOCA analyses.

## LOOP DESCRIPTION

5. The layout of the RD-4 loop is shown in (Fig. 1) and component dimensions are given in Table 1. It is a small scale recirculating loop with two pumps, two sets of inlet and outlet headers, two horizontal tubular heated sections and two heat exchangers arranged in a "figure-of-eight" geometry. The heated sections are at the lowest elevation and other component elevations relative to the heated sections are: headers = 3.88 m; heat exchanger inlet/outlet = 4.54 m; pump inlet/outlet = 4.12 m. The specifications of the major components follow.

6. The heated sections were 2 m long Inconel tubes of 13.4 mm* inside diameter and 1.24 mm* wall thickness. Heat was generated by the Joule effect and up to 110 kW of direct current electrical power was available for each tube. Fifteen thermocouples were brazed into each tube wall at approximately 130 mm intervals for temperature measurement.

7. The single stage centrifugal pumps (impeller diameter = 158 mm; specific speed = 760 rpm $gpm^{1/2}/ft^{3/4}$) had a rated flow and head of 4.7 ℓ/s and 43 m respectively. Alternating current electrical motors were used to drive the pumps and hence, as long as electrical power is supplied, the impeller speed remains essentially constant.

8. The heat exchangers consisted of a single 6.23 m long Inconel U-tube (inside diameter = 18.8 mm; wall thickness = 3.9 mm). Heat was rejected to an organic coolant maintained at constant temperature during blowdown.

9. The loop instrumentation included the following:

| Quantity | Instrument |
|---|---|
| Absolute/differential pressure | strain gauge transducers |
| Fluid temperature | resistance temperature detectors; thermocouples |
| Pipe/heater wall temperature | 1 mm diameter chromel-alumel thermocouples |
| Flow rate | turbine flow meters; pitot tubes |
| Flow density | attenuation of a single γ-beam from a $Cs^{137}$ source (ref. 4) |

* In some experiments Inconel tubes of 14.1 mm inside diameter and 0.89 mm wall thickness were used.

Transducer voltage signals were recorded on magnetic tape via a digital voltmeter and a PDP-8 digital computer. The sampling period was 8 ms for each of 55 channels.

## EXPERIMENT DESCRIPTION

10. All blowdowns started from the following initial subcooled state:

| | | |
|---|---|---|
| flow rate (g/s) | | 685 |
| heat exchanger secondary side temperature (°C) | | 150 |
| heat section inlet | pressure (MPa) | 4.7 |
| | temperature (°C) | 236 |
| heated section power (kW/tube) | | 54 |

11. Heat was rejected, expressed as a fraction of the total heat input, as follows:

| | |
|---|---|
| loop piping | 0.54 |
| bus bar cooling at heated section ends | 0.09 |
| heat exchanger | 0.22 |
| pressure tap cooling (various locations) | 0.15 |

12. Blowdown was initiated by opening the fast acting valve (Fig. 1) located at inlet header 1. Simultaneous closing of a second valve isolated the surge tank. The break area was determined by an appropriately sized orifice plate; specific break areas covered in our experiments were 2.5, 7.5, 10, 25, 50, and 100% of the piping cross-section area (284 $mm^2$). In experiments done with the pumps bypassed, the bypass lines were opened by fast acting valves which were activated simultaneous with the blowdown valve. All valve opening and closing times were approximately 5 ms.

13. In each experiment the power to the heated sections was increased to 70 kW/tube just prior to opening the blowdown valve. It was maintained at 70 kW for 7 seconds and then abruptly dropped to 5 kW and held constant for the duration of blowdown.

## DESCRIPTION OF RODFLOW

14. The RODFLOW code is based on the one-dimensional conservation equations assuming thermal equilibrium between vapour and liquid. The governing equations were solved, assuming homogeneous flow, in conservation law form as follows:

$$A\frac{\partial \rho}{\partial t} + \frac{\partial}{\partial z}(GA) = 0 \qquad (1)$$

$$A\frac{\partial E}{\partial t} + \frac{\partial}{\partial z} GA\left(h + \frac{G^2v^2}{2}\right) = q + GAW \qquad (2)$$

$$A\frac{\partial G}{\partial t} + \frac{\partial}{\partial z} G^2vA + A\frac{\partial p}{\partial z} + A\rho g\frac{dZ}{dz} + F + \rho W = 0 \qquad (3)$$

where A = flow area ($m^2$), e = internal energy (J/kg), $E \equiv \rho(e + G^2v^2/2)$, F = friction force per unit volume ($N/m^3$), G = mass velocity ($kg/m^2.s$), h = enthalpy (J/kg), p = pressure (Pa), q = external heating rate per unit volume ($W/m^3$), t = time (s), v = mixture specific volume ($m^3/kg$), W = shaft work per unit mass per unit length (J/kg.m), Z = elevation (m), z = space coordinate (m), and $\rho$ = mixture density. The equation of state was used in the form $p = p(e,\rho)$.

15. The two-phase friction factor was calculated using the empirical equation of Jones and Dight (ref. 5). Heat addition at the bounding walls (heated sections, loop piping and heat exchanger tubes) was determined from a finite difference solution to the heat conduction equation for the wall material, along with appropriate surface heat transfer coefficients (ref. 3). Because heat loss from the loop was significant, heat transfer coefficients for outer surfaces were calculated from steady state fluid temperature measurements. These heat transfer coefficients were held constant during the transient.

16. The pump shaft work (W) is related to the head by the following equation:

$$W = \frac{gH^*}{L} H_r \qquad (4)$$

where g = gravitational acceleration ($m/s^2$), $H_r$ = rated pump head, H* = ratio of the actual head to the rated head, and L = length of the segment to which the work is added. The head developed by the pump in two-phase flow is defined as (ref. 6):

$$H^* = H^*_1 - M(H^*_1 - H^*_2) \qquad (5)$$

where $H^*_1$ and $H^*_2$ are the single-phase and fully graded head respectively, and M is the following function of void fraction ($\alpha$):

$$M = 5 \qquad 0 < \alpha < 0.2$$
$$= 1 \qquad 0.2 < \alpha < 0.95$$
$$= 20(1-\alpha) \qquad 0.95 < \alpha < 1.$$

For these experiments, the pump impeller speed was maintained constant throughout the blowdown and hence a torque-speed equation is not required. The single-phase and fully degraded head are functions of the volumetric flow rate ($Q^*$) only; a partial tabulation of the functions used follows:

| Q* | $H^*_1$ | $H^*_2$ | Q* | $H^*_1$ | $H^*_2$ |
|---|---|---|---|---|---|
| - 1 | 2 | 4.5 | 0.4 | 1.23 | 0.15 |
| - 0.75 | 1.63 | 3.70 | 0.6 | 1.2 | 0.14 |
| - 0.5 | 1.43 | 2.89 | 0.8 | 1.10 | 0.13 |
| - 0.25 | 1.30 | 2.09 | 1.0 | 1.0 | 0. |
| 0 | 1.28 | 1.28 | 1.5 | 0.59 | -0.72 |
| 0.2 | 1.26 | 0.16 | 2.0 | 0. | -1.25 |

17. To solve equations (1) to (3) the loop piping is divided into control volumes with fluid properties defined at the center and mass velocities at the boundaries as shown below

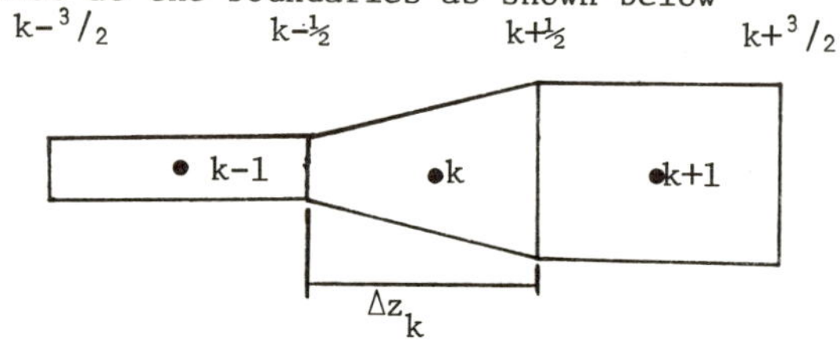


The conservation equations are approximated by

$$\left(\frac{d\rho}{dt}\right)_k = \frac{(AG)_{k-\frac{1}{2}} - (AG)_{k+\frac{1}{2}}}{A_k \Delta z_k} \qquad (6)$$

for $G_{k+\frac{1}{2}} > 0$.

$$\left(\frac{dE}{dt}\right)_k = \frac{1}{A_k \Delta z_k} \left[ (AG)_{k-\frac{1}{2}} \left(h_{k-1} + \frac{G^2_{k+\frac{1}{2}} v^2_{k-1}}{2}\right) - (AG)_{k+\frac{1}{2}} \left(h_k + \frac{G^2_{k+\frac{1}{2}} v^2_k}{2}\right) \right] + q_k + (AG)_{k+\frac{1}{2}} W_k \quad (7)$$

$$\left(\frac{dG}{dt}\right)_{k+\frac{1}{2}} = \frac{(AG^2)_{k-\frac{1}{2}} v_{k-1} - (AG^2)_{k+\frac{1}{2}} v_k}{A_{k+\frac{1}{2}} \Delta z_k} - \frac{P_k - P_{k+1} - \rho_k g \Delta Z_{k+\frac{1}{2}}}{\frac{1}{2}(\Delta z_k + \Delta z_{k+1})} - F_{k+\frac{1}{2}} - \rho_k W_k \quad (8)$$

for $G_{k+\frac{1}{2}} < 0$

$$\left(\frac{dE}{dt}\right)_k = \frac{1}{A_k \Delta z_k} \left[ (AG)_{k-\frac{1}{2}} \left(h_{k-1} + \frac{G^2_{k-\frac{1}{2}} v^2_{k-1}}{2}\right) - (AG)_{k+\frac{1}{2}} \left(h_{k+1} + \frac{G^2_{k+\frac{1}{2}} v^2_{k+1}}{2}\right) \right] + q_k + (AG)_{k+\frac{1}{2}} W_k \quad (9)$$

$$\left(\frac{dG}{dt}\right)_{k+\frac{1}{2}} = \frac{(AG^2)_{k+\frac{1}{2}} v_{k+1} - (AG^2)_{k+3/2} v_{k+2}}{A_{k+\frac{1}{2}} \Delta z_{k+1}} + \frac{P_k - P_{k+1} - \rho_k g \Delta Z_{k+\frac{1}{2}}}{\frac{1}{2}(\Delta z_k + \Delta z_{k+1})} - F_{k+\frac{1}{2}} - \rho_k W_k \quad (10)$$

For a system of K discrete volumes, this procedure results in a system of 3K ordinary differential equations that can be solved by a variety of integration techniques. We have used the simplest integration technique by approximating the time derivatives by

$$\left(\frac{df}{dt}\right)_k = \frac{f_k^{n+1} - f_k^n}{\Delta t} \quad (11)$$

where superscript n denotes the value at the nth time step.

COMPARISON OF EXPERIMENT AND PREDICTION

18. To illustrate the important features of the results we have selected two experiments:

- a break area equal to 50% of the feeder cross-section with pumps bypassed, and

- a break area equal to 7.5% of the feeder cross-section with pumps operating during blowdown.

19. Figure 2* shows selected experimental and predicted absolute pressure, differential pressure and heated section temperature histories for a 50% break with pumps bypassed during blowdown. As can be seen, the agreement between experiment and prediction is good. Note the abrupt drop in differential pressure across heated section 1 during the first 7 seconds of the blowdown. Due to the attendant low flow during this period there is a rapid rise in heated section 1 temperature until (at 7 seconds) the electric power is dropped abruptly to 5 kW. Beyond 7 seconds we used a simple rewetting model; the post dryout heat transfer coefficient was imposed until the surface temperature reached 400°C at which time the surface was assumed to rewet.

20. Figure 3+ shows results obtained for a 7.5% break with pumps operating. Again the agreement between experiment and prediction is good. There is an extended period of low flow through heated section 2 and thus an attendant rapid rise of heater temperature. This case gave the highest heater temperature in both experiment and prediction. As may be seen from Fig. 3, pump 1 develops full head for approximately 12 seconds into blowdown thus locating the region of low flow, associated with flow reversal, in heated section 2.

21. Maximum heated section temperature is shown as a function of break size, for the complete series of experiments, in Fig. 4. The RODFLOW predictions and the experiments are in excellent agreement with regard to the break size (7.5%) which yields the highest heater temperature. However, RODFLOW predictions of heater temperatures are in every case higher than in the experiment. The conservatism is primarily due to the onset of dryout being predicted earlier in the blowdown than actually observed. This is because the thermal equilibrium model predicts an earlier sharp increase in density than occurs in the experiment. We believe that a thermal non-equilibrium model is needed to reduce the degree conservatism. Changes to the empirical heat transfer coefficients have little effect.

CONCLUSIONS

22. These experiments for blowdown of a small scale recirculating water loop, RD-4, provide a sensitive basis for determining the performance of LOCA analysis codes. Accurate predictions of heater temperature demand that rates of depressurization, mass flow and heat transfer throughout the loop be correctly predicted. Because the loop was well instrumented and geometrically simple, we believe that the data are of fundamental interest.

23. Predictions of the RODFLOW code are shown to be in good agreement with the experimental results. The predicted results are relatively insensitive to the choice of empirical heat transfer coefficients and hence fluid dynamic effects dominate. We conclude that the homogeneous equilibrium model for flow-boiling is adequate for blowdown prediction. The conservatism in temperature predictions could be reduced by introducing thermal non-equilibrium effects into the model.

REFERENCES

1. EDWARDS A.R. and O'BRIEN T.P. Studies of phenomena connected with depressurization of water reactors. Journal of the British Nuclear Energy Society, April 1970. pp. 125-135.

* heated tube inside diameter = 13.4 mm; wall thickness = 1.24 mm

\+ heated tube inside diameter = 14.1 mm; wall thickness = 0.89 mm

2. PREMOLI A. and HANCOX W.T. An experimental investigation of subcooled blowdown with heat addition. Proceedings of CSNI Specialists Meeting on Transient Two-Phase Flow, Toronto, 1976, to be published.

3. BANERJEE S., HANCOX W.T., JEFFRIES R.B. and SULATISKY M.T. Transient two-phase flow and heat transfer during blowdown from subcooled conditions with heat addition. AIChE paper 24, 15th National Heat Transfer Conference, San Francisco, 1975.

4. HEIDRICK T.R., SALTVOLD J.R. and BANERJEE S. Application of a 3-beam γ-densitometer to two-phase flow regime and density measurements. Nuclear, Solar and Process Heat Transfer, St. Louis, AIChE, 1976.

5. JONES A.B. and DIGHT D.G. Hydrodynamic stability of a boiling channel, Part 2, Knolls Atomic Power Laboratory report KAPL-2208, 1962.

6. OLSON D.J. Personal communication, Idaho National Engineering Laboratory, Idaho Falls I.D., 1974.

| | SECTION | LENGTH (m) | DIAMETER (mm) | WALL THICKNESS (mm) | ELEVATION CHANGE (m) | MEASURED* ΔP (kPa) |
|---|---|---|---|---|---|---|
| HEADER | A,E,J,N | 0.914 | 73.7 | 7.62 | 0 | - |
| | B | 10.58 | 20.9 | 2.87 | -3.88 | 25.5 |
| HEATED SECTION | C,L | 2 | 13.4/14.1 | 1.24/0.89 | 0 | 79.2 |
| | D | 11.66 | 20.9 | 2.87 | 3.88 | 104.1 |
| | F | 3.14 | 20.9 | 2.87 | .67 | 48.2 |
| HEAT EXCHANGER U-TUBE | G,P | 6.23 | 18.8 | 3.9 | 3.12+ | 20.7 |
| | H | 9.73 | 20.9 | 2.87 | -.43 | 43.4 |
| | I | 12.75 | 26.6 | 3.38 | -.24 | 24.1 |
| | K | 11.94 | 20.9 | 2.87 | -3.88 | 25.5 |
| | M | 10.71 | 20.9 | 2.87 | 3.88 | 104.1 |
| | O | 3.38 | 20.9 | 2.87 | .67 | 48.2 |
| | Q | 2.94 | 20.9 | 2.87 | -.43 | 43.4 |
| | R | 7.3 | 26.6 | 3.38 | -.24 | 24.1 |
| | | 104.25 | | | | |

Total Surface Area = 7.634 $m^2$

Total Volume = .05275 $m^3$

Table 1: Specifications of loop components

* At a flow rate of 685 g/s and water temperature $\simeq$ 236°C.
\+ Elevation of U-tube.

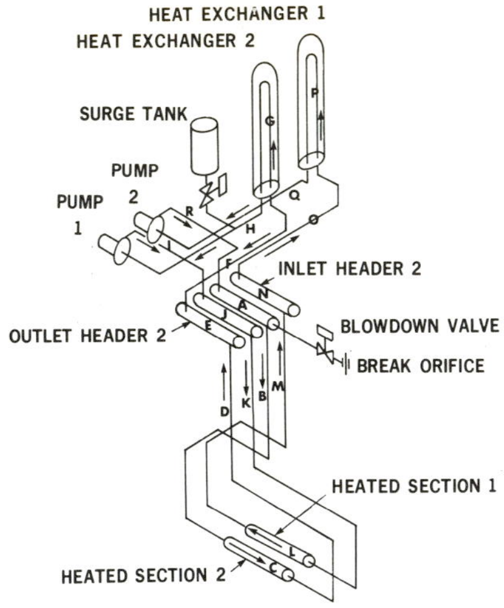


**Fig. 1: RD–4 loop schematic**

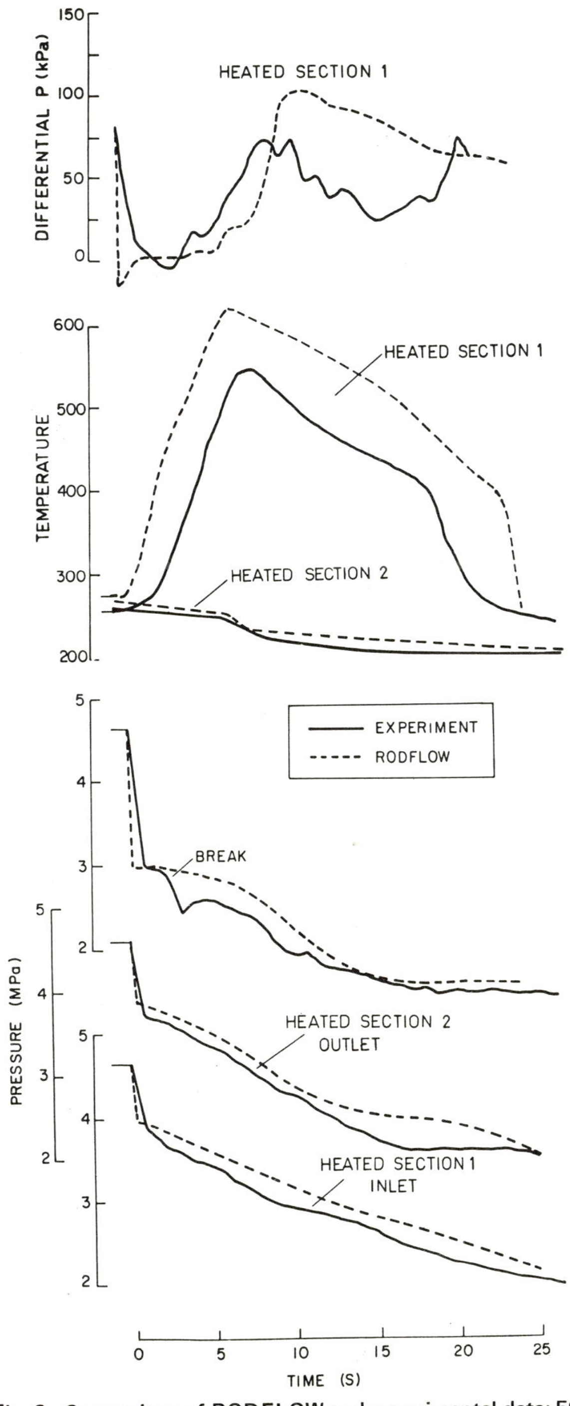


**Fig. 2: Comparison of RODFLOW and experimental data; 50% break with pumps bypassed**

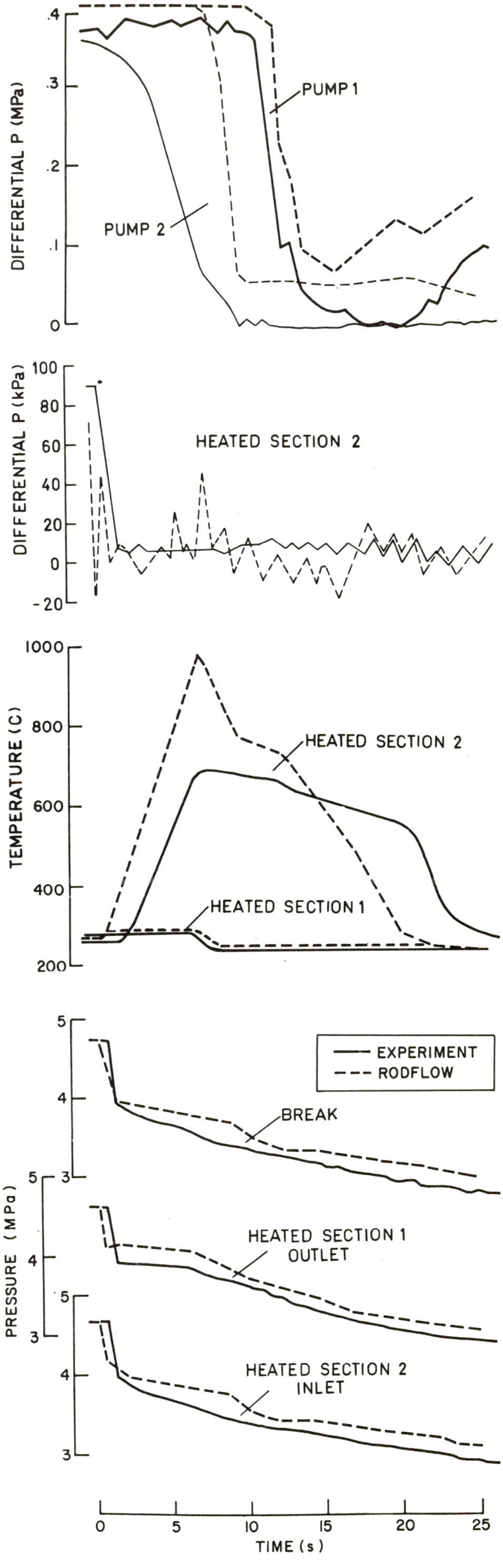


**Fig. 3: Comparison of RODFLOW and experimental data; 7.5% break with pumps operating**

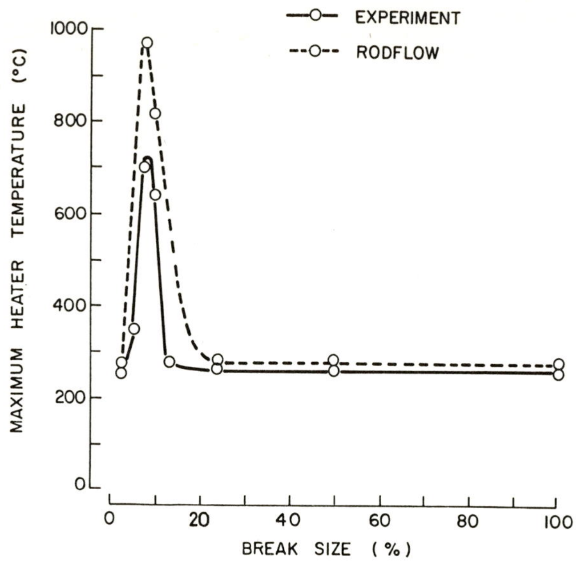


**Fig. 4: Maximum heater temperature as a function of break size**

# C203/77

# TWO-PHASE PERFORMANCE CHARACTERISTICS OF A PWR PRIMARY PUMP UNDER LOCA CONDITIONS

P. GRISON and J. F. LAURO
Electricité de France, Direction des Etudes et Recherches,
Service E. P., 6, Quai Watier, 78400 Chatou, France

SYNOPSIS A mathematical model, based on the Euler's theory and a limited flashing, is presented for the flow calculation through a pump working in two-phase conditions. Similarity criteria for representative experimental conditions are studied. Experimental test loop and the first experimental results are described.

## INTRODUCTION

The discharge rate in connection with critical flow through a pipe or a pump, is a key parameter in safety analyses of the loss of coolant accident (L.O.C.A.) in a P.W.R. reactor. The ability to calculate accurate parameters for the flow through the pump is of fundamental importance for the calculation of the loss of coolant rate because of the large pressure drop through the pump, and, as a consequence, of the important flashing rate and critical flow occuring at this point. In order to test the mechanical and hydraulic behaviour of the P.W.R. primary pump in such conditions, Electricité de France decided to build a test loop at CHATOU. In this paper we firstly present the theoretical studies, and the flow calculation based on a non-equilibrium thermal-hydraulic model. Then we present some experimental results in single phase flow and in two phase flow.

## MATHEMATICAL MODEL

### 1.1 Description of the model.

In order to calculate the pump characteristics during the L.O.C.A., we need to know at all times the pressure, temperature, density, velocity of the fluid along each part of the pump.

As these performances largely depend on the density we have to correctly describe all the phenomena which can occur in the pump such as vaporization, developed cavitation, waves propagation, critical flow etc ....

Analysis of the flow through a pump shows that the pressure gradient is very high and in case of accident the fluid may pass through the saturation line inside the pump. But as the pressure change is very rapid, the fluid can not flash immediatly and is in a non-equilibrium state. So, the fluid properties depend not only on local conditions, (pressure, temperature) but also on its "history". An axial model describing the fluid evolution along each part of the pump (inlet, impeller, diffuser and volute casing) as a function of the section-averaged properties of the fluid may take into account such a behaviour. All the functions only depend on time t and a space coordinate Z.

### 1.2 Equations.

This model is established using fundamental conservation equations of mass, momentum and energy (ref. 1). A 4th equation describes the fluid evolution giving a correlation for the flashing rate. This equation has been established using various experimental results about flashing in pipes.

Conservation of mass

$$\frac{\delta A\rho}{\delta t} + \frac{\delta A\rho V}{\delta z} = 0 \qquad (1)$$

Conservation of momentum

$$\frac{\delta A\rho V}{\delta t} + \frac{\delta A\rho V^2}{\delta z} + A\frac{\delta P}{\delta z} = -\tau X + A\rho g \qquad (2)$$

Conservation of energy

$$\frac{\delta A\rho\ (H + V^2/2)}{\delta t} + \frac{\delta A\rho V\ (H + V^2/2)}{\delta z} - A\frac{\delta P}{\delta t} = A\rho g V \qquad (3)$$

Flashing rate

$$\frac{\delta x}{\delta t} + V\frac{\delta x}{\delta z} = \frac{x^* - x}{\theta} \qquad (4)$$

and the state equation for the fluid.

In the runner these equations are written in a moving coordinate system, that is we add centrifugal and coriolis effect in equation of momentum and energy (ref. 2).

This one-dimensional model supposes the flow remains parallel to the vanes in the runner. But, as a matter of fact, this hypothesis is only an approximation. So, all three-dimensional effects as losses of momentum or energy by separation or eddies effect and recirculation are introduced by correlations at the inlet and at the outlet of the impeller. Three different kinds of losses may be distinguished (ref. 3).

a) Friction losses due to viscosity effect, introduced by a classical correlation (such as $\lambda\ \rho\ \frac{|V| \cdot V}{2}$) and distributed all along the pump. ($\lambda$ may be a function of quality for the two phase flow).

b) Enlargement effect due to separation or eddy effect in the non uniform section part or due to an abrupt change of velocity at the inlet or at outlet of the impeller for off-design flow rate.

c) Recirculation effect due to a non uniform velocity in the channel of the impeller and to a recirculation in the diffuser and in the casing. This loss is important only for small flow rate.

The formulation of these losses is indicated in Fig. 1, as a function of the flow rate.

## SIMILARITY CONSIDERATIONS

In order to have a good description of the actual phenomena the flow conditions have to be similar for the model and for the actual pump. So, some criteria are necessary to define experimental conditions.

### 2.1 Mechanical considerations

Similarity of the flow in hydrodynamic machines requires a constant ratio between the fluid velocity and peripheral velocities for all geometrically similar points of the machines compared. This condition is therefore expressed by the relation :

$$\frac{V}{U} = \frac{V}{\omega r} = \text{constant} \qquad (5)$$

As we know, in two-phase flow we may have different configurations for the flow pattern (slug flow, annular flow ...) and to have a similar behaviour, we have to stay in the same flow regime. Taking the values for the minimum velocity and according to the different maps of flow pattern (ref. 4), the flow regime should be the same (dispersed flow in both cases) with the condition $V_m = V_p$.

But it is obvious that in our case we neglect some aspects of the problem :

a) Friction effects : the Reynolds'number will not be the same :

$$R = \frac{VD}{\nu} \rightarrow R_m = 1/10\ R_p$$

$$R_m = 10^6 \quad R_p = 10^7$$

But for so high a Reynolds number the flow is quite the same and the formulation for friction losses or singular pressure drop are the same in single phase as in two-phase flow if it is independant of the pressure gradient.

b) Centrifugal effect. In two-phase flow, rotation of the runner may be responsible for a separation between vapor and water by centrifugal effect. If we calculate in both extreme cases (vapor and water) we find (ref. 6) :

x ≠ 1 for little drops of water in vapor : the separation effect will be approximately the same

x ≠ 0 for little bubbles of vapor in water : the separation effect will be multiplied by 10.

But more investigations are necessary to determine how important is the separation effect.

### 2.2 Thermodynamic considerations

The behaviour of the fluid will be similar only under a lot of conditions.

The state equation has to be the same. For this reason the experimental conditions for the pressure and temperature will be the same in both cases. And if we add the fact velocities are identical, the pressure line along the pump will be the same for a fluid in thermodynamical equilibrium.

But if the flashing rate is limited, as it seems to be, (ref. 7, 8, 9) we may have some difference between the two cases. This difference will depend on the flashing rate law.

As we can see, it will be impossible to have a complete similarity for all the phenomena. Power consideration and the fact that we want a steady state during the test have led us to choose a geometrical scale of 1/10. According to (5), the corresponding rotation speed will be $\omega_m = 10\ \omega_p$. But the fact that some phenomena will not be the same implies that on the one hand we have to carry out complementary experiments in order to precise the notion of relaxation time. On the other hand we will be able to give realistic predictions for the actual pump only by means of a performant code calculation.

## THEORETICAL PREDICTIONS OF THE PUMP BEHAVIOUR DURING L.O.C.A.

Depending on the location of the break (see Fig.2) we have two possibilities for the pump behaviour (ref. 10) :

a) if the break occurs between E and D, the pressure falls from 120 bars to 1 bar and the fluid flow from the reactor through the pump in the C.D. direction : the impeller speed increases as the flow rate increases. The pump becomes a centrifugal turbine. The evolution of the pump may be represented in diagram (3) by line (a).

b) if the break occurs between A and C, the pressure of the core is very much higher than the head of the pump, so the flow rate becomes zero and then negative. Due to the inertia, the speed of pump remains positive and then becomes negative. The pump becomes a turbine. Evolution of the pump may be represented by line (b) in diagram (3).

But as we do not know exactly the location of the two curves (a) and (b), the first step is to investigate the pump behaviour, in steady state, for all points of the regions I, II and III of the diagram. The second step is to introduce all the actual parameters of the pump (inertia, torque...) in the computer code to describe the pump behaviour during the transient phenomenon.

## EXPERIMENTAL TEST FACILITY

### 4.1 Description

Fig. 4 shows a diagram of the test facility (a detailed description is given in ref. 11). Schematically we have two pressure vessels B1 and B2 producing 6 m$^3$ of pressurized hot water with constant conditions of pressure and temperature. This water is partially vaporized in a convergent-divergent section. By means of various sections we may adjust the pressure drop between the vessel and the pump, and consequently the quality. Then the fluid passes through the pump. The pump may be connected in both directions : the inlet connected to the high pressure for the direct flow tests and outlet connected to the high pressure for the reverse flow tests. The pump is drawen by a 2 MW D.C. motor and a high speed shaft. The pressure downstream the pump is regulated either to have a constant value for this downstream pressure or to hold the flow rate at a constant value. Then the fluid goes to the atmosphere through a silencer.

### 4.2 Basic test loop data

| | |
|---|---|
| Upstrean pressure ........................ | 120 bars - 20 bars |
| Downstream pressure .................... | 130 bars - 1 bar |
| Temperature max. ......................... | 320°C |

Quality .................................... 0 - 40 %
Test duration ......................... 30 s - 180 s
Max. flow rate ........................ 200 l/s
Pump velocity ......................... 45000 rpm - 350 rpm
Electrical power of the motor 2 MW
Time to reach the maxi. velocity 8 s.

4.3 Measurement techniques

The determination of the flow rate and the initial thermodynamic conditions upstream the pump is very important for the calculations. Most of the instrumentation is therefore focussed on this task. A lay-out of the instrumentation is shown in Fig. 5.

a) Mass flow rate. The mass flow rate is determined by a venturi flowmeter located just beneath the pressure vessel, so it works in single phase flow conditions. An indirect method can be used by means of the variation of the level in the vessel.

b) Pressure. In order to calculate the quality of the fluid upstream the pump we need to know precisely all the pressures before the pump. For this purpose we use 20 differential pressure transducers and 2 absolute pressure transducers with a time-response about 0.2 s. Fast pressure fluctuations are measured by two additional piezotransducers. The pressure downstream the pump is determined by an absolute pressure transducer.

c) Temperature. We measure the temperature only in 3 different points : in single phase flow before the vaporization to know the initial conditions, and upstream and downstream the pump. But in two-phase flow it seems very difficult to measure the temperature of a non equilibrium flow and we probably always measure the saturation temperature.

d) Void fraction. Currently we use only a gamma-densitometry method. The system utilizes a single gamma beam crossing the pipe by two berylium windows, and passing through the center of the pipe. The characteristics of the gamma source are :

NATURE .................................... Cobalt 57
ACTIVITY .................................. 80 mci
ENERGY .................................... 122 kev
CONTRAST .................................. 2,5

The two void fraction measurements are located upstream and downstream the pump.

To complete the average result given by the gamma densitometry method we will use two miniature conductivity probes to know the flow pattern and the void fraction distribution.

EXPERIMENTAL RESULTS

5.1 Single phase flow

The first step is to verify the code calculations in single phase flow for all the cases ($\omega > 0$, $\omega < 0$, $Q > 0$, $Q < 0$). Experimental results show a quite good agreement with the theoretical predictions for a flow rate varying from - 200 % up to + 300 % and a rotation speed varying between - 100 % and + 300 % (Fig. 6).

5.2 Two-phase flow

Experimental results are not yet available and will be presented directly at the meeting.

NOTATION

A : cross-section area
P : pressure
$\rho$ : density
V : velocity
z : space coordinate
H : enthalpy
$x^*$ : theoretical quality (assuming equilibrium)
x : actual quality
$\theta$ : relaxation time
U : peripherical velocity
$\omega$ : rotation speed
r : radius
$\mathcal{R}$ : Reynolds number
$\nu$ : viscosity
g : gravity projection on the z axis
X : perimeter
$\tau$ : shear stress

SUBSCRIPT

M : model
P : actual pump

REFERENCES


1. HOUDAYER, SUREAU and BAUER A non equilibrium axial flow model, Toronto 1976.

2. HOUDAYER and REGNIER Le code pompe, rapport E.D.F. 1976.

3. STEPANOF A.J. Centrifugal and axial flow pumps, J. Wiley, U.S.A.

4. MANHANE and AL. INT.J. Multiphase flow, Oct. 1974.

5. POLOMIK E.E. Nuclear Eng. and design 35, 1975.

6. GRISON P. Note sur les similitudes, rapport E.D.F., 1975.

7. REOCREUX M. Débit critiques en diphasique, Thèse, Grenoble 1974

8. FRIZ G. and RIEBOLD W. Studies on thermodynamic non equilibrium, Toronto 1976.

9. KROEGER P.G. Non-equilibrium drift flux model, Toronto 1976.

10. LAURO J.F. Essai sur maquette d'une pompe, rapport E.D.F. 1975.

11. BOIVIN J.Y. and CHAPUZY J.J. Test loop INDIRA, Manchester 1977.

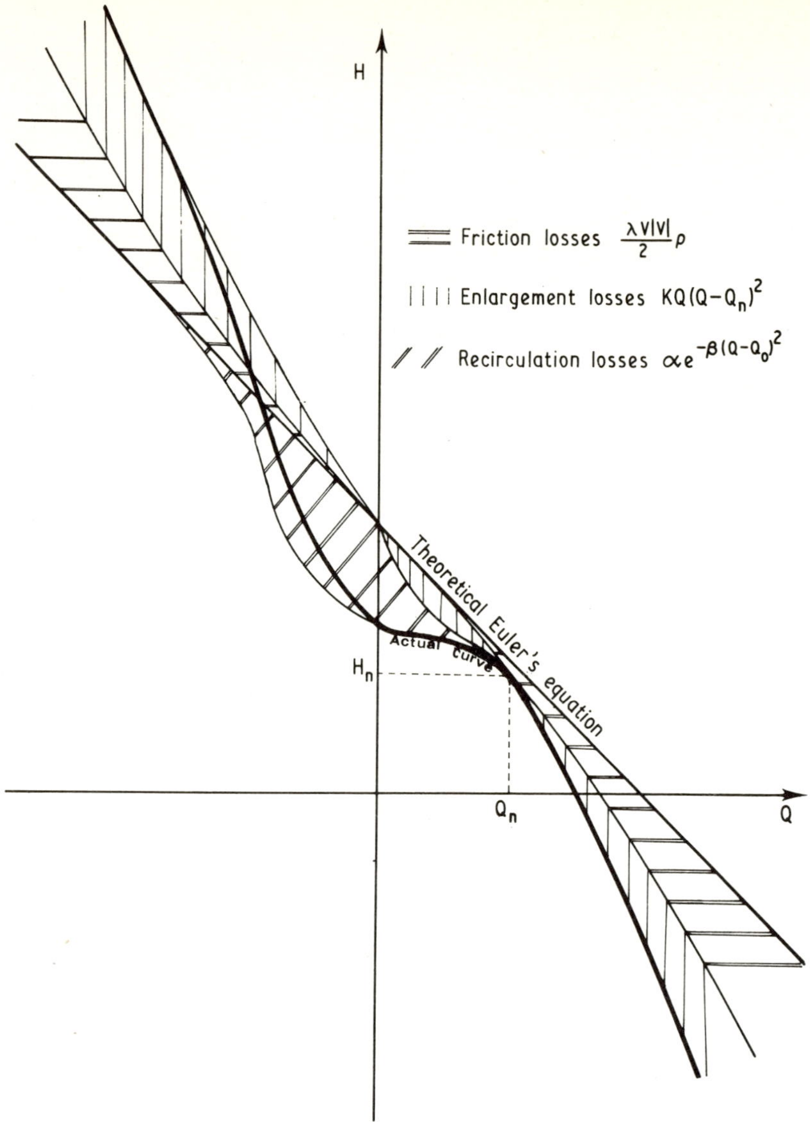


Fig. 1: Pump performance curve

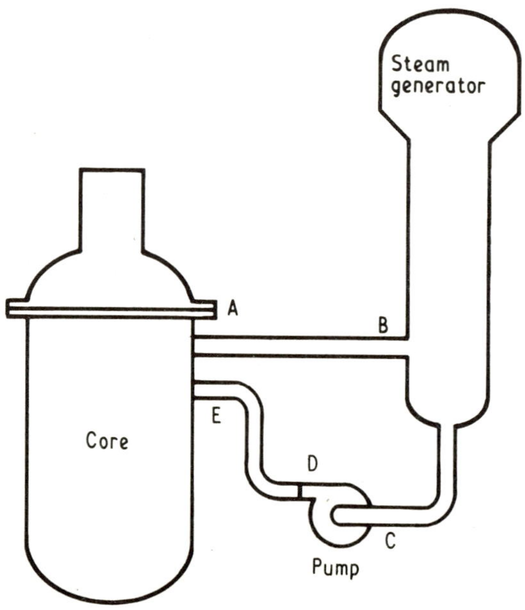


Fig. 2: PWR primary loop

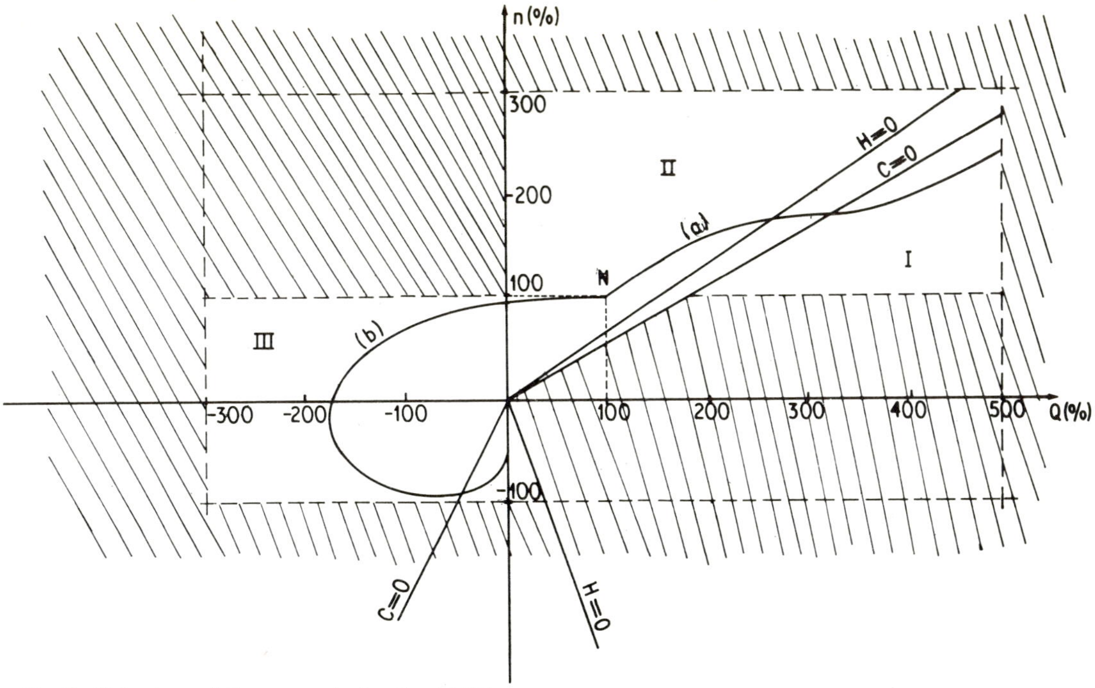


Fig. 3: Behaviour of the pump during the LOCA

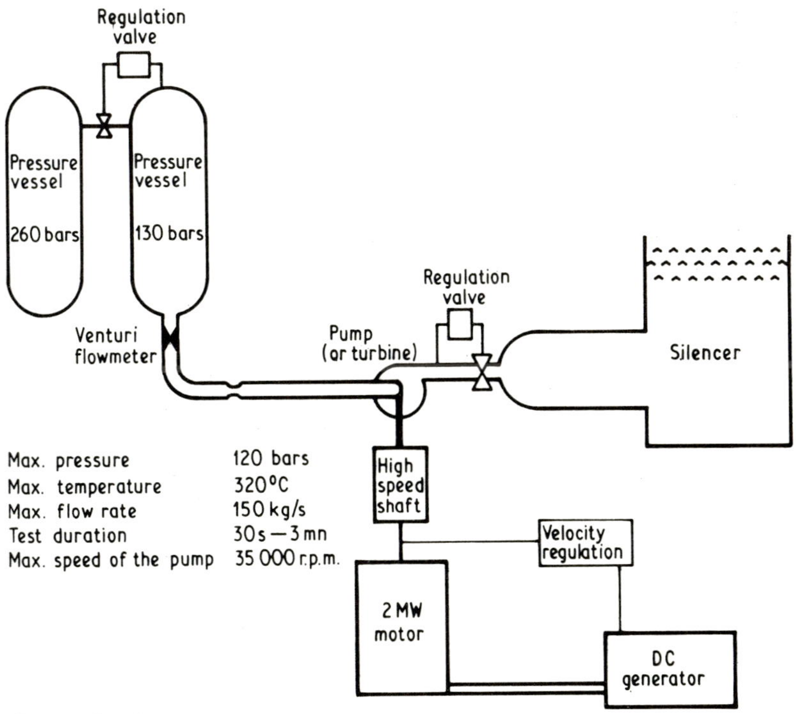


Fig. 4: Test loop

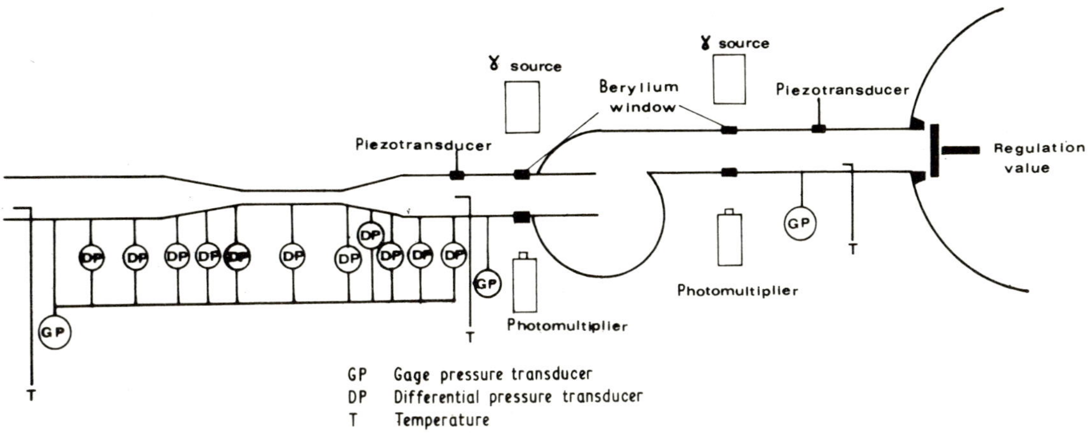


Fig. 5: Instrumentation

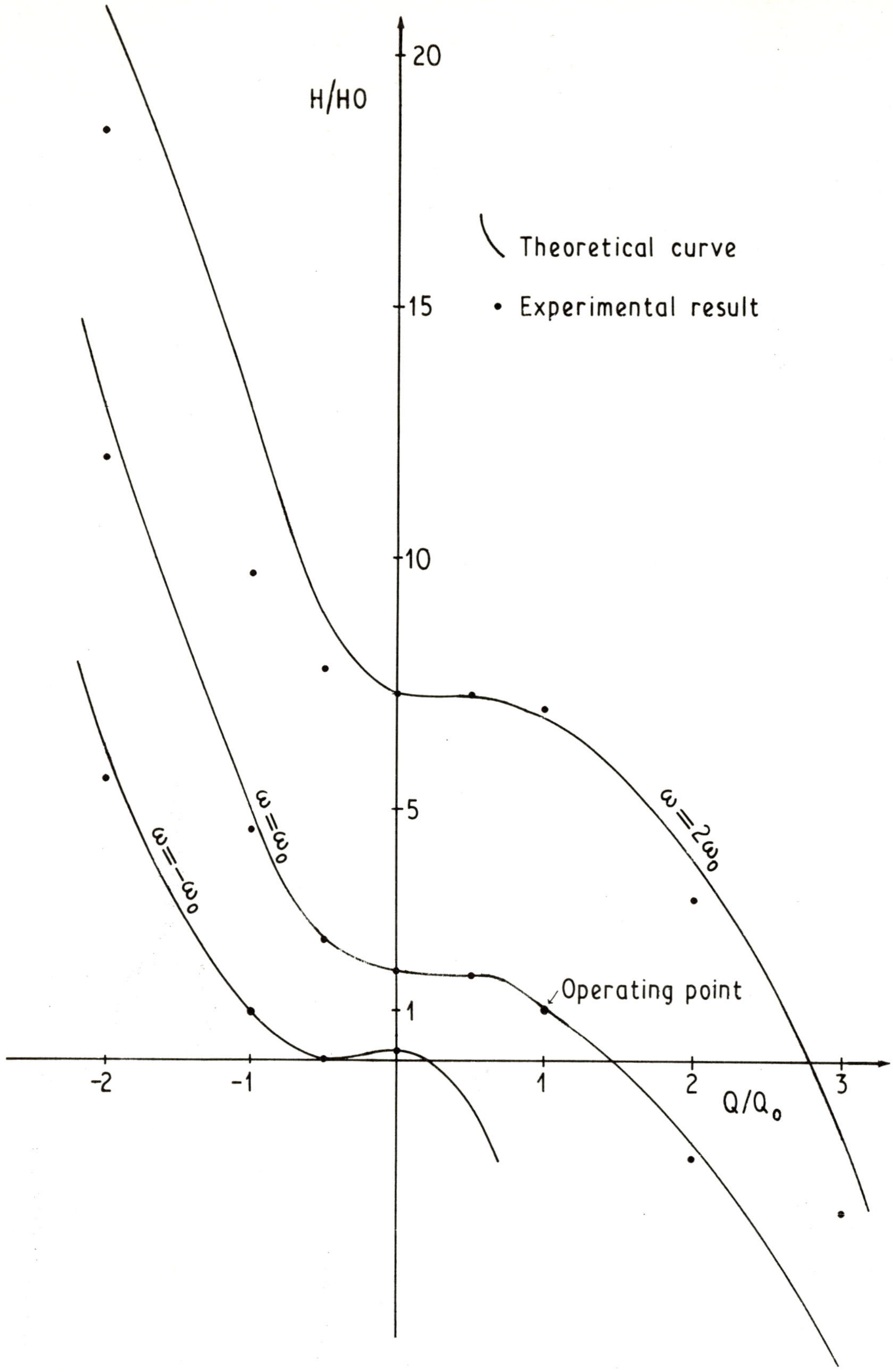


**Fig. 6: Head-flow characteristics in single phase**

C204/77

# STUDIES OF THE PHYSICAL MODELS USED IN TRANSIENT TWO-PHASE FLOW ANALYSIS

K.H. ARDRON, BA, MSc, R.B. DUFFEY, BSc, PhD, P.C. HALL, BSc
Central Electricity Generating Board, Berkeley Nuclear Laboratories, Berkeley, Glos.

SYNOPSIS We examine two-fluid and 'drift-flux' models of the form proposed for calculation of two-phase flow phenomena important in reactor safety analysis.

Two-fluid theory is used to develop new models for acoustic wave propagation in a vapour-liquid mixture, and for calculating critical vapour-liquid flow rates in a pipe. Predictions agree reasonably with available data.

Drift flux correlations are compared with new measurements of void fraction in a vertical pipe with and without a pin-bundle. Purely empirical correlations give reasonable agreement with data. Physically based models usually over-predict the effective mixture height in the channel, and fail to represent the observed trends with pressure.

NOTATION

| | |
|---|---|
| $a_i$ | interfacial area per unit mixture volume |
| $A_z$ | flow area at position z |
| c | sound speed |
| D | pipe diameter |
| $D_\ell$ | liquid diffusivity |
| $G_c$ | gas constant |
| $G^*$ | critical mass velocity |
| h | enthalpy per unit mass |
| j | volumetric flux |
| $j_k^*$ | non-dimensional volumetric flux $(=j_k[\rho_k/gD\rho_{\ell g}]^{1/2})$ |
| J | dimensional group $(J = [s\, g\rho_{\ell g}/\rho_\ell^2]^{1/4})$ |
| k | wavenumber |
| Ku | Kutateladze Number $(=j_g/J)$ |
| $N_b$ | bubble number per unit liquid mass |
| p | pressure |
| $q_{ik}$ | heat flux from interface to phase k |
| r | distance from bubble centre |
| $R_b$ | bubble radius |
| s | surface tension |
| t | time |
| T | temperature |
| U | velocity |
| $V_{gj}$ | drift velocity $(U_g-j)$ |
| z | axial co-ordinate |
| $D/D_kt$ | substantive derivative for phase k $(\equiv \partial/\partial t + U_k\, \partial/\partial z)$ |
| $\alpha_k$ | volumetric concentration of phase k $(\alpha_g = (1 - \alpha_\ell))$ |
| $\gamma$ | specific heat ratio of vapour |
| $\Gamma_k$ | mass generation rate of phase k per unit volume |
| $\varepsilon_k$ | conductivity of phase k |
| $\zeta$ | time co-ordinate $(=\int_o^t R_b^4(t')dt')$ |
| $\lambda$ | spatial co-ordinate $(= \frac{1}{3}(r^3-R_b^3(t))$ |
| $\rho$ | density |
| $\tau_k^d(\tau_k^w)$ | drag force per unit mixture volume due to interphase (wall) shear |
| $\xi$ | force per unit volme $(= \tau_g^d + (U_g-U_\ell)\frac{\partial\tau_g^d}{\partial U_g})$ |

 

| | |
|---|---|
| $\Phi$ | energy dissipation rate per unit volume due to friction |
| $\omega$ | angular frequency |
| $\omega_q$ | characteristic frequency = $\omega^{1/2} D_\ell^{-1/2} \varepsilon_\ell a_i \, (dT/dp)^{SAT} e^{i\pi/4}$ |

Subscripts

| | |
|---|---|
| g | gas property |
| $\ell$ | liquid property |
| i | interface property |
| $g\ell$ | difference between gas and liquid property |
| o | property of undisturbed flow or reservoir property |

Superscripts

| | |
|---|---|
| SAT | saturation property |
| ' | small perturbation |
| ^ | maximum value in channel |

## GENERAL INTRODUCTION

1. The need for more rigorous models for water reactor safety analysis has stimulated an intensive development of techniques for evaluating transient vapour-liquid flows. Very general two-fluid models have been proposed, using separate conservation equations for the vapour and liquid phases (see, e.g. references (1-3)). Simpler models reduce the number of conservation equations to be solved by allowing for interphase relative motion through a slip or 'drift-flux' correlation (4-6).

2. This paper examines the ability of the two-fluid and drift-flux models to predict two-phase flow phenomena important in reactor safety analysis. The studies reported comprise:

(1) calculations of sound-speeds and critical flow rates from the two-fluid theory of vapour-liquid flow, and comparison with data;

(2) comparison between predictions of drift-flux correlations and new void fraction measurements for a vertical boiling channel.

## TWO FLUID MODEL OF VAPOUR LIQUID FLOW

3. The two-fluid equations are used to calculate sound speeds and critical flow rates for a vapour liquid mixture. The three-dimensional conservation equations given in (7) are simplified by considering an adiabatic uni-directional flow, and averaging the resultant equations across the duct area. Mass, momentum and energy conservation give, respectively,

$$\frac{\partial}{\partial t}(\alpha_k \rho_k) + \frac{1}{A_z}\frac{\partial}{\partial z}(A_z \alpha_k \rho_k U_k) = \Gamma_k \qquad (1)$$

$$\alpha_k \rho_k \frac{DU_k}{D_k t} + \alpha_k \frac{\partial p}{\partial z} = (U_{ki} - U_k) + \tau_k^d + \tau_k^w \qquad (2)$$

$$\alpha_k \rho_k \frac{Dh_k}{D_k t} - \alpha_k \frac{Dp}{D_k t} = \Phi_k + \Gamma_k (h_{ki} - h_k) + q_{ik} a_i + \tau_k^d (U_{ki} - U_k) - \frac{\partial}{\partial z}\left(\alpha_k \varepsilon_k \frac{\partial T_k}{\partial z}\right) \qquad (3)$$

where the phase properties $\alpha$, T, $\rho$ etc, are duct-averaged values.

4. Equations (1)-(3) assume that (i) phase pressures are equal, (ii) surface tension effects are negligible, and (iii) the average of products of dependent variables across the duct is identical to the product of averages. These are conventional simplifications.

5. Interphase transfer terms must satisfy the interfacial balance conditions (7):

$$\left.\begin{aligned} &\Sigma_k \Gamma_k = 0 \\ &\Sigma_k [\tau_k^d + U_{ki} \Gamma_k] = 0 \\ &\Sigma_k \tau_k^d = 0 \\ &\Sigma_k [q_{ik} a_i + \tau_k^d U_{ki} + \Gamma_k h_{ki}] = 0 \end{aligned}\right\} \qquad (4)$$

Small kinetic energy terms and drag force fluctuations have been neglected in the energy balance.

## PROPAGATION OF SOUND WAVES IN VAPOUR-LIQUID FLOW

6. Equations (1)-(4) can be linearised to provide a dispersion relation for acoustic waves. The calculated sound speed should be physically applicable for wavelengths long compared with bubble sizes, film thicknesses, etc., so that the fluid behaves like a continuum.

7. Mecredy and Hamilton (8) were first to use the two-fluid equations for similar calculations. They assumed that for a vapour-liquid mixture the $\Gamma_k$'s were controlled by kinetic theory limitations, concluding that c depended strongly on the accommodation coefficient for molecular transfer, $\sigma$. The very large uncertainty in $\sigma$ implied that sound speeds could not be confidently predicted.

8. Ardron and Duffey (9) argued that for frequencies of practical interest ($\omega \lesssim 10^6$ Hz for steam and water) mass transfer rates are independent of $\sigma$, being limited instead by conduction in the liquid. On this basis c can be determined in terms only of measurable flow properties. This calculation is outlined below. Sound speeds are predicted for low velocity bubbly and annular steam water flows, and results compared with data.

## Calculation of Dispersion Relations

9. Consider the propagation of acoustic waves in the z direction. The dependent variables and transfer terms in equations (1-4) vary as

$$\nu = \nu_o + \nu' \exp i\,(\omega t - kz) \qquad (5)$$

in the disturbed flow. It is assumed that in the undisturbed flow the phase temperatures are equal, although the phase velocities are not necessarily equal (i.e. $\Gamma_{ko} = q_{iko} = 0$).

10. Ardron and Duffey (9) show that provided the signal frequency is in the range $D_\ell a_i^2/(1-\alpha_g)^2 << \omega << 2\sigma^2\rho_g^2 h_{g\ell}^4 D_\ell/[\pi(2-\sigma)^2 G_c \varepsilon_\ell^2 T_o^3]$ (for steam/water typically implying $1 \lesssim \omega \lesssim 10^6$Hz), then the interface liquid remains saturated. The transient heat flux from the liquid to the interface is then given by,

$$q_{i\ell} a_i \simeq \omega_q\, p' \exp i\,(\omega t - kz) \qquad (6)$$

obtained by solving the conduction equation, neglecting interface curvature. Using equation (6) and the interfacial energy balance [eq(4)], the mass transfer relation is [cf(9)]:

$$\Gamma_\ell = -\Gamma_g \doteq \omega_q h_{g\ell}^{-1}\, p' \exp i\,(\omega t - kz) \qquad (7)$$

11. Effects of wall shear on wave propagation are usually small and thus we take $\tau_k^w = \Phi_k = 0$ (9). Interphase drag terms depend on $(U_g - U_\ell)$ as primary variable, so for the disturbed flow field a linear expansion is used of the form

$$\tau_k^d = \tau_{ko}^d + \left(\frac{\partial \tau_g^d}{\partial U_g}\right)(U_g' - U_\ell') \exp i\,(\omega t - kz) \qquad (8)$$

12. Substituting equations (5)-(8) into the conservation equations (1)-(4), and using an ideal gas state equation for the vapour, and a compressible liquid state equation, $dp/d\rho_\ell = c_\ell^2$, gives a set of linear homogeneous equations in the primed quantities. The secular equation of this set is the sound-wave dispersion relation. For flow velocities much smaller than the sound velocity this can be written (9),

$$\left\{ i\omega^2 \rho_\ell \rho_g \alpha_g (1-\alpha_g)[\rho_\ell \alpha_g + \rho_g (1-\alpha_g)] - \omega \rho_\ell \rho_g \left(\frac{\partial \tau_g^d}{\partial U_g}\right)\right\} k^2$$

$$+ \left\{ \frac{(\gamma-1)}{c_g^2} \xi \omega^2 (1-\alpha_g)\rho_\ell \rho_{g\ell} - \frac{\rho_\ell \rho_g \omega \omega_q}{h_{g\ell}} (U_g - U_\ell)[(1-\alpha_g)\alpha_g \rho_\ell \omega + i \left(\frac{\partial \tau_g^d}{\partial U_g}\right)] \right\} k$$

$$- i \frac{(\gamma-1)}{h_{g\ell}\, c_g^2} \xi \omega^2 (1-\alpha_g)\rho_\ell \omega_q (U_g - U_\ell)$$

$$- i\omega^3 \left[ \frac{\rho_g (1-\alpha_g)}{c_\ell^2} + \frac{\rho_\ell \alpha_g}{c_g^2} - \frac{\omega_q \rho_{g\ell}}{i h_{g\ell} \omega} \right] \left[ i \left(\frac{\partial \tau_g^d}{\partial U_g}\right)(\rho_\ell + \alpha_g \rho_{g\ell}) + (1-\alpha_g)\alpha_g \rho_\ell \rho_g \omega \right] = 0 \qquad (9)$$

## DISCUSSION

13. Equation (9) has two complex solutions in k of nearly equal magnitude, being waves propagating in directions ± z. Frequency dependent wave speeds, $c = \omega / \{\text{Real part of } k\}$, are plotted in Fig. 1 for sound waves in bubbly and annular flows at 70 bar, using standard expressions for interfacial area and interphase drag (9). Note that, for bubbly flow, the sound speed increases with frequency between approximate limits obtained assuming homogeneous flow with interphase thermal equilibrium, and homogeneous flow with zero mass transfer. The upper limit in annular flow is the 'stratified' sound speed of Grolmes and Fauske (10).

14. Fig. 2 shows comparisons with Freon-113 data of reference (11). Predictions assume bubbly flow relations for $\tau_k^d$ and $a_i$. Bubble sizes [not reported in (11)] have been calculated as $R_b \simeq 1$ mm from a textbook correlation (12). The good agreement, obtained without the use of adjustable parameters, supports the applicability of the two-fluid model.

## VAPOUR-LIQUID CRITICAL FLOW RATES

15. Estimation of the critical mass velocity ($G^*$) of a steam-water mixture discharging from a pressure vessel through a pipe is very important for reactor blowdown analysis. Measured values of $G^*$ are inconsistent with thermal equilibrium flow models (13) (14), and available non-equilibrium flow theories do not describe the observed effects of pipe size on flowrate. This leads to substantial uncertainties in flow rates for the wide bore pipes relevant to reactor calculations (D > 200 mm), for which no data are available (14).

16. Determination of $G^*$ requires a representation of the entire non-equilibrium flow field upstream of the pipe exit. This calculation has been attempted in references (15) and (16) assuming honogeneous flow. We use the two-fluid equations (1)-(4) to calculate the exit flow field: the main differences of this work from that described in (15) and (16) are: (i) the use of more general equations; (ii) allowance for interphase slip; and (iii) the introduction of a representation of boiling upstream of the pipe entrance. The calculations, described in detail (17) are outlined below. The theory is compared with measured critical flow-rates of saturated water in small diameter tubes, and predictions made of discharge flowrates in wide bore pipes (100 mm < D < 900 mm).

#### Theoretical Model

17. The flow geometry is idealised to allow one-dimensional calculations. Field equations (1)-(3) are solved for steady flow in the regions I, II and III shown in Fig. 3. The main assumption is that of purely radial inviscid flow in the entry region I; flow contraction effects are included using an entrance discharge coefficient, $C_D$. For sharp-edged entries, separated flow is assumed in a region II, downstream of the vena contracta, with an inviscid two-phase jet expanding within a vapour annulus. This pattern was observed by Bailey (18), and was represented in Edwards's model (16). The conditions in the jet at the pipe entrance (ABODE) are taken as those on a hemisphere (BCD) in region I with surface equalling the minimum flow area, thus ensuring continuity of $\alpha$, $\rho$, U etc. between regions I and II. As in (16) it is assumed that in region II friction and evaporation at the surface of the jet are negligible, so that the area expands only because of bubble growth in the liquid.

18. We consider low void fraction bubbly flow; following Edwards (16) we assume a constant bubble number density which is retained as an adjustable parameter.

19. As in the sound speed calculation we assume a saturated interface, obtaining the evaporation rate from the energy balance equation (4), as $\Gamma_g = - q_{i\ell}a_i/h_{g\ell}$. Again $q_{i\ell}$ is assumed to be dominated by conduction, and individual bubble thermal boundary layers are assumed not to interact. The conduction equation has been solved for a moving spherical boundary with a time-varying surface temperature in the thin thermal boundary layer approximation (19). This solution leads to an equation for the heat flux:

$$q_{i\ell}(t) = \underset{\lambda\to o}{\mathrm{Lt}} \left\{ - \frac{\varepsilon_\ell R_b^2(t)}{\sqrt{(4\pi D_\ell)}} \frac{\partial}{\partial \lambda} \left[ \lambda \int_o^\zeta \frac{T(\lambda,\zeta')-T_o}{(\zeta - \zeta')^{3/2}} \exp \left( \frac{-\lambda^2}{4D_\ell(\zeta-\zeta')} \right) d\zeta' \right] \right\} \quad (10)$$

where $\lambda(r,t)$ and $\zeta(t)$ are transformed spatial and time co-ordinates used in (19).

If the interface temperature variation is approximated by a series of small steps equation (10) can be replaced by a finite difference formula:

$$q_{i\ell}^{(j)} = \varepsilon_\ell R_b^{(j)2} \sum_{k=1}^{k=j} \frac{T_{i\ell}^{(k-1)} - T_{i\ell}^{(k)}}{\sqrt{\pi D_\ell \left(\zeta^{(j)} - \zeta^{(k-1)}\right)}} \quad (10')$$

where the superscript j denotes conditions at the $j^{th}$ time step. In the numerical calculations the time t is introduced as a dependent variable, related to bubble position by

$$dt/dz = U_g^{-1} \quad (11)$$

The origin in time is taken as a point in the reservoir where the fluid velocity is negligible.

20. The interphase drag $\tau_k^d$ is obtained by generalising the standard expression for the drag on a small spherical gas bubble in a liquid. The form used is identical to that adopted for the bubbly flow sound-speed calculations described above and is given in reference (9). Pipe wall friction is introduced using a conventional homogeneous flow approximation (12). The Reynolds number dependent friction factor is obtained from a standard correlation for turbulent liquid flow in smooth tubes.

21. Equations (1) (2) (3) and (11) are solved numerically in the three regions using the programme CRACKPOT (17). $q_{i\ell}$, and hence the evaporation rate, are calculated from equation (10') at a series of mesh points.. $z_{i-1}$, $z_i$, $z_{i+1}$.. In the region between points $z_i$ and $z_{i+1}$ the equations are integrated assuming the evaporation rate is constant, and equal to the value at point $z_i$. As in (16) an initial mass flowrate is postulated and the distance down the pipe at which dp/dz becomes effectively infinite determined: this is the pipe length consistent with this steady choked flowrate.

## RESULTS

22. In Fig. 4(a) the theory is compared with Fauske's data (20), for the flow of saturated water in sharp-edged entry tubes (D = 6.35 mm). A value of $C_D$ = 0.8 has been used to give a best fit to data. Agreement can be improved if $N_b$ is increased with pipe length. This may be due to the failure to describe finite nucleation rates, and this possibility is being explored in further calculations. Predictions for wide bore pipes are given in Fig. 4(b) in a form that shows the independent effects of length and diameter. Strong effects of pipe length, and weak effects of pipe diameter are indicated by the model. Predictions approach homogeneous equilibrium flowrates for large L/D.

## DRIFT FLUX MODELS

23. Several models for reactor depressurisation analysis use drift flux correlations relating individual phase velocities to the total flow-rate and void fraction (4-6). This avoids solving momentum equations for each phase. Unlike a slip model, a drift flux model can describe gravitational separation of steam from the two-phase mixture. Steam separation rates determine the height of coolant in a channel during blow-down and reflooding (21, 22), and can strongly influence depressurisation behaviour (23).

24. Void fraction data for vertical boiling channels, suitable for testing drift flux models, are scarce. New data are presented below for pipes and pin-bundles under steady and falling pressures for the range $\alpha_g$<0.6, p < 4.0 bar; measurements are compared with predictions of conventional drift flux correlations.

#### Experiments and Observations

25. The closed loop apparatus (see Fig. 5a) consisted of a vertical glass vessel (105 mm i.d) containing 19 x 1 KW electrical heater pins (10 mm o.d) immersed in distilled water. The

liquid was effectively stagnant in all tests ($j_\ell$=0). Differential pressure measurements within and above the bundle (regions (a) and (b) respectively) were used to calculated average void fractions through the equation $\Delta p \simeq \rho_\ell g z(1-\bar{\alpha}_g)$.

26. Except for lowest void fractions ($\alpha_g$ < 0.05) the flow was unsteady and turbulent, with upflow at the vessel centre and downflow at the walls. High-speed photographs of region(b) showed large bubbles (∿ 10 mm dia.). Differential pressure measurement in region (b) oscillated with period 0.1 - 0.05 sec, of order of the rise time of the large bubbles through this region.

27. Figs. 5(a, b) show plots of $\bar{\alpha}_g$ against the maximum non-dimensional steam flowrate in the channel containing a heated bundle ($j_g$ is obtained from bundle power). Fig. 5(c) shows results of depressurisation tests without the bundle. Depressurisation was effected by steam venting. $\hat{j}_g$ was deduced from $\dot{p}$, and $\bar{\alpha}_g$ derived from mixture height observations.

Comparison with Predictions of Drift Flux Models

28. Continuity considerations lead to drift flux models for bubbly and slug flows of the general form (25):

$$\alpha_g = j_g/(C_o j + V_{gj}) \qquad (12)$$

$V_{gj}$ allows for gravitational separation and $C_o$, the profile parameter, accounts for the effect of non-uniform concentration and velocity profiles ($C_o$ = 1 for uniform flows). Proposed values for $C_o$ and $V_{gj}$ are listed in Table 1. Alternative empirical correlations are available which are derived from dimensional analysis of data. These have the form

$$\alpha_g = K_1 \left(\frac{\rho_g}{\rho_{\ell g}}\right)^{K_2} \left(\frac{j_g}{j}\right)^{0.6} (Ku)^{K_3} \left(\frac{\sqrt{(s/g\rho_{\ell g}}}{D}\right)^{K_4} \qquad (13)$$

Values of $K_1$, etc are listed in Table 1, (see Fig. 5).

29. Correlations are compared with data in Fig. 5. Physically based models [equ. (12)] tend to over-estimate $\bar{\alpha}_g$ and do not describe the tendency of $\bar{\alpha}_g$ to increase with pressure. Agreement with the slug flow correlations of references (25) (26) can be improved by increasing $C_o$ to the maximum values reported in (26) for low liquid flow ($C_o$ ∿ 2). Use of $C_o$ = 2 raises the predicted gas velocity from 2.7 to 3.6 $ms^{-1}$, in better agreement with the observed velocity of large bubbles (4 $ms^{-1}$). Empirical correlations [eq. (13)] give reasonable agreement with data, and show the correct trend with pressure.

30. Whilst it is desirable to use models based on physical arguments for reactor safety calculations, the present experiments show deficiencies in the available physically based drift-flux models. Effects of flow and geometry on $C_o$ are particularly uncertain.

CONCLUSIONS

31. A two-fluid model with simple constitutive equations has been used to calculate sound speeds and critical vapour liquid flow-rates in a pipe. Reasonable agreement of the results with data supports the use of a simple two-fluid theory for transient two-phase flow analysis.

32. Drift flux models for interphase relative motion have been tested against void fraction measurements in vertical boiling channels. Purely empirical correlations give reasonable agreement with data. Models with a physical basis do not represent the observed effects of pressure. Significant over-prediction of void fraction obtained using physically-based models highlights the need to improve the representation of effects of flow and geometry on concentration and velocity profiles.

ACKNOWLEDGMENT

This paper is published by permission of the Central Electricity Generating Board.

REFERENCES


(1) Solbrig, C.W., Mortensen, G.A. and Lyczkowski, R.W. "An Unequal Phase Velocity Unequal Phase Temperature Theory Applied to Two-Phase Blowdown from a Horizontal Pipe". Proc. 1976 Heat Transfer and Fluid Mech. Inst., Stanford University Press, 1976.

(2) Hancox, W.T., Mathers, W.G. and Kawa, D. "Analysis of Transient Flow Boiling; Application of the Method of Characteristics". AIChE Paper No. 42, Two-Phase Flow Session, 15th National Heat Transfer Conf., San Francisco, Aug. 1975.

(3) Harlow, F.H. and Amsden, A.A. "Numerical Calculation of Multiphase Flow". J. Comp. Phys., 17, 19, 1975.

(4) Kelly, R.D., Dougall, R.S. and Cantineau, B.J. "Application of Drift Flux to Loss of Coolant Analysis". ASME Paper 74-WA/HT-40, 1974.

(5) Brittain, I. and Fayers, F.J. "A Review of U.K. Developments in Thermal-Hydraulic Methods for Loss-of-Coolant Accidents. Presented at OECD CSNI Specialist Meeting on Transient Two-Phase Flow, Toronto, 1976.

(6) Kroeger, P.G. "Application of a Non-Equilibrium Drift Flux Model to Two-Phase Blowdown Experiments". ibid.

(7) Ishii, M. "Thermo-Fluid Dynamic Theory of Two-Phase Flow". Eyrolles, Paris, 1975.

(8) Mecredy, R.C. and Hamilton, L.J. "The Effects of Non-equilibrium Heat, Mass and Momentum Transfer on Two-Phase Sound Speed". Int. J. Heat Mass Transfer, 15, 61, 1972.

(9) Ardron, K.H. and Duffey, R.B. "A Two-Fluid Continuum Model for Acoustic Wave Propagation in a Non-Equilibrium Liquid Vapour Flow", CEGB Report RD/B/N3915, 1977.

(10) Grolmes, M.A. and Fauske, H.K. "Propagation Characteristics of Compression and Rarefaction Pressure Pulses in One-Component Vapour-Liquid Mixtures". Nucl. Eng and Des., 11, 137, 1969.

(11) Weisman, J., Ake, T. and Knott, R. "Two-Phase Pressure Drop Across Abrupt Area Changes in Oscillatory Flow". Nucl. Sci and Eng., 61, 297, 1976.

(12) Wallis, G.B. "One-Dimensional Two-Phase Flow". McGraw-Hill Book Co., New York, 318-322, 1969.

(13) Simon, U. "Blowdown Flow Rates of Initially Saturated Water". ANS Topl. Mtg. Water Reactor Safety, Salt Lake City, 172, 1973.

(14) Ardron, K.H. and Furness, R.A. "A Study of Critical Flow Models Used in Reactor Blow-down Analysis". Nucl. Eng and Des, 39,257, 1976.

(15) Simpson, H.C. and Silver, R.S. "Theory of One-Dimensional Two-Phase Homogeneous Non-Equilibrium Flow", Proc. Instn. Mech Engrs. Symp. Two-Phase Flow, Paper No. 6, 1962.

(16) Edwards, A.R. "Conduction Controlled Flashing of a Fluid, and the Prediction of Critical Flow Rates in One-Dimensional Systems". UKAEA Report AHSB(S) R147, 1968.

(17) Ardron, K.H. "A One-Dimensional Two-Fluid Model for Critical Vapour-Liquid Flow in a Pipe", to be submitted to Nucl. Sci Eng., 1977.

(18) Bailey, J.F. "Metastable Flow of Saturated Water". Trans. ASME, 73, 1109, 1951.

(19) Plesset, M.S. and Zwick, S.A. "A Nonsteady Heat Diffusion Problem with Spherical Symmetry", J. Appl. Phys., 23, 95, 1952.

(20) Fauske, H.K. "The Discharge of Saturated Water Through Tubes", Chem. Eng. Prog. Symp. Ser. 61, 210, 1965.

(21) Cunningham, J.P. and Yeh, H.C. "Experiments and Void Correlations for PWR Small-Break LOCA Conditions". Trans. ANS, 17, 269, 1973.

(22) Sozzi, G.L., Muzzy. R.J. and Burnette, G.W. "Blowdown Heat Transfer Phenomena in the Scaled BWR Test System". GE Report GEAP-21126, 1976.

(23) Ardron, K.H., Furness, R.A. and Hall. P.C. "Experimental and Theoretical Studies of Transient Boiling and Two-Phase Flow During the Depressurisation of a Simple Vessel". OECD CSNI Specialist Mtg. Two-Phase Flow, Toronto, 1976.

(24) Wilson, J.F. "Primary Separation of Steam from Water by Natural Separation". US/EURATOM Report ACNP-65002, 1965.

(25) Zuber, N. and Findlay, J.A. "Average Volumetric Concentration in Two-Phase Systems". Trans. ASME J. Ht. Transfer, 87, 453, 1965.

(26) Nicklin, D.J., Wilkes, J.O. and Davidson, J.F. "Two-Phase Flow in Vertical Tubes". Trans. Instn. Chem. Engrs. 40, 61, 1962.

(27) Griffith, P. "Prediction of Low Quality Boiling Voids". Trans. ASME J. Ht. Transfer, 85, 327, 1964.

| Ref. | Conditions | $C_o$ | $V_{gj}$ |
|---|---|---|---|
| (26) | Slug flow (pipes) Re > 8000 | 1.2 | $0.35\sqrt{gD}$ |
| (27) | Rod Bundles Re > 8000 | Dependent on geometry | Dependent on geometry$^{+}$ |
| (25) | Churn Turbulent Flow | < 1.5 | 1.53 J |
| (12) | Bubbly Flow | 1.0 | $1.53\,(1-\alpha_g)^2 J$ |
| (5) | No restrictions | $\dfrac{1}{a + (1-a)\alpha^b}$ | $\dfrac{C_o\,(1-\alpha C_o)\,Ku^{*}J}{\left(\frac{\rho_g}{\rho_\ell}\right)^{\frac{1}{2}}\alpha C_o + (1-\alpha C_o)}$ $^{++}$ |

(a) Parameters for correlations of form of equation (12)

$^{+}$Multiplier of 1.6 used in "developing" flows.

$^{++}$a, b and $Ku^{*}$ are functions of pressure and geometry defined in ref. (5)

| Ref. | Conditions | | $K_1$ | $K_2$ | $K_3$ | $K_4$ |
|---|---|---|---|---|---|---|
| (24) | 0 < Ku < 1.5 | Pipes D < 0.91m | .564 | .12 | .67 | .1 |
| (24) | 1.5 ≤ Ku < 10 | 0 < p < 136bars | .619 | .12 | .47 | .1 |
| (21) | Ku < 1.53 | PWR bundle | .696 | .24 | .67 | 0 |
| (21) | Ku ≤ 1.53 | 7 < p < 26 bars | .757 | .24 | .47 | 0 |

(b) Parameters for correlations of form of equation (13)

Table 1. Parameters used in drift flux correlations

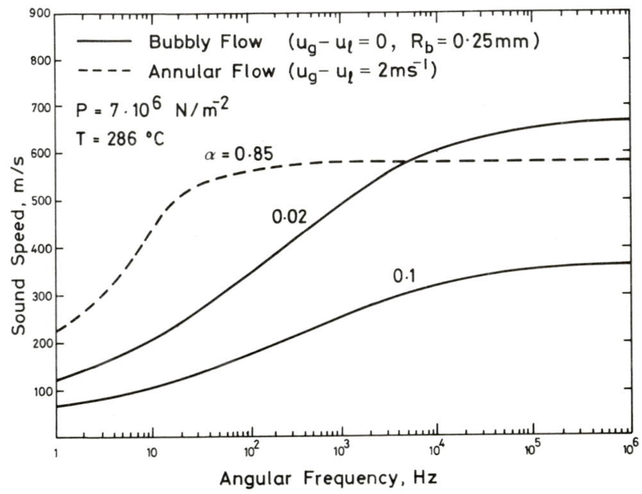


Fig. 1: Sound speed as a function of frequency for bubbly and annular steam water flow

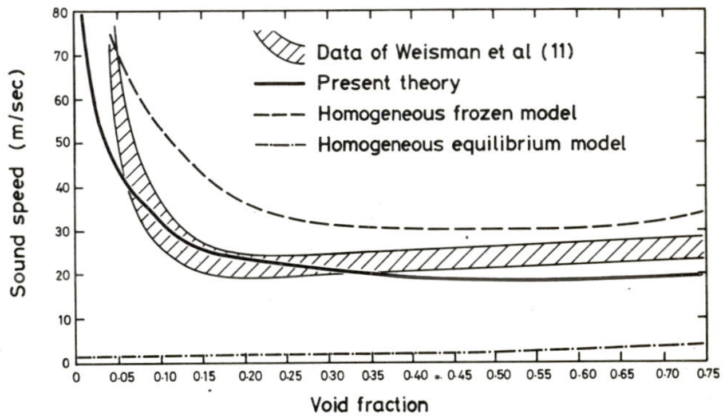


Fig. 2: Comparison of theory with experimental data

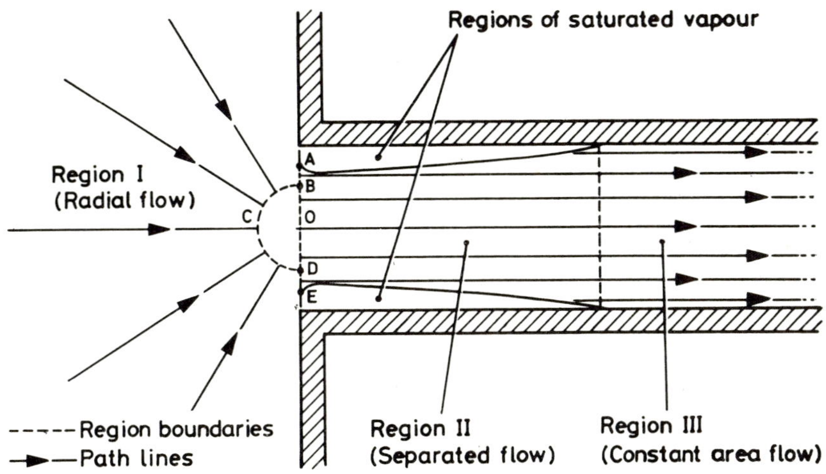


Fig. 3: Idealized flow geometry used as a basis for one-dimensional critical flow calculations

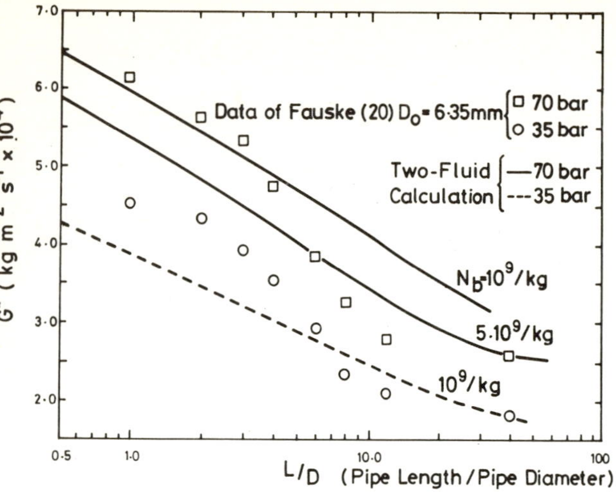


Fig. 4(a): Comparison of two fluid critical flow calculation with data

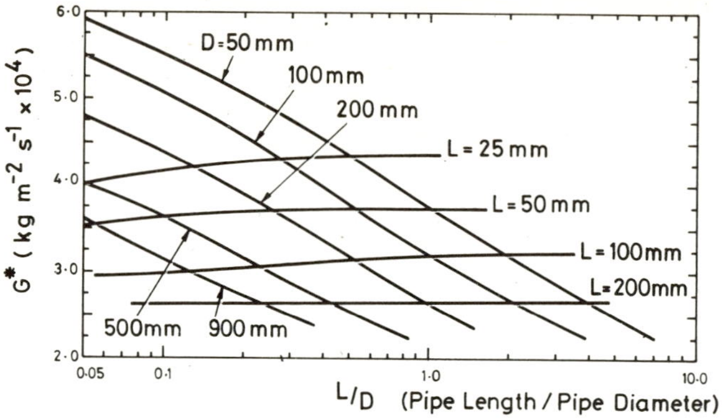


Fig. 4(b): Predicted critical flow rates for large sharp-edged entry pipes p=70 bar, ($C_D$ =0.8, $N_b$ =5.10$^9$ /kg)

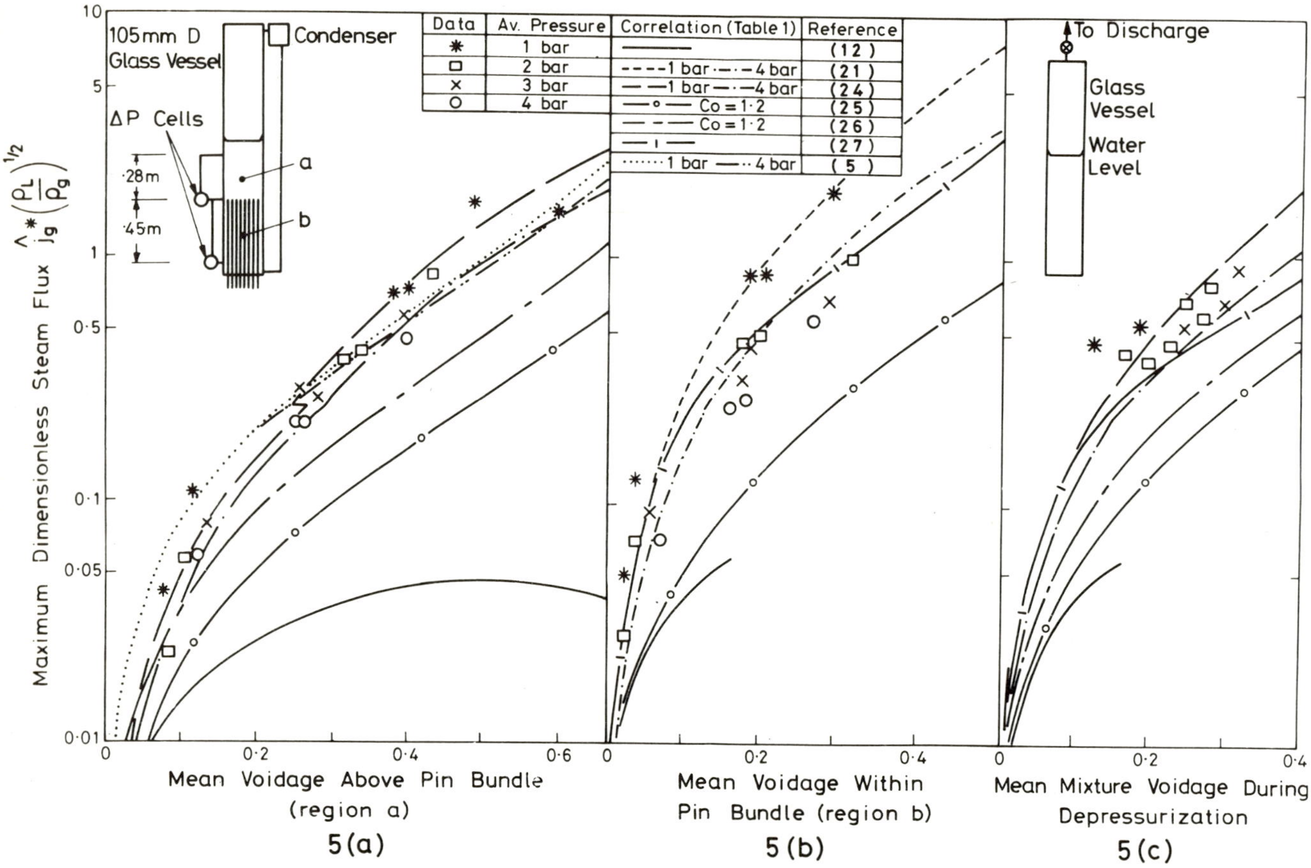


Fig. 5: Comparison of drift-flux correlations with voidage measurements

C205/77

# FLUID DYNAMICS AND HEAT TRANSFER METHODS FOR THE TRAC CODE

W. H. REED, PhD and W. L. KIRCHNER, PhD
Los Alamos Scientific Laboratory, Los Alamos, New Mexico, USA

SYNOPSIS A computer code called TRAC is being developed for analysis of loss-of-coolant accidents and other transients in light water reactors. This code involves a detailed, multidimensional description of two-phase flow coupled implicitly through appropriate heat transfer coefficients with a simulation of the temperature field in fuel and structural material. Because TRAC utilizes about 1000 fluid mesh cells to describe an LWR system, whereas existing lumped parameter codes typically involve fewer than 100 fluid cells, we have developed new highly implicit difference techniques that yield acceptable computing times on modern computers. Several test problems for which experimental data are available, including blowdown of single pipe and loop configurations with and without heated walls, have been computed with TRAC. Excellent agreement with experimental results has been obtained.

## INTRODUCTION

1. It is widely recognized that so-called "conservative" computer codes for analysis of loss-of-coolant accidents in light water reactors, which involve the most pessimistic data and modeling assumptions at every stage, are difficult if not impossible to verify against experimental data. Because we rely on computer models of these plants to extrapolate small scale data to the full size geometry of interest and to synthesize data for individual components into an integral description of the entire system, "best-estimate" computer codes are also being developed. These latter codes attempt to provide neither pessimistic nor optimistic predictions of the events occurring during a reactor accident but instead strive for realistic predictions of accident consequences. Among the several "best-estimate" codes currently under development, the TRAC code utilizes the most detailed and complex models of plant components. This paper presents a brief description of some of these models and the numerical techniques used to solve them. Finally, some preliminary comparisons between experimental results and TRAC predictions are presented.

## TRAC MODELS AND METHODS

2. The TRAC code is highly modular in structure, both by function and by system component. For example, there are sections of the code that are devoted to modeling PWR components like pipes, pumps, steam generators, etc., and there are other sections of the code that carry out the basic fluid dynamics and heat transfer solution processes. This very flexible code structure allows a more complicated and detailed modeling of important components, with cheaper and less detailed models for less important components. At present, the code contains models for most PWR components, including pipes, pumps, pressurizer, steam generators, tee's, and vessel (including upper and lower plenum, core, downcomer, and upper head). BWR component models will be added later.

### Fluid dynamics and heat transfer models

3. The fluid dynamics models in the TRAC code offer several capabilities not present in existing LOCA codes. First, relative motion between the liquid and vapor phases is accounted for by a drift-flux formulation and associated slip model in some code modules and by a fully two-field model in others. Second, nonequilibrium thermodynamics is included, in the sense that liquid and vapor may be at different temperatures and a nonequilibrium phase change model is used. Finally, a three-dimensional representation of the fluid flow inside the reactor vessel has been incorporated in TRAC (space limitations preclude a detailed discussion of this feature).

4. Drift-flux models of two-phase flow are used in all one-dimensional TRAC components. A fully two-field model is used only for components inside the reactor vessel, where counter-current flow is expected to occur. Both of these models incorporate conservation equations for the mass and energy of the individual liquid and vapor phases. They differ only in their representation of conservation of momentum. The TRAC drift-flux model uses a single equation representing conservation of the momentum of the two-phase mixture. The two-field model involves separate momentum conservation equations for each phase.

5. The "nonequilibrium" drift flux model that is used in one-dimensional TRAC components involves the following partial differential equations:

Conservation of vapor mass

$$\frac{\partial}{\partial t}(\alpha\rho_v) + \frac{\partial}{\partial z}(\alpha\rho_v v_m) + \frac{\partial}{\partial z}\left[\frac{\alpha\rho_v(1-\alpha)\rho_\ell v_r}{\rho_m}\right] = \Gamma \quad (1)$$

Conservation of mixture mass

$$\frac{\partial \rho_m}{\partial t} + \frac{\partial}{\partial z}(\rho_m v_m) = 0 \qquad (2)$$

Conservation of mixture momentum

$$\frac{\partial}{\partial t}(\rho_m v_m) + \frac{\partial}{\partial z}(\rho_m v_m^2) + \frac{\partial}{\partial z}\left[\frac{(1-\alpha)\rho_\ell \alpha \rho_v v_r^2}{\rho_m}\right] + \frac{\partial p}{\partial z} + \rho_m g_z + \tau = 0 \qquad (3)$$

Conservation of vapor specific internal energy

$$\frac{\partial}{\partial t}(\alpha \rho_v e_v) + \frac{\partial}{\partial z}(\alpha \rho_v e_v v_m) + \frac{\partial}{\partial z}\left[\frac{\alpha \rho_v (1-\alpha)\rho_\ell e_v v_r}{\rho_m}\right] + p\frac{\partial}{\partial z}(\alpha v_m) + p\frac{\partial}{\partial z}\frac{\alpha(1-\alpha)\rho_\ell v_r}{\rho_m} = q_v'' + q_{\ell v}'' \qquad (4)$$

Conservation of mixture specific internal energy

$$\frac{\partial}{\partial t}(\rho_m e_m) + \frac{\partial}{\partial z}(\rho_m e_m v_m) + \frac{\partial}{\partial z}\left[\frac{(1-\alpha)\rho_\ell \alpha \rho_v v_r (e_v - e_\ell)}{\rho_m}\right] + p\frac{\partial v_m}{\partial z} + p\frac{\partial}{\partial z}\left[\frac{\alpha(1-\alpha)v_r(\rho_\ell - \rho_v)}{\rho_m}\right] = q'' \quad . \qquad (5)$$

Variables appearing in the above equations have the following meanings:

| | |
|---|---|
| $\alpha$ | vapor volume fraction |
| $\rho_v$ | vapor microscopic density |
| $\rho_\ell$ | liquid microscopic density |
| $\rho_m$ | mixture density |
| $v_m$ | mixture velocity |
| $v_r$ | relative velocity $(v_v - v_\ell)$ |
| $v_v$ | vapor velocity |
| $v_\ell$ | liquid velocity |
| $p$ | pressure |
| $g_z$ | gravitational force |
| $\tau$ | wall shear |
| $e_v$ | vapor specific internal energy |
| $e_\ell$ | liquid specific internal energy |
| $e_m$ | mixture specific internal energy |
| $q_v''$ | heat flux to vapor from wall |
| $q''$ | heat flux to mixture from wall |
| $\Gamma$ | vapor production rate due to flashing |
| $q_{\ell v}''$ | interfacial heat flux from liquid to vapor |

6. In addition to Eqs. (1)-(5), several additional equations are required to form a closed set. The simplest of these are definitions of mixture properties:

$$\rho_m = \alpha\rho_v + (1-\alpha)\rho_\ell \qquad (6)$$

$$\rho_m e_m = \alpha\rho_v e_v + (1-\alpha)\rho_\ell e_\ell \qquad (7)$$

$$\rho_m v_m = \alpha\rho_v v_v + (1-\alpha)\rho_\ell v_\ell \qquad (8)$$

The remaining equations are called "constitutive" relationships. The particular forms of these relationships presently incorporated in TRAC are presented below.

## Equations of state

7. The equations of state in TRAC are represented by low-order polynomial fits to the steam tables. These relationships apply for both superheated and subcooled liquid and vapor and are evaluated in the following forms:

$$e_\ell = e_\ell(p, T_\ell) \qquad (9)$$

$$e_v = e_v(p, T_v) \qquad (10)$$

$$\rho_\ell = \rho_\ell(p, e_\ell) \qquad (11)$$

$$\rho_v = \rho_v(p, e_v) \qquad (12)$$

## Wall friction

8. The wall friction term takes the form

$$\tau = c\frac{\rho_m v_m |v_m|}{2D} \qquad (13)$$

where D is the hydraulic diameter and c is a friction multiplier depending strongly upon the void fraction $\alpha$ and other fluid properties. Several options are available in TRAC for computing $\tau$, including an annular flow multiplier [1], a homogeneous factor [2], and others.

## Slip model

9. The Zuber-Findlay [3] slip model is currently used in TRAC. Because this model can lead to infinite slip for large void fractions, necessitating a void fraction cutoff, a revised regime dependent slip package is being incorporated in TRAC.

## Phase change

10. A Nigmatulin phase change model is used in TRAC. This model is presented in Ref. 4. The phase change term is composed of an evaporation and a condensation term as indicated below:

$$\Gamma_e = \lambda_e A\rho_\ell\,\alpha(1-\alpha)\sqrt{R_{gas}T_{sat}}\,\frac{(T_\ell - T_{sat})}{T_{sat}} \qquad (14)$$

$$\Gamma_c = \lambda_c A\,\rho_v\alpha(1-\alpha)\sqrt{R_{gas}T_{sat}}\,\frac{(T_v - T_{sat})}{T_{sat}} \qquad (15)$$

where $\lambda$ is a constant, A is an interfacial area, and $R_{gas}$ is a gas constant.

## Interfacial heat transfer

11. The interfacial heat flux $q_{\ell v}''$ is now set as

$$q_{\ell v}'' = c\alpha(1-\alpha)(T_\ell - T_v) + \Gamma\,h_{sat} \qquad (16)$$

where c is a constant and $h_{sat}$ is the saturation enthalpy for the vapor or liquid, depending on whether $\Gamma$ is positive or negative, respectively. At present, the coefficient c is set to a very large number to produce temperature equality between the phases.

## Wall heat transfer

12. The wall heat transfer coefficient is based on construction of a generalized boiling curve. For each coolant cell with adjoining structural material, local coolant conditions, structural material properties, and surface temperature are used to identify the heat transfer regime and to determine an appropriate heat transfer coefficient from the stored correlations which comprise the generalized boiling curve. The particular correlations used are as follows:

- i Critical heat flux (CHF) - Bowring correlation [5]
- ii Minimum film boiling temperature - Henry-Berenson formulation [6]
- iii Single phase flow - turbulent flow - Dittus-Boelter correlation [7]
- iv Nucleate boiling - Chen correlation [8]
- v CHF point - intersection of Chen and Bowring correlations
- vi Transition boiling - log-log interpolation from CHF point to minimum film boiling heat flux point on boiling curve
- vii Film boiling - void fraction $\leq 0.5$ - modified Bromley correlation [9] void fraction $>0.5$ - Dougall-Rohsenow correlation [10]

## Solution techniques

13. The TRAC code utilizes strongly implicit numerical techniques for solving the partial differential equations that describe heat conduction in the metal walls and fluid flow in the hydraulic system. These two processes are coupled by a semi-implicit treatment of the convective heat transfer between wall and fluid. To advance all system variables by a single time step, we first determine heat transfer coefficients using existing fluid and wall conditions. We then update the fluid conditions throughout the system using an iterative technique for solving the implicit difference equations. During this iteration the heat flux between wall and fluid is represented as

$$q'' = h^n A(T_w^n - T_f^{n+1}) \quad , \tag{17}$$

where h is the heat transfer coefficient, A is the wall area, $T_w$ is the wall temperature, $T_f$ is some representative fluid temperature, and the superscripts n and n+1 refer to the two successive time levels. We note that the heat transfer coefficient and wall temperature are constant during the fluid iteration, but that $q''$ is changing because $T_f$ is being repeatedly updated. Following the fluid dynamics iteration, the wall heat flux $q''$ is determined by the final value $T_f^{n+1}$, and the wall temperatures are determined everywhere by solving the wall heat conduction equations with $q''$ as a fixed boundary condition. The above method for handling heat transfer between wall and fluid guarantees strict energy conservation and has been found to be quite stable. We conjecture, however, that the method may perform poorly for systems with walls of lower heat capacity than the adjacent fluid. In this case one might wish to represent the wall heat flux as

$$q'' = h^n A(T_w^{n+1} - T_f^n) \tag{18}$$

14. We will now discuss the manner in which the drift-flux equations (1) - (5) are solved by the TRAC code. The three-dimensional, two-field equations used to model the vessel are substantially more complicated and will not be discussed here. The difference equations corresponding to Eqs. (1-5) are:

$$\frac{(\alpha\rho_v)_i^{n+1} - (\alpha\rho_v)_i^n}{\Delta t} + \frac{(\alpha\rho_v)^n_{i+\frac12}(v_m)^{n+1}_{i+\frac12} - (\alpha\rho_v)^n_{i-\frac12}(v_m)^{n+1}_{i-\frac12}}{\Delta z}$$
$$+ \frac{1}{\Delta z}\left\{\left[\frac{(1-\alpha)\rho_\ell\alpha\rho_v v_r}{\rho_m}\right]^n_{i+\frac12} - \left[\frac{(1-\alpha)\rho_\ell\alpha\rho_v v_r}{\rho_m}\right]^n_{i-\frac12}\right\}$$
$$= \dot{\Gamma}^{n+1} \tag{19}$$

$$\frac{(\rho_m)_i^{n+1} - (\rho_m)_i^n}{\Delta t} + \frac{(\rho_m)^n_{i+\frac12}(v_m)^{n+1}_{i+\frac12} - (\rho_m)^n_{i-\frac12}(v_m)^{n+1}_{i-\frac12}}{\Delta z}$$
$$= 0 \tag{20}$$

$$\frac{(v_m)^{n+1}_{i+\frac12} - (v_m)^n_{i+\frac12}}{\Delta t} + (v_m)^n_{i+\frac12}\left[\frac{(v_m)^n_{i+1} - (v_m)^n_i}{\Delta z}\right]$$
$$+ \frac{1}{(\rho_m)^n_{i+\frac12}\Delta z}\left\{\left[\frac{(1-\alpha)\rho_\ell\alpha\rho_v v_r^2}{\rho_m}\right]^n_{i+1} - \left[\frac{(1-\alpha)\rho_\ell\alpha\rho_v v_r^2}{\rho_m}\right]^n_i\right\}$$
$$+ \frac{1}{(\rho_m)^n_{i+\frac12}}\left(\frac{P^{n+1}_{i+1} - P^{n+1}_i}{\Delta z}\right) + (g_z)_{i+\frac12} + \left(\frac{\tau}{\rho_m}\right)_{i+\frac12} = 0 \tag{21}$$

$$\frac{(\alpha\rho_v e_v)^{n+1}_i - (\alpha\rho_v e_v)^n_i}{\Delta t} + \frac{1}{\Delta z}\left\{\left[(\alpha\rho_v e_v)^n_{i+\frac12}(v_m)^{n+1}_{i+\frac12}\right]\right.$$
$$\left.- \left[(\alpha\rho_v e_v)^n_{i-\frac12}(v_m)^{n+1}_{i-\frac12}\right]\right\} + \frac{1}{\Delta z}\left\{\left[\frac{\alpha\rho_v(1-\alpha)\rho_\ell e_v v_r}{\rho_m}\right]^n_{i+\frac12}\right.$$
$$\left.- \left[\frac{\alpha\rho_v(1-\alpha)\rho_\ell e_v v_r}{\rho_m}\right]^n_{i-\frac12}\right\} + P^{n+1}_i\left[\frac{\alpha^n (v_m)^{n+1}_{i+\frac12} - \alpha^n (v_m)^{n+1}_{i-\frac12}}{\Delta z}\right]$$
$$+ \frac{P^n_i}{\Delta z}\left\{\left[\frac{\alpha(1-\alpha)\rho_\ell v_r}{\rho_m}\right]^n_{i+\frac12} - \left[\frac{\alpha(1-\alpha)\rho_\ell v_r}{\rho_m}\right]^n_{i-\frac12}\right\}$$
$$= q''_v + q''^{n+1}_{\ell v} \tag{22}$$

$$\frac{(\rho_m e_m)_i^{n+1} - (\rho_m e_m)_i^{n}}{\Delta t} + \frac{1}{\Delta z}\Big\{(\rho_m e_m)^n_{i+\frac{1}{2}}(v_m)^{n+1}_{i+\frac{1}{2}}$$

$$(\rho_m e_m)^n_{i-\frac{1}{2}}(v_m)^{n+1}_{i-\frac{1}{2}}\Big\} + \frac{1}{\Delta z}\left\{\left[\frac{(1-\alpha)\rho_\ell \alpha \rho_v (e_v - e_\ell)}{\rho_m} v_r\right]^n_{i+\frac{1}{2}}\right.$$

$$\left. - \left[\frac{(1-\alpha)\rho_\ell \alpha \rho_v (e_v - e_\ell)}{\rho_m}\right]^n_{i-\frac{1}{2}}\right\} + p_i^{n+1}\left[\frac{(v_m)^{n+1}_{i+\frac{1}{2}} - (v_m)^{n+1}_{i-\frac{1}{2}}}{\Delta z}\right]$$

$$+ \frac{p_i^n}{\Delta z}\left\{\left[\frac{\alpha(1-\alpha)(\rho_\ell - \rho_v)}{\rho_m} v_r\right]^n_{i+\frac{1}{2}}\right.$$

$$\left. - \left[\frac{\alpha(1-\alpha)(\rho_\ell - \rho_v)}{\rho_m} v_r\right]^n_{i-\frac{1}{2}}\right\} = q'' \qquad (23)$$

In the above equations the superscripts n and n+1 refer to time levels, i and $i+\frac{1}{2}$ to space points, and $\Delta t$ and $\Delta z$ are mesh spacings in time and space, respectively. The mesh spacing $\Delta z$ is assumed to be a constant, although this is not a fundamental restriction of the method. Note that the mixture momentum equation is differenced in nonconservative form. We particularly emphasize that several of the terms in the above equations are evaluated at the new time level, but that all drift terms are evaluated at the old time level. The particular semi-implicit difference technique presented here was chosen to achieve a maximum of stability with a minimum complication in the equations that must be solved to obtain the fluid state at the new time level.

## RESULTS

15. Calculational results obtained with the TRAC code for two different experiments are presented in this section. Figure 1 illustrates a TRAC prediction of the measured pressure history at gauge station 5 of the Edwards and O'Brien blowdown experiment [11]. This result was obtained with a noding of 40 cells for the 4.10 meter pipe used in the experiment. A modified exponent (1.75) in the two-phase multiplier of the annular flow friction factor [1] was used for all predictions presented in this section.

16. Figures 2, 3, and 4 illustrate TRAC predictions of experimental measurements of pressure history, mass hold-up, and wall temperature for the experimental data of Premoli and Hancox [12]. This experiment was a heated pipe blowdown from initially subcooled conditions. The three pipe sections, feeder, heater, and riser, were each represented with five fluid cells. Gravitational effects and flow area changes were included in the modeling. Computational running times were (on a CDC 7600 computer) approximately 1 minute for the Edwards problem (about 1000 time steps) and 4 minutes for the Premoli problem (about 9000 time steps).

17. Results presented above give confidence in the modeling and numerical techniques of TRAC for two-phase flow applications.

## REFERENCES


1. HIRT, C.W. and ROMERO, N.C. "Application of a Drift-Flux Model to Flashing in Straight Pipes." Los Alamos Scientific Laboratory report LA-6005-MS, July 1975.

2. WALLIS, G. "One-dimensional Two-phase Flow." McGraw-Hill Inc., New York, 1969.

3. ZUBER, N. and FINDLAY, J. A. "Average Volumetric Concentration in Two-phase Flow Systems." (in J. of Heat Trans. 87, No. 4, 453-468, 1965).

4. RIVARD, W.C. and TORREY, M.D. "Numerical Calculation of Flashing from Long Pipes Using a Two-field Model." Los Alamos Scientific Laboratory report LA-6104, November 1975.

5. BOWRING, R.W. "A Simple but Accurate Round Tube, Uniform Heat Flux Dryout Correlation over the Pressure Range 0.7-17 $MN/m^2$ (100-2500 psia)." Atomic Energy Establishment, Winfrith, England report AEEW-R789, 1972.

6. HENRY, R.E. "A Correlation for the Minimum Film Boiling Temperature." AIChE Symposium Series 138 No. 70, 81-90, 1974.

7. DITTUS, F.W. and BOELTER, L.M.K. University of California Publ. Eng. 2, p. 443, 1930.

8. CHEN, J.C. "A Correlation for Boiling Heat Transfer of Saturated Fluids in Convective Flow." ASME, 63-HT-34, 1963.

9. BAILEY, N.A. "Film Boiling on Submerged Vertical Cylinders." Atomic Energy Establishment, Winfrith, England report AEEW-M1051, 1971.

10. DOUGALL, R.S. and ROHSENOW, W.M. "Film Boiling on the Inside of Vertical Tubes with Upward Flow of the Fluid at Low Qualities." MIT Mechanical Engineering report 9079-26, 1963.

11. EDWARDS, A.R. and O'BRIEN, T.P. "Studies of Phenomena Connected with the Depressurization of Water Reactors." (in. J. British Nucl. Energy Society 9, 125-135, April 1970).

12. PREMOLI, A. and HANCOX, W.T. "An Experimental Investigation of Subcooled Blowdown with Heat Addition." (in Proceedings, Specialists Meeting on Transient Two-Phase Flow, Toronto, Canada 1976, to be published).


This work was performed under the auspices of the USNRC.

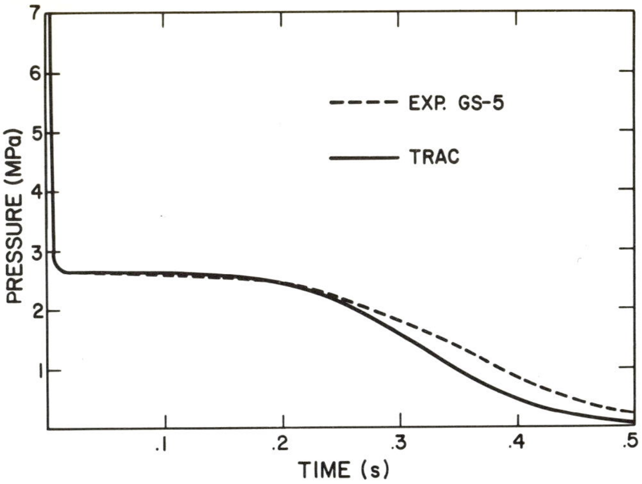


Fig. 1: Pressure history — Edwards pipe blowdown [11]

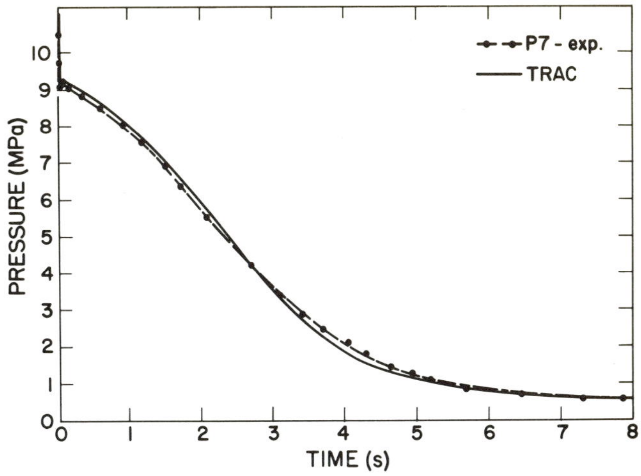


Fig. 2: Pressure history — Premoli and Hancox heated pipe blowdown [12]

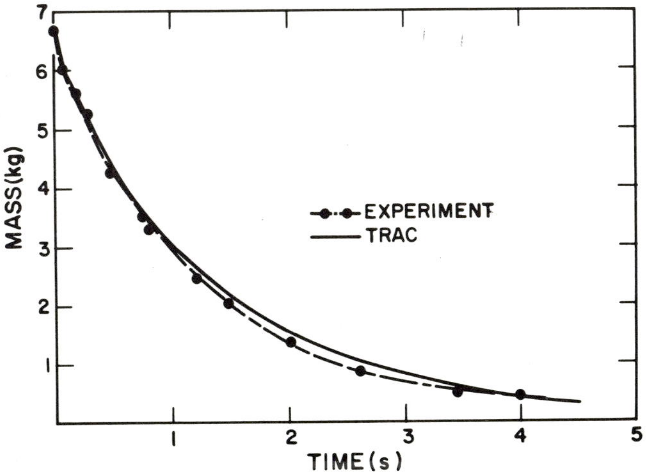


Fig. 3: Mass hold up — Premoli and Hancox heated pipe blowdown [12]

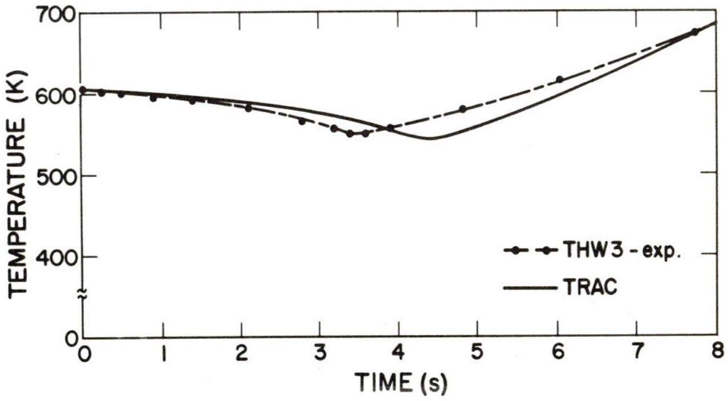


Fig. 4: Temperature history — Premoli and Hancox heated pipe blowdown [12]

# C206/77

# DESCRIPTION OF THE DRIFT FLUX MODEL IN THE LOCA CODE RELAP-UK

J.A. HOLMES, MA, DPhil
UKAEA, Atomic Energy Establishment, Winfrith

SYNOPSIS A numerical technique of modelling the effects of drift flux in vertical channels is described, which has been included in the LOCA code RELAP-UK. A drift flux correlation resolves the net flow rate of fluid of specified thermodynamic properties into the individual flow rates of the gas and liquid phases. It is based on the assumption that the difference in the average velocities of the two phases is a result of the linear superposition of the profile slip and the local slip. The profile slip may be obtained from a choice of profile slip correlations, which includes the flow-dependent Bryce correlation designed to preserve the hyperbolicity of the fluid conservation equations at high flow rates. The local slip is calculated from the Wallis flooding correlation, modified at low void fractions to be consistent with the analysis of Zuber and Findlay. Comparisons are given between the drift flux correlation, using the Bryce profile slip correlation, and certain sets of published experimental data.

NOTATION

| | |
|---|---|
| $C_o$ | = profile parameter |
| $D$ | = pipe diameter |
| $D^*$ | = dimensionless pipe diameter, defined by equation (10) |
| $G_d$ | = mass drift flux |
| $G_g, G_l$ | = superficial mass velocities of gas and liquid phases |
| $G$ | = net mass flow rate per unit area = $G_g + G_l$ |
| $g$ | = gravitational acceleration |
| $h_g, h_l$ | = specific enthalpies of gas and liquid phases |
| $j_g, j_l$ | = superficial velocities of gas and liquid phases |
| $j$ | = volumetric flux of mixture = $j_g + j_l$ |
| $K(\alpha)$ | = interpolation function defined by equation (18) |
| Ku | = critical Kutateladze number |
| $P$ | = absolute pressure |
| $Q$ | = rate of heat input |
| $U$ | = thermodynamic energy of fluid |
| $v_c$ | = characteristic velocity defined by equation (8) |
| $v_d$ | = mean drift velocity of the gas relative to the mixture |
| $v_g, v_l$ | = weighted mean velocities of gas and liquid phases |
| $v_g(v_l=0)$ | = weighted mean rise velocity of gas through a stagnant liquid |
| $W_g, W_l$ | = mass flow rates of gas and liquid phases |
| $W$ | = net mass flow rate = $W_g + W_l$ |
| $X^*$ | = flow quality in absence of local slip |
| $\alpha$ | = average void fraction |
| $\alpha_1, \alpha_2$ | = limits of interpolation function $K(\alpha)$ |
| $\Delta t$ | = time step size |
| $\rho_g, \rho_l$ | = densities of gas and liquid phases |
| $\sigma$ | = surface tension |

INTRODUCTION

1. RELAP-UK is a computer code designed to calculate the effects of a loss of coolant accident (LOCA) in a water cooled nulcear reactor (Ref 1). The coolant circuit is represented by a series of nodal volumes which are interconnected by a set of junctions. The thermodynamic properties of the coolant are derived from three main variables: the mass and the energy of the fluid in each node and the mass flow rate of the fluid in each junction. The solution is advanced at each time step by numerical integration of the one-dimensional conservation equations for the main variables in each node or junction.

2. In order to improve the code's capability of modelling low-flow situations, it has recently been adapted to include the effects of drift flux in vertical junctions where the fluid is two-phase. The energy conservation equation accounts for the energy transported across each junction by the two individual phases, and the flow rates of the two phases are obtained from the net flow rate of the mixture by a drift flux correlation. In this paper we derive the drift flux correlation and compare its predictions with some published experimental data.

THE ENERGY CONSERVATION EQUATION

3. The finite difference form of the energy conservation equation for a given node is:

$$U^{n+1} - U^{n} = \Delta t^{n+1} \left[ \sum_{\substack{\text{inflow}\\ \text{junctions}}} (W_g h_g + W_l h_l) - \sum_{\substack{\text{outflow}\\ \text{junctions}}} (W_g h_g + W_l h_l) + Q \right] \qquad (1)$$

where U is the internal energy of the fluid in the node, $W_g$ and $W_l$ represent the mass flow rate of the gas and liquid phases within the adjacent junctions, $h_g$ and $h_l$ are the specific enthalpies of the two phases, and Q represents the rate of heat input from external resources (eg fuel rods). $\Delta t^{n+1} = t^{n+1} - t^n$ is the size of the time step. The summations on the right hand side are taken over all junctions which connect with the given node.

4. The motion of the fluid is described by one momentum equation for each junction in the circuit, and numerical integration of these gives the net mass flow rate in each junction, $W = W_g + W_l$. The net flow rate is subsequently resolved into the flow rates of the two individual phases by a drift flux correlation, which takes account of the geometry and orientation of the junction and the thermodynamic properties of the fluid within it. The need for such a correlation arises from the process of describing the motions of the two individual phases by one momentum equation, which is computationally more economical than solving a separate momentum equation for each phase.

5. In RELAP-UK the fluid conservation equations are solved by a linearised implicit method, in which certain terms on the right hand sides are calculated at the time $t^{n+1}$ by means of a first order Taylor expansion about their values at time $t^n$ (Ref 1). Consequently the terms in the conservation equations, and their first derivatives, must vary in a continuous manner with respect to the main variables. Certain discontinuities are unavoidable (such as those produced by a change of phase), and the code reduces the time step size to a minimum value whenever these are likely to be encountered. Accordingly the drift flux correlation should be a smooth function of its constituent variables, and moreover it should be algebraically as simple as possible since it must be calculated for each junction at every time step. In the next section we derive a correlation which is satisfactory in this context and is versatile enough to allow easy adjustment should this be required.

THE DRIFT FLUX CORRELATION

6. The RELAP-UK drift flux correlation is derived from the results of Zuber and Findlay (Ref 2), who established a linear relation between the weighted mean velocity of the gas phase $v_g$ and the average volumetric flux of the mixture $j \equiv \alpha v_g + (1-\alpha)\, v_l$:

$$v_g = C_o\, j + v_d \qquad (2)$$

The 'profile parameter' $C_o$ depends upon the form of the velocity and local void fraction profiles in the mixture, and the 'mean drift velocity' $v_d$ represents the weighted mean relative velocity between the gas and the mixture. Since RELAP-UK uses the mass flow rate as its main flow variable, we rewrite Equation (2) in terms of the superficial mass velocity of the gas phase $G_g$ and the mass flux of the mixture G,

$$G_g = X^* G + G_d \qquad (3)$$

X* represents the flow quality ($W_g/W$) which would exist in the absence of drift flux, and depends on the average void fraction and the profile parameter,

$$X^* = \frac{\rho_g \alpha C_o}{\rho_g \alpha C_o + \rho_l (1-\alpha C_o)} \qquad (4)$$

$\rho_g$ and $\rho_l$ are the densities of the gas and liquid phases. The 'mass drift flux' $G_d$ is related to the mean drift velocity by

$$G_d = \frac{X^*}{C_o} \rho_l v_d \qquad (5)$$

The Profile Parameter

7. Zuber and Findlay derived a relation for $C_o$ in terms of power-law profiles of the local velocity and void fraction for axially-symmetric flow (Ref 2). When the void fraction is greatest at the centre of the pipe, the value of $C_o$ lies between 1.5 and 1.0. They also observed that, as the flow rate increases, the profiles become flatter and the value of $C_o$ approaches 1.0. Bryce showed that the use of flow-independent profile slip correlations can cause the fluid conservation equations of RELAP-UK to become non-hyperbolic at high flow rates, which may lead to numerical instability in their solution (Ref 3). He derived a correlation which preserves the hyperbolicity of the RELAP-UK conservation equations by reducing the profile parameter in the appropriate manner as the flow rate increases (Ref 4):

$$C_o = \left[ A_o + (1-A_o)\, \alpha^{B_o} \right]^{-1} \qquad (6)$$

where

$$A_o = 1 - 0.328\,(1 - P/3206.2)\, F^2$$

$$B_o = B_1 B_2 / \left[ 1 + F\,(B_2 - 1) \right]$$

$$B_1 = 2.94 + 0.159\,(P/1000) + 0.407\,(P/1000)^2$$

$$B_2 = (\rho_l / \rho_g)^{1/2}$$

$$F = \exp\,(-3\,|G|/G_{HS})$$

$$G_{HS} = \left[ \bar{X}\, G_2^{-2} + (1-\bar{X})\, G_1^{-2} \right]^{-1/2}$$

$$G_1 = -251.7 + 105.3\sqrt{P} + 3.766\,P$$

$$G_2 = -4.8 + 6650\,\rho_g/\rho_l + 3.45\sqrt{P} + 2.909\,P$$

G is the mass flux of the mixture in lb $s^{-1}$ $ft^{-2}$, $\bar{X}$ is the thermodynamic quality, and P is the absolute pressure in lbf $in^{-2}$. This correlation results in values of $C_o$ which lie between 1.488 and 1.0, which are within the limits derived for power-low profiles. The above correlation is available in RELAP-UK as the recommended choice, along with the options of Smith slip (Ref 5), Bankoff-Jones slip (Ref 6) and homogeneous flow ($C_o = 1$).

The Drift Velocity

8. A correlation for $v_d$ may be constructed from experimental measurements of the rise velocity of gas through a stagnant liquid. We obtain from Equation (2) a relation for the weighted mean rise velocity of the gas when the liquid velocity is zero,

$$v_g\ (v_l = 0) = \frac{v_d}{1 - \alpha C_o} \qquad (7)$$

As $\alpha$ approaches zero, the velocity $v_g$ ($v_l = 0$) should approach the rise velocity of a single bubble, for which Harmathy (Ref 7) recommends the value $1.53 v_c$, where $v_c$ is the characteristic velocity,

$$v_c = \left[\frac{\sigma g\ (\rho_l - \rho_g)}{\rho_l^{\,2}}\right]^{1/4} \qquad (8)$$

$\sigma$ is the surface tension and g is the gravitational acceleration. The data considered by Zuber and Findlay (Ref 2) correlate well with the relation $v_d = 1.53 v_c$ for low void fractions ($0 < \alpha < 0.3$). In their experiments on downward co-current flow Arosio et al obtain the relation $v_d = 1.5 v_c$ over the range ($0 < \alpha < 0.2$), which is in close agreement (Ref 8).

9. In the annular flow regime, at high values of $\alpha$, we consider the results of Wallis and Makkenchery (Ref 9), who measured the upward flow rate of air required to support a thin annular film of water in a vertical pipe of circular cross-section. Describing the behaviour of the liquid film in terms of the Kutateladze number $v_g \rho_g^{1/2} \left[\sigma g\ (\rho_l - \rho_g)\right]^{-1/4}$, they obtained the relation

$$v_g\ (v_l = 0) = \mathrm{Ku}\left[\frac{\rho_l}{\rho_g}\right]^{\frac{1}{2}} v_c \qquad (9)$$

where Ku, the 'critical Kutateladze number', is related to the dimensionless pipe diameter,

$$D^* = \left[\frac{g\ (\rho_l - \rho_g)}{\sigma}\right]^{1/2} D \qquad (10)$$

by a function which is tabulated in Table 1. Subsequent experiments involving a variety of interface shapes resulted in values of Ku ranging from 1.8 to 3.2 (Ref 10). Part of this variation may be due to void-fraction dependent effects, and until more data on the value of Ku are available we shall assume that the relation defined by Equations (9) and (10) and Table 1 represents the behaviour of $v_g$ ($v_l = 0$) as $\alpha$ approaches 1.0.

Table 1

Values of the critical Kutateladze number

| Dimensionless pipe diameter $D^*$ | Critical Kutateladze number Ku |
|---|---|
| $\leq 2$ | 0 |
| 4 | 1.0 |
| 10 | 2.1 |
| 14 | 2.5 |
| 20 | 2.8 |
| 28 | 3.0 |
| $\geq 50$ | 3.2 |

10. For intermediate values of the void fraction, we take an approach similar to that which Kelly et al chose for the SATAN code (Ref 11), using the Wallis flooding curve to define the limit of the countercurrent flow regime (Refs 12, 13). Wallis observed that the gas and liquid flow rates which initiate the disruption of steady countercurrent flow lie on the curve

$$\left[\frac{\rho_g}{\rho_l}\right]^{1/4} (j_g)^{1/2} + (-j_l)^{1/2} = c\left[\frac{(\rho_l - \rho_g)}{\rho_l}\ gD\right]^{1/4} \qquad (11)$$

where c is a constant which depends upon the pipe geometry, $j_g$ and $j_l$ are the superficial velocities of the two phases, and we have used the sign convention that upward flow is positive. In accordance with the results described in the previous paragraph, which were obtained over a wider range of pipe diameters, we renormalise the flooding curve to agree with Equation (9) when $j_l = 0$ by rewriting the right hand side of Equation (11) as $(\mathrm{Ku}\ v_c)^{1/2}$. We now note that Equation (3) may be written

$$G_g\ (1 - X^*) - G_l\ X^* = G_d \qquad (12)$$

If we assume that, at a constant pressure, $G_d$ is a function of $X^*$ alone, then Equation (12) represents a family of straight lines of constant $X^*$ on a graph of $G_g$ vs $G_l$. The flooding curve may be expressed in terms of mass fluxes:

$$\left[\frac{\rho_l}{\rho_g}\right]^{1/4} (G_g)^{1/2} + (-G_l)^{1/2} = (\rho_l\ \mathrm{Ku}\ v_c)^{1/2} \qquad (13)$$

The condition that the lines of constant $X^*$ (Equation (12)) are tangential to the flooding curve (Equation (13)) defines the mass drift flux:

$$G_d = \frac{X^*\ (1 - X^*)}{\left[\frac{\rho_l}{\rho_g}\right]^{1/2} X^* + 1 - X^*}\ \rho_l\ \mathrm{Ku}\ v_c \qquad (14)$$

Substitution of Equations (4) and (5) into Equation (14) gives the drift velocity:

$$v_d = \frac{(1 - \alpha C_o)\ C_o\ \mathrm{Ku}\ v_c}{\left[\frac{\rho_g}{\rho_l}\right]^{1/2} \alpha C_o + 1 - \alpha C_o} \qquad (15)$$

The above analysis yields the same expression for the drift velocity $v_d$ as that of Kelly et al (Ref 11). However, Kelly's derivation of $v_d$ contains the implied assumption that $C_o$ is independent of the total flow rate, and it is therefore not valid when $C_o$ is obtained from the Bryce G-dependent slip correlation. The derivation outlined above does not rely on this limiting assumption, since any dependence of $C_o$ on the value of G is already accounted for in the value of the parameter X*.

11. We may now combine the arguments of this section to produce a correlation for $v_d$ valid for the complete range of void fractions:

$$v_d = \frac{(1 - \alpha C_o)\, C_o\, K(\alpha)\, v_c}{\left[\frac{\rho_g}{\rho_l}\right]^{1/2} \alpha C_o + 1 - \alpha C_o} \quad (16)$$

which gives a mass drift flux of:

$$G_d = \frac{X^* (1 - X^*) \rho_l\, K(\alpha)\, v_c}{\left[\frac{\rho_l}{\rho_g}\right]^{1/2} X^* + 1 - X^*} \quad (17)$$

The function $K(\alpha)$ is defined as:

$$K(\alpha) = 1.53/C_o \quad \text{when } 0 \leq \alpha \leq \alpha_1 \quad (18a)$$

$$K(\alpha) = \frac{(1.53/C_o)\,(\alpha_2 - \alpha)^2 + Ku\,(\alpha - \alpha_1)^2}{(\alpha_2 - \alpha)^2 + (\alpha - \alpha_1)^2} \quad \text{when } \alpha_1 < \alpha < \alpha_2 \quad (18b)$$

$$K(\alpha) = Ku \quad \text{when } \alpha_2 \leq \alpha \leq 1.0 \quad (18c)$$

This provides a smooth interpolation between the correlations for low and high values of $\alpha$. When the void fraction is greater than a specified value $\alpha_2$, $K(\alpha)$ is set equal to Ku, which makes the drift flux correlation consistent with the Wallis flooding curve. At lower values of $\alpha$, where Wallis reported appreciable scatter in his flooding measurements (Ref 12), a value of $K(\alpha) = 1.53/C_o$ gives agreement between Equation (16) and the Zuber and Findlay correlation for bubbly flow, provided that $(\rho_g/\rho_l)^{1/2} \alpha C_o \ll (1 - \alpha C_o)$. The rise velocity of steam through stagnant water $v_g$ ($v_l$=0) at atmospheric pressure (14.7 lbf in$^{-2}$) is plotted as a function of $\alpha$ in Figure 1, using the two values 3.2 and $1.53/C_o$ for $K(\alpha)$, with $C_o$ obtained from the Bryce slip correlation. The limits of the interpolation function, $\alpha_1$ and $\alpha_2$, are left as free parameters for the user to decide. Comparison with the data of Bailey et al (Ref 14), which Zuber and Findlay use to demonstrate the transition from the bubbly flow regime (Ref 2), gives the best agreement when $\alpha_1 = 0.18$ and $\alpha_2 = 0.45$. This is demonstrated in Figure 1, in which the crosses mark points which lie on a curve drawn by eye through the centre of Bailey's data. However, the appropriate values of $\alpha_1$ and $\alpha_2$ will in general depend upon the manner in which the gas phase is introduced, and accordingly we must emphasise the need to conduct drift flux experiments in which steam is generated by boiling in geometries similar to that of a fuel rod cluster.

12. Although the RELAP-UK drift flux correlation is derived principally from observations of gas upflow through a stationary liquid, it is intended to be valid in the co-current downflow regime. Arosio et al (Ref 8) correlate their results with the values $C_o = 1.03$ and $v_d = 1.5\, v_c$. In their experiments the void fraction remains less than 0.18, and the RELAP-UK drift flux correlation accordingly sets $v_d = 1.53\, v_c$. At the values of G associated with their measured downflows of water the Bryce slip correlation predicts values of $C_o$ between 1.023 and 1.002. The variation of the measured superficial gas velocity $j_g$ with void fraction, at specified values of the superficial liquid velocity $j_l$, is shown in Figure 2. The continuous lines represent the predictions of the RELAP-UK drift flux correlation, using the Bryce profile slip correlation, at the corresponding values of the liquid superficial velocity. Agreement between the drift flux correlation and the experimental results is excellent at void fractions below 0.14 but the correlation slightly underestimates the gas flow at higher void fractions. At void fractions higher than 0.2 the effects of the change in flow regime can be seen in the drift flux correlation, increasing the amount of local slip between the phases and thereby decreasing the downward velocity of the gas. It would be interesting to extend Arosio's experiment to higher void fractions, to determine whether this effect can be seen.

## SUMMARY

13. RELAP-UK has been adapted to model the effects of drift flux in vertical channels. The net mass flow rate of the coolant, obtained from the solution of the momentum conservation equation, is resolved into the respective flow rates of the gas and liquid phases by a drift flux correlation. This assumes that the difference in the average velocities of the two phases in vertical pipes is a result of the linear superposition of the profile slip and the local slip. The profile slip may be obtained from a choice of profile slip correlations, which includes the flow-dependent Bryce slip correlation designed to preserve the hyperbolicity of the RELAP-UK conservation equations at high flow rates. The local slip is calculated from the Wallis flooding correlation, modified at low void fractions to agree with the analysis of Zuber and Findlay.

14. The user is able to adjust the values of three parameters in the drift flux correlation: Ku, $\alpha_1$ and $\alpha_2$. For each junction in the circuit, the user may either specify a particular value of the critical Kutateladze number Ku, or the code can determine a value from the hydraulic diameter and the data in Table 1. If the latter option is chosen, it must be noted that Table 1 is derived from the results of experiments with unheated pipes of circular cross-section. The appropriate values of $\alpha_1$ and $\alpha_2$, which govern the interpolation between the Wallis flooding correlation and the Zuber and Findlay correlation for bubbly flow, will in general depend upon the manner in which the gas phase is introduced. Comparison with an experiment in which air was passed upwards through stagnant water gives the best agreement when $\alpha_1 = 0.18$ and $\alpha_2 = 0.45$.

## REFERENCES


1. BRITTAIN, I. et al. "The status of RELAP-UK Mk III at July 1976 - A computer program for

transient thermal hydraulic analysis", AEEW - R1083, 1976.

2. ZUBER, N., FINDLAY, J. A. "Average volumetric concentration in two-phase flow systems", Trans ASME, Series C, 1965, Vol 87, p 453.

3. BRYCE, W. M. "Determination of the hyperbolic regime of the hydrodynamic equations modelled in the LOCA code RELAP-UK." CSNI Conference on Two-Phase Flow, Toronto, August 1976.

4. BRYCE, W. M. "A new G-dependent slip correlation which gives hyperbolic steam-water mixture flow equations." AEEW - R1099 (1977).

5. SMITH, S. L. "Void fraction in two-phase flow: a correlation based upon an equal velocity head model." Proc IME, 1970, Vol 184, pt 1, no 36, p 647.

6. JONES, A. B., DIGHT, D. G., "Hydrodynamic stability of a boiling channel". KAPL-2208, 1962.

7. HARMATHY, T. "Velocity of large drops and bubbles in media of infinite and restricted extent". AIChE J, 1960, Vol 6, p 281.

8. AROSIO, S. et al. "Two-phase downward flow in large diameter ducts - Preliminary results". Energia Nucleare, 1976, Vol 23, p 224.

9. WALLIS, G. B., MAKKENCHERY, S. "The hanging film phenomenon in vertical annular two-phase flow". Trans ASME, Series I, 1974, Vol 96, p 297.

10. WALLIS, G. B., KUO, J. T. "The behaviour of gas-liquid interfaces in vertical tubes". Int J Multiphase Flow, 1976, Vol 2, p 521.

11. KELLY, R. D. et al. "Application of drift flux to loss of coolant analysis". ASME Publication 74-WA/HT-40, 1974.

12. WALLIS, G. B. "Flooding velocities for air and water in vertical tubes". AEEW - R123, 1961.

13. WALLIS, G. B. "One-dimensional two-phase flow". New York, McGraw-Hill, 1969.

14. BAILEY, R. V. et al. "Transport of gases through liquid-gas mixtures." Paper presented at the AIChE New Orleans Meeting, 1956.

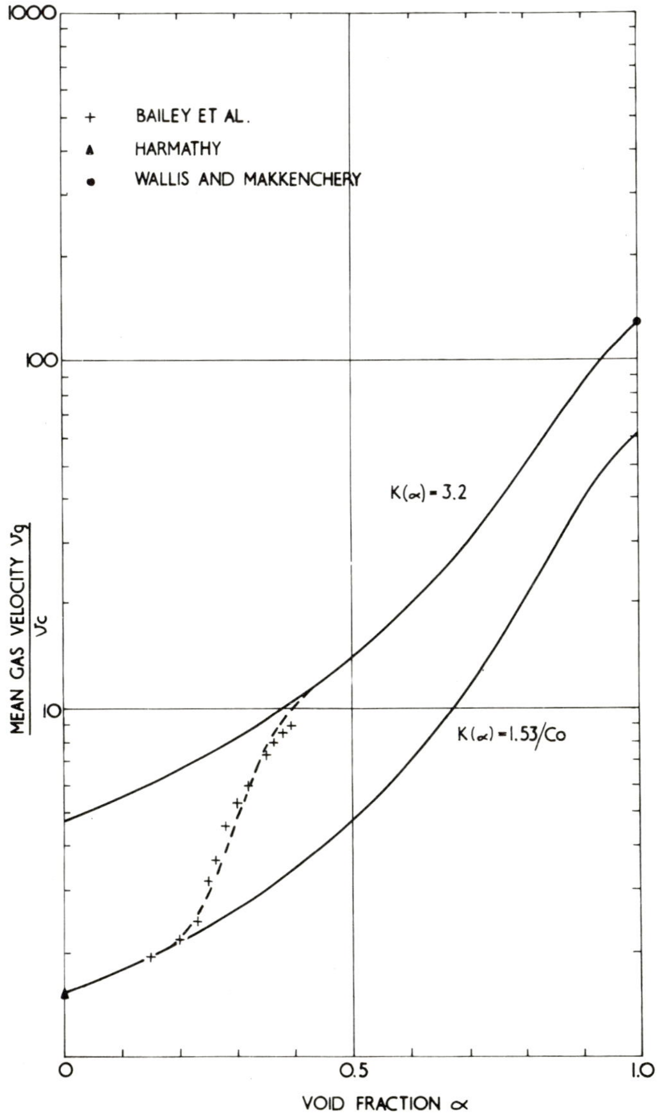


Fig. 1: Weighted mean rise velocity of steam through stagnant water at atmospheric pressure

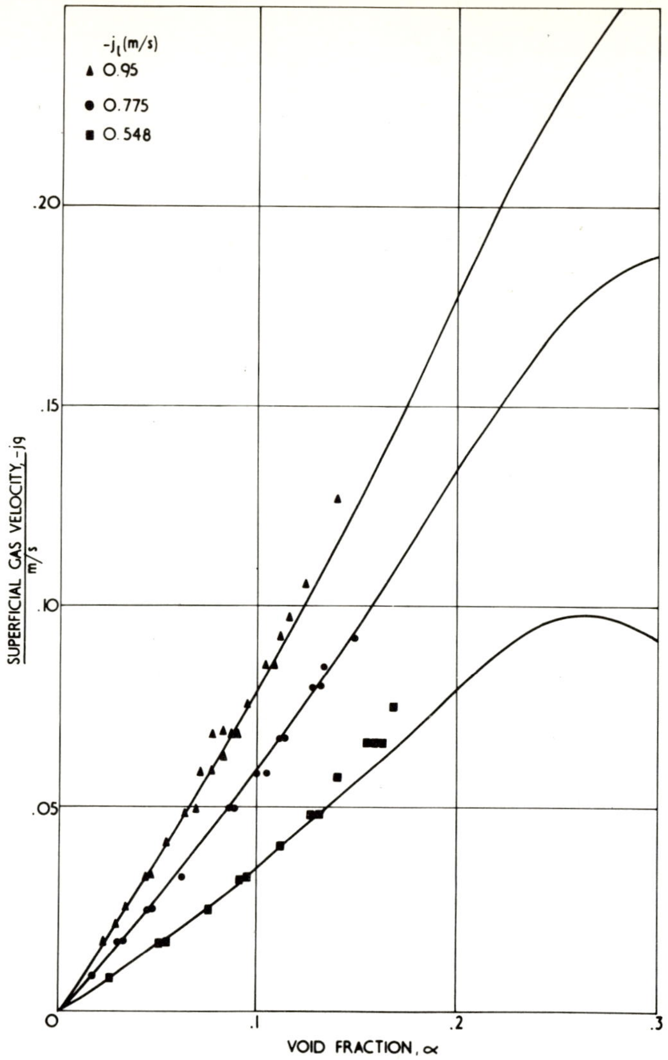


**Fig. 2: Variation of the superficial gas velocity with void fraction in downward concurrent flow**

C207/77

# EMERGENCY CORE COOLANT DELIVERY IN THE LOFT NON-NUCLEAR EXPERIMENT SERIES

S. A. NAFF, BS(Physics), MS(Eng)
EG&G Idaho, Inc., Idaho Falls, Id 83401, USA
W. F. SCHWARZ, BSEE, MBA
US Energy Research and Development Administration, Idaho Falls, Id 83401, USA

SYNOPSIS The performance of Emergency Core Cooling Systems during the injection following the hypothetical rupture of a primary system pipe in a pressurized water reactor has been the subject of extensive investigation, but has never been observed under actual accident conditions, since no such accident has ever occurred. The first experiments of a program designed to study such behavior under operating reactor plant conditions in a facility constructed for that purpose have been performed. The facility and the completed experiments are described, and the results discussed. Comparisons with predictions from the RELAP 4 computer code and other analytical techniques are included.

## INTRODUCTION

1. The Loss of Fluid Test (LOFT) experiments are part of a program managed by the Energy Research and Development Administration for the Nuclear Regulatory Commission (Ref. 1). These experiments are being conducted at the Idaho National Engineering Laboratory (INEL). The purpose of the LOFT experiments is to provide data for verifying and improving reactor safety computer codes designed to predict the response of a large pressurized water reactor (LPWR) during a hypothetical loss of coolant accident (LOCA).

2. LOFT (Figure 1) is a complete but scaled down (~1/50 LPWR Volume) pressurized water reactor with a system volume of 7.7 $m^3$ (Ref. 2). Its single intact loop contains a steam generator and two paralleled coolant pumps, and is sized relative to the remainder of the system to simulate the three unbroken loops of a four loop LPWR. Instead of a nuclear core in the reactor vessel, a core simulator is temporarily installed to provide flow resistance representative of the LOFT core for the non-nuclear experiments. No core heat simulation is provided and the system is heated to pre-blowdown conditions by mechanical energy addition from the primary coolant pumps.

3. Unique to LOFT is the "broken" loop, sized to simulate the loop in a four loop LPWR in which a pipe rupture occurs. The broken loop contains coolant pump and steam generator simulators. These are passive devices which simulate the flow resistances of the actual components. Normally closed Quick Opening Blowdown Valves (QOBV's) in this loop are opened to initiate system decompression, called a Loss of Coolant Experiment (LOCE).

4. When the QOBV's are opened, the decompressing effluent is directed through a header and downcomers into the Suppression Tank, a vessel partially filled with water and designed to simulate containment back pressures.

5. LOFT is equipped with redundant Emergency Core Cooling (ECC) systems scaled to those in a typical LPWR. There are High Pressure Injection Systems (HPIS), gas-driven accumulator systems, and Low Pressure Injection Systems (LPIS).

6. The LOFT program will ultimately examine the combined effects of nuclear, thermal, hydraulic, and structural processes which would occur during a LOCA. The non-nuclear (L-1) isothermal experiments discussed herein are the first phase of the program.

7. The LOFT L1 experiments are five LOCE's in which the parameters to be varied are:

break location, done by placing the broken loop simulators on either the hot or cold side of the QOBV's;

ECC system injection location and time, done by pre-aligning the injection path and using pre-set timers; and

operating loop flow resistance, done by orificing at the steam generator.

8. The principal objectives of the L1 series are:

a. To evaluate overall ECC performance.
b. To compare break flow data with predictions
c. To provide overall system thermal hydraulic information for use in code verification

This paper addresses only objective (a).

## THE NON-NUCLEAR EXPERIMENT SERIES

9. Prior to beginning the L1 series, the transit time of ECC water from the cold leg injection point to the reactor vessel lower plenum with the primary system empty and at ambient conditions was determined. This "cold wall delay" test was done by injecting ECC accumulator water at a typical rate into the cold leg, and observing the plenum fill rate.

10. The first LOFT LOCE, L1-1, simulated a 100% double-ended off-set shear break (50% of the flow area on each side of the break) in the hot leg of one loop of a LPWR (Ref. 3). This experiment was conducted from a reduced initial

pressure of 9.1 MPa to minimize subcooled blowdown-induced mechanical forces. Water from all three ECC systems was injected into the intact loop cold leg.

11. The second LOCE, L1-2, simulated a full 200% (100% of the flow area in each leg) double-ended off-set shear in the cold leg of one loop of a LPWR and blowdown was initiated from 15.5 MPa initial system pressure (Ref. 4). Intact loop hydraulic resistance was higher than in other tests to determine effects of intact loop resistance during the ECC portion of decompression. ECC was injected into the intact loop cold leg but was delayed until approximately 13 minutes after blowdown was complete. This LOCE effectively consisted of two separate experiments. The "blowdown test" gave thermal-hydraulic baseline data during decompression independent of phenomena which would have been caused by ECC injection. The follow-on "hot wall test" yielded data concerning the effects of heated reactor vessel walls on ECC delivery independent of complications which would have been caused by decompression-induced countercurrent flow.

12. The third LOCE, L1-3, simulated a cold leg 200% double-ended off-set shear at 15.5 MPa, with undelayed ECC water being directed horizontally to the reactor vessel lower plenum side near the bottom (Ref. 5). The HPIS and LPIS functioned as planned, but due to a procedural error, accumulator injection did not occur. The experiment was repeated identically as L1-3A, this time with all ECC systems injecting to the lower plenum as planned (Ref. 6).

13. The next LOCE, L1-4, is at this writing scheduled to be run in May of 1977. It will again be a 200% cold leg break with decompression from 15.5 MPa, without the programmed ECC delay of the L1-2 LOCE, and with ECC injected to the intact loop cold leg. Following L1-4 the nuclear core will be loaded in preparation for the nuclear experiment series.

POTENTIAL CAUSES OF ECC DELAY

14. In addition to the normal transport time of ECC water during cold conditions (cold wall delay), there are several physical conditions that have the potential to further delay or prevent ECC delivery to the reactor vessel lower plenum. These are:

bypass: ECC water could be swept around the inlet annulus and out the break by primary system fluid flowing up the downcomer; or, the inlet annulus configuration and ECC injection momentum could encourage loss of some ECC out the break.

Hot walls: Hot reactor vessel downcomer walls could vaporize some ECC water, adding to the countercurrent steam flow up the downcomer and decreasing downflow; or, the hot walls could sputter the ECC water into the upflowing steam which could carry it out.

Entrainment: ECC water which has collected in the lower plenum could be entrained in the primary system fluid flowing from the core barrel, and swept up the downcomer.

Lower plenum boiling: ECC water in the lower plenum could be boiled away by stored wall heat.

15. The following table identifies which of these potential causes of ECC delay are possible in each test.

| | Bypass | Hot Wall | Entrainment | Lower Plenum Boiling | Cold Wall Delay |
|---|---|---|---|---|---|
| L1-1 | X | X | X | X | X |
| L1-2 (Hot Wall) | | X | | X | |
| L1-2 (Blowdown) | | | | X | |
| L1-3 | | | | X | |
| L1-3A | | | X | X | |
| L1-4 | X | X | X | X | X |
| Cold Wall Delay | | | | | X |

L1 EXPERIMENT RESULTS

16. Before final conclusions can be drawn concerning ECC delivery behavior in the L1 series, the results from L1-4 must be evaluated. However, results to date provide a number of encouraging indications worth reporting.

17. LOCE L1-1 results indicate that contrary to predictions, the lower plenum did not void (Ref. 7). The model also predicted significant bypass and ECC loss, but none was observed. ECC water reached the downcomer in four seconds and the lower plenum in seven seconds after ECC initiation, and the entire vessel was filled in 46 seconds.

18. Non-axisymmetric behavior in the downcomer was observed (Figure 2). ECC water tended to flow down the intact loop side, and escaping primary fluid tended to flow up the broken loop side.

19. LOCE L1-1 is not directly comparable to the other L1 experiments because the others are cold leg breaks. However, the trends observed in L1-1 provide insight into expected behavior in the other L1 experiments.

20. The "hot wall" portion of L1-2 was primarily to study hot wall delay. By the start of delayed ECC injection, the lower plenum had emptied due to lower plenum boiling, thus the "hot wall" experiment started with a dry reactor vessel (Figure 3). ECC water was first detected in the lower plenum 2.5 $\pm$ 0.5 seconds after injection initiation, compared to the 1.9 $\pm$ 0.5 second cold wall delay, indicating that additional delay from hot wall effects was only 0.6 second. It was also noted that condensation produced a net gain of liquid starting at 20 seconds after ECC injection initiation (Figure 3). Hot wall delay predictions for these experiments were obtained from Creare and Semiscale experimental work.

21. Application of the Creare empirical correlations resulted in predictions of delay effects much longer (12 seconds) than the data indicated (Ref. 8). However, the LOFT system piping and downcomer walls were initially at

28$^{o}$C and the system absolute pressure was 0.34 MPa. This wall temperature was above the Leidenfrost temperature, so the Creare correlations are apparently not applicable for LOFT. In general, hot wall delay is expected to increase with increasing wall temperature until, at around the Leidenfrost temperature, sputtering of liquid from the wall allows it to fall free of wall friction. Beyond this temperature, the hot wall delay time can decrease with temperature.

22. Application of the INEL-produced Semiscale correlations gave a delivery time prediction of 5.5 seconds, compared with a measured value of 2.5 seconds for the sum of hot wall, cold wall and lower plenum boiling delays (Ref. 9). However, this correlation is based on data from the more one-dimensional Semiscale facility, and because of non-axisymmetric LOFT downcomer behavior, would bear further study before it is used for LOFT.

23. During L1-3, ECC was injected at a low flow rate into the lower plenum. Accumulation was sufficiently slow to maintain a liquid level below the core barrel bottom during blowdown, reducing losses due to entrainment. The only potential cause for loss of ECC water in this test was lower plenum boiling. Figure 4 shows ECC water loss by this mode was small.

24. In L1-3A, the potential causes of ECC water loss were entrainment and lower plenum boiling. However, condensation produced a net gain of liquid greater than the losses from these two sources, as is seen in Figure 5. The prediction indicates a more rapid fill rate due to its assumption of instantaneous mixing of mass and energy in each volume, producing complete and massive condensation in the lower plenum from cold ECC water.

## CONCLUSION

25. The conclusions are based on data from the LOFT non-nuclear test series. All of these tests were initiated from isothermal conditions and no core heat existed. The effects of energy addition from a nuclear core may alter the results of the reported data during the reflood portion of a LOCA. However, the reported information provides valuable insight into the refill portion of a LOCA, and will assist in predicting the beginning of reflood. It will also form a data base for comparison with other experiments. Specific conclusions drawn so far are:

a. ECC is delivered rapidly to the core inlet regions in all cases investigated.

b. The lower plenum is not completely voided for conditions studied.

c. Evidence exists indicating nonaxisymmetric downcomer behavior minimizing countercurrent flow effects.

d. Hot wall effects are small when downcomer wall temperatures are approximately 280$^{o}$C.

e. Loss of lower plenum liquid through boiling and entrainment is small.

26. The results of the LOFT non-nuclear experiment series are expected to provide better understanding of ECC scaling and delivery phenomena, and when considered in conjunction with other reactor safety experiments, should provide an improved basis for verification of computer codes used to predict ECC response in full-sized PWR's.

## REFERENCES


1. LOFT program description, LPD-1, Rev. 0, prepared by Aerojet Nuclear Company for the Energy Research and Development Administration, January, 1975.

2. ROBINSON, H.C., LOFT system and test description (Loss of coolant experiments using a core simulator), TREE-NUREG-1019, November, 1976.

3. NAFF, S.A. Quick-look report-LOFT non-nuclear experiment L1-1, Idaho National Engineering Laboratory, March 11, 1976.

4. BERTA, V.T. Quick-look report on LOFT non-nuclear experiment L1-2 and the hot wall delay experiment, QLR-76-3. Idaho National Engineering Laboratory, May, 1976.

5. CHAPPELL, J.R. Quick-look report on LOFT non-nuclear experiment L1-2 and the hot wall delay experiment, QLR-76-4. Idaho National Engineering Laboratory, July, 1976.

6. CHAPPEL, J.R. Quick-look report on LOFT non-nuclear experiment L1-3A, QLR-76-5. Idaho National Engineering Laboratory, July, 1976.

7. HOWE, T.M. LOFT Experiment Test Predictions for Non-nuclear Test Series L1, Experiment 1, ETP L1-1, Idaho National Engineering Laboratory, June, 1975.

8. BLOCK, James A. et al Analysis of ECC delivery, Topical Report CREARE TN-231, Creare, Inc., Hanover, New Hampshire, April, 1976.

9. BERTA, V.T. et al Experiment prediction for LOFT experiment L1-2, EP L1-2, Idaho National Engineering Laboratory, December, 1975.

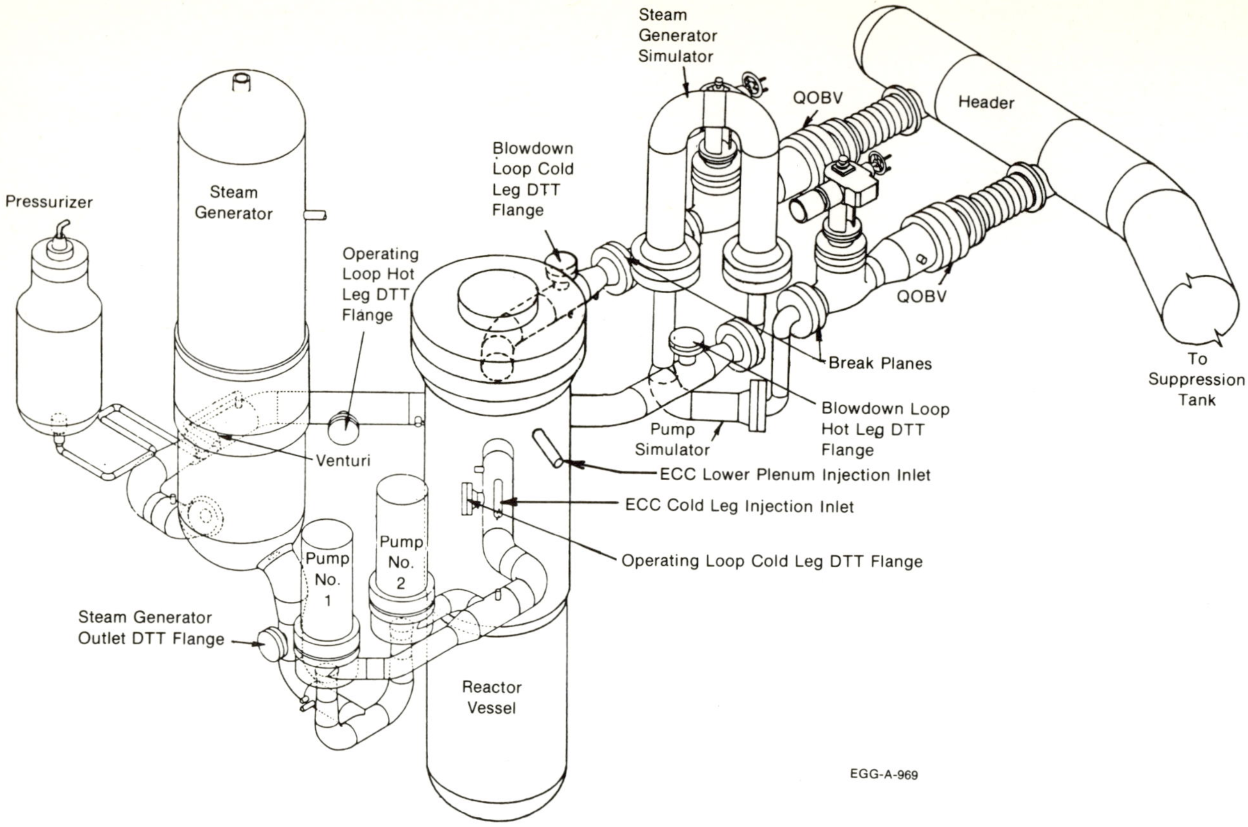


**Fig. 1: LOFT major components**

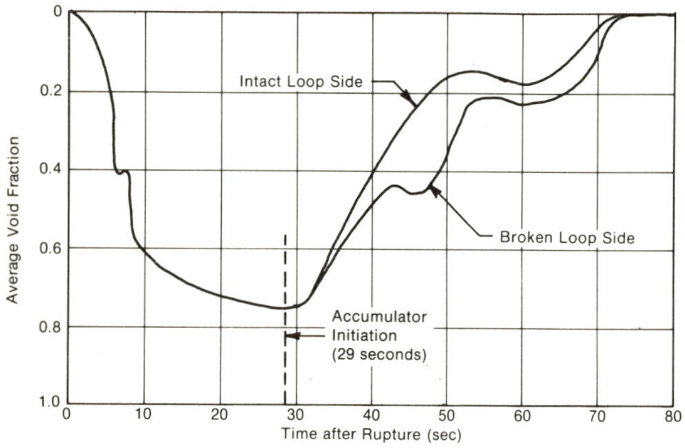


**Fig. 2: Average void fraction in reactor vessel, LOFT experiment L1-1**

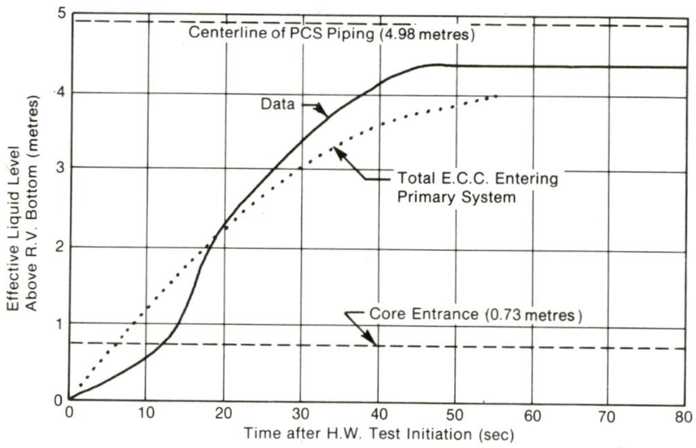


**Fig. 3: Effective liquid level in reactor vessel, LOFT experiment L1-2 (hotwall)**

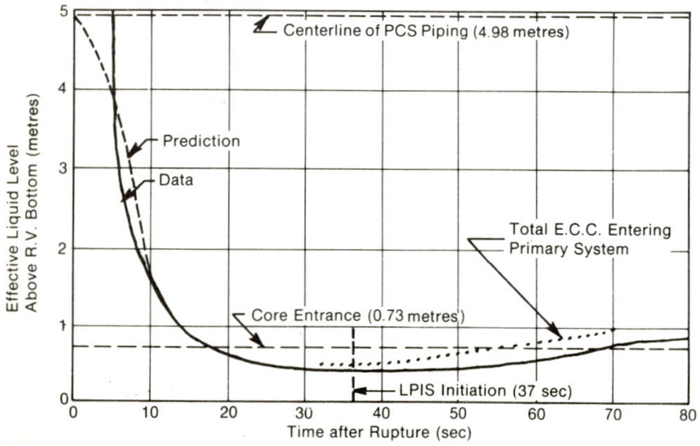


**Fig. 4: Effective liquid level in reactor vessel, LOFT experiment L1-3**

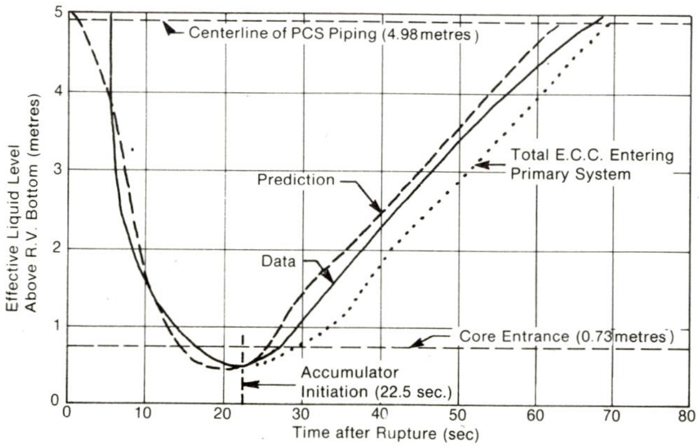


**Fig. 5: Effective liquid level in reactor vessel, LOFT experiment L1-3A**

C208/77

# THERMO-HYDRAULIC BEHAVIOUR OF FLUID IN PRESSURE VESSEL DURING BLOWDOWN

H. NARIAI, DrEngr
Ship Research Institute, Tokyo, Japan

SYNOPSIS Single volume blowdown experiments were conducted at the initial pressures of 5 MPa and 3 MPa. Pressures and void fractions at 6 points vertically in pressure vessel were measured during blowdown. The steam distributions in water were studied and compared with the distributions by the bubble gradient model used in analytical models for blowdown phenomena. Several transient characteristics during the blowdown were explained from the steam distributions experimentally derived and the input parameters in analytical models were discussed.

## INTRODUCTION

1. Investigation on the blowdown process is one of the most important problems for the development of the computer codes to analyse the transient behaviours of fluid at the loss-of-coolant accident of light water reactors, and many experimental and analytical works have been performed over more than 10 years throughout the world. This report presents the results of blowdown experiments conducted at the Ship Research Institute in Japan. The void fractions at 6 points vertically in pressure vessel were measured during blowdown. Analytical models on blowdown process and also the input parameters in the models were discussed.

## EXPERIMENT

2. Experimental apparatus consists of a pressure vessel, a discharge pipe, and a quench tank to condense the discharged steam-water mixture. The dimensions of pressure vessel are 284 mm ID and about 1700 mm height. The discharge pipe is installed at either the upper, middle or lower nozzle of the pressure vessel, simulating the upper, middle or lower blowdown respectively. The positions of three nozzles are at 1412 mm (upper), 862 mm (middle) and 312 mm (lower) from the bottom of the pressure vessel.

3. Pressures in pressure vessel, differential pressures between upper and lower nozzles, and void fractions at 6 points vertically in pressure vessel were measured during blowdown. Pressures and differential pressures were measured by force balance type transducers and void fractions by probe type void meters. The void meters were set at any six of the positions of 1412 mm (V-1), 1274 mm (V-2), 1137 mm (V-3), 1000 mm (V-4), 862 mm (V-5), 725 mm (V-6), 587 mm (V-7), and 312 mm (V-9) from the bottom of pressure vessel.

4. Three break positions, three orifice diameters, two initial liquid levels and two values of initial saturation pressure of fluid (5 MPa and 3 MPa) were selected as the parameters for experimental conditions. 27 experiments were conducted with various combinations of the parameters.

### Experimental results

5. Transient behaviours of pressure are summarized as follows. In the case of upper blowdown, an undershoot of the pressure is observed at the beginning of the blowdown, and then comparatively rapid pressure decrease followed showing that the flow is mainly in the form of steam. In the case of lower blowdown, two stages of pressure transients are observed. The first stage shows the slow pressure decrease corresponding two-phase steam water flow. At the second stage the pressure decreases rapidly, since the discharge changes to steam flow. In the case of middle blowdown, the pressure transients are in the intermediate band between those of upper and lower blowdowns and the changes from two-phase to steam flow occurs earlier than in the case of lower blowdown.

6. Figure 1 shows an example of the records of electro-magnetic oscillograph for steam and water in pressure vessel measured by void meters, just after the initiation of upper blowdown. The initial liquid level was between the positions of void meters V-3 and V-5. The generation of steam bubbles in water is observed from the records of V-5, and the liquid level moves upward, crossing the void meters of V-3, V-2 and V-1 successively. The same records of oscillograph were obtained in the whole time span of each experiment. The void fractions at time t were determined from the records as the time fractions indicating steam within a time interval $\Delta t$ around time t. The time interval $\Delta t$ was chosen between 0.2 sec and 1.0 sec.

7. The void fraction at each point of void meters in pressure vessel changes with time. In the case of upper blowdown, the generation of steam by the depressurization boiling occurs in the whole of the pressure vessel for the first several seconds. After this period, the values of void fractions begin to increase and distribute to take the larger values at the positions higher in the pressure vessel. The vertical distributions of the void fractions from the experiments coincided very well with those calculated from the so-called bubble gradient model with the value of 0.8 for the bubble distribution parameter Cx (refer to section 10). In the case of lower blow-

 

down, the values of void fractions are very small compared with those in the case of upper blowdown. And the void fractions in water take relatively large values only in the vicinity of the liquid level. This means that the depressurization boiling is occuring only in the vicinity of the liquid level, and the distributions of the void fractions take steeper gradients than those estimated from the bubble gradient model with Cx = 1.0 and the gradients are steeper with the decrease of the orifice diameter. And so the void fractions at the lower point in water take smaller values than those calculated from the bubble gradient model. In the case of middle blowdown, the void distributions take steeper gradients than those predicted from bubble gradient model for the first stage of blowdown which is similar to the case of lower blowdown. When the liquid level falls to the break position, the distribution changes to that predicted from the bubble gradient model with Cx = 0.8, which is similar to the case of upper blowdown. Figure 2 is an example of the changes of void distributions with time in the case of middle blowdown. The ordinate shows the normalized height (the positions of void meters divided by the position of liquid level), and the abscissa shows the void fractions at each void meter. The void fractions were derived from the time mean curve of the experimental data. At the time of 3.5 sec, the liquid level has reached at the break point. Before that time, the distributions of void fractions differ from the straight lines which show the distributions from the bubble gradient model with three Cx values. At the time of 12 sec, the distribution coincides very well with the straight line of Cx = 0.8.

## Discussion of the experimental results

8. When the initial fluid conditions and the break position are the same, the pressure in the experiment with large orifice decreases faster than that with small one. But as the blowdown time would be almost inversely proportional to the cross-sectional area of orifice, the time scales of the diagrams for the transient characteristics were changed so that the product of the cross-sectional area of orifice and the time may take the same value, and the comparisons of transient characteristics were made with the converted time scales for three orifice diameters. Figure 3 shows an example of the comparisons for the middle blowdown, in which the transient behaviour of pressure and void fractions are simultaneously compared. As shown in the figure, time scales for the blowdowns of 12.7 mm and 6.3 mm orifice diameters are 1/4 and 1/16 of those for the blowdown of 25.4 mm orifice diameter, respectively. When the transient behaviour of pressure is concerned, the pressure in the experiment with small orifice decreases relatively faster than that with large one. The same results were obtained in the either case of upper and lower blowdowns.

9. The comparisons for the void fractions are also made in the figure. The comparisons show that the falling speeds of liquid level increase and the void fractions in water decrease with the decrease of orifice diameters. The same results were also obtained in the either case of upper and lower blowdowns. These results are explained from the strength of depressurization boiling. When the orifice diameter is small, the depressurization boiling is not so strong and the steam can not be generated so much in water, resulting in low void fractions in water and increasing the falling speed of liquid level. In the case of upper blowdown, the discharging fluid will contain more water since the liquid level goes up with the increase of orifice diameter, resulting in relatively slow transients of pressure. In the case of lower blowdown, the discharging fluid will be in the form of relatively lower quality of two-phase steam-water flow since the gradients of void distributions become steeper with the decrease of orifice diameters, resulting in relatively short blowdown time. The explanations for both upper and lower blowdowns will also be applicable to the case of middle blowdown. In the middle blowdown, the behaviour is almost the same as in the case of lower blowdown for the first stage, and also the same as in the case of upper blowdown for the later stage.

## ANALYSIS

10. A computer code is developed for the analysis of the present results of single volume blowdown experiments. The code includes the bubble gradient model for the steam distribution in water as shown in equation (1).

$$\alpha(Z) = a\frac{Z}{Zm} + b \qquad (1)$$

$$a = \begin{cases} 2Cx\,\alpha m & 0 \leq \alpha m \leq 0.5 \\ 2Cx(1-\alpha m) & 0.5 \leq \alpha m \leq 1.0 \end{cases}$$

$$b = \begin{cases} (1-Cx)\alpha m & 0 \leq \alpha m \leq 0.5 \\ (1+Cx)\alpha m - Cx & 0.5 \leq \alpha m \leq 1.0 \end{cases}$$

where

$\alpha(Z)$: void fraction at elevation Z
$\alpha m$: average void fraction in water
Z: elevation from the bottom of pressure vessel (m)
Zm: height of liquid level from the bottom of pressure vessel (m)
Cx: bubble distribution parameter

In equation (1), the bubble distribution parameter Cx is contained as an input parameter. The bubble rise velocity Vb (m/s) from the water surface to the steam dome has also be contained in the code as another input parameter. The flow rate of discharging fluid through the orifice was calculated as either the steam flow of single phase with flow coefficient Cd1 as shown in equation (2), or the Moody critical flow of two phase steam-water mixture with the flow coefficient Cd2 as shown in equation (3).

$$W = Cd1\ Ws \qquad (2)$$

$$W = Cd2\ Wmc \qquad (3)$$

where

W: mass rate of flow (kg/s)
Ws: mass rate of single phase steam flow (kg/s)
Wmc: mass rate of flow by Moody two-phase critical flow model (kg/s)
Cd1: flow coefficient of single phase
Cd2: flow coefficient of two phase

The value of 0.6 was adopted for Cd1 in this study and Cd2 remained as the other input parameter. Some calculations showed that the results by this code gave almost the same results from the blowdown code RELAP 3 (Ref. 1) or 4 (Ref. 2) for the single volume blowdown calculations. So the discussions on the input parameters Cx, Cd2 and Vb are in general for other blowdown codes containing the similar models as described above.

## Discussion of the analytical results

11. As shown in section 7, the steam distribution can be estimated very well by the bubble gradient model with the value of 0.8 for Cx in the case of upper blowdown. Though the steam distributions by the bubble gradient model are not exact for the lower and middle blowdowns, the comparisons between the experimental results and the calculational results by the code were also made with the value of 0.8 for Cx in the case of lower and middle blowdowns. Many comparisons were carried out with the various combinations of Cd2 and Vb values. The ranges of the combinations of Cd2 and Vb are found so that the results of the calculations for pressure, void fractions and falling speed of liquid level fit fairly well the respective experimental results. The ranges are given in diagrams in which Cd2 is in the ordinate and Vb is in the abscissa. The diagrams show that Cd2 decreases with Vb for pressure and falling speed of liquid level, and Cd2 increases with Vb for the void fractions. The result that Cd2 decreases with Vb for pressure and liquid level speed can be explained as follows. The falling speed of liquid level increases and the blowdown becomes more rapid with the increase of Vb, and so Cd2 has to be changed to small value for the blowdown to be slower. The result that Cd2 increases with Vb for void fractions can be explained as follows. As the void fraction in water decreases with the increase of Vb, Cd2 has to be larger so that the flow rate and the strength of depressurization boiling may increase. The limited ranges of Cd2 and Vb combinations are obtained with which the calculational results fit fairly well the experimental data of pressure, void fraction and liquid level speed simultaneously. Figure 4 shows an example of the diagrams for the limited ranges of Cd2 and Vb. The following results are obtained from the figure.

(a) Cd2 increases, the lower the break position with the same orifice diameter. In the case of upper blowdown the most part of the discharge process is in the form of steam flow and the calculational curve does not depend so much on Cd2 values, though the best value of Cd2 is near 0.6. In the case of lower blowdown, Cd2 takes the values of about 0.8 - 0.9 for the blowdown with the orifice diameter of 25.4 mm, and about 1.3 with the orifice diameters of 12.7 mm and 6.3 mm.

(b) In the same break position, Cd2 value decreases and Vb value increases with the increase of orifice diameter.

12. The result that Cd2 values increase with the decrease of orifice diameter in the case of lower blowdown, can be explained as follows. As shown in section 7, the gradient of the void distribution is steeper than that predicted by the bubble gradient model in the lower blowdown with the same orifice diameter. And so the real void fractions are smaller compared with those predicted by the analytical calculations. So the void fractions of discharging fluid take larger values in the analysis than the real values. And Cd2 has to be put larger so that the discharge flow rate from Moody critical flow model may fit the real one. The larger Cd2 values with small orifice diameter are sometimes explained from the non-equilibrium effects. However, the estimation of the void distribution in water is also seen to be important from the above mentioned results. The analysis of lower blowdown has to be done either with the bubble gradient model with the experimentally determined Cd2 values, or with the another model with real void fractions.

13. Results of the analysis with a variable parameter model are compared with the present experimental results. The model includes Cx = 0.8, Vb values from the experimental correlations by Wilson et al (Ref. 3), and Cd2 values from the correlation by Sobajima (Ref. 4). As the Cd2 value from the correlation is larger than 1.0 for the discharging fluid with small quality, the calculational results from the model agree fairly well with the experimental data over a wide range of pressure and void fractions.

## CONCLUSION

14. Many data on pressure and void fractions are obtained in the single volume blowdown experiments. The void distributions for upper blowdown are predictable very well by the bubble gradient model with Cx = 0.8. But the void distributions for lower blowdown take much steeper gradients than those predicted by the bubble gradient model. The comparisons of three kinds of blowdown characteristics are made with the converted time scales so that the products of the cross-sectional area of orifice and the time may take the same value. The result that the blowdown time is relatively longer with the larger orifice diameter, is explained from the strength of depressurization boiling. An analytical model with some input parameters is developed. The combinations of the input parameters Cd2 and Vb are obtained so that the analytical results may fit fairly well the experimental results. The large Cd2 values with small orifice diameter are explained as the void fractions at break position are overestimated in the analysis for small orifice diameter. The calculational results by a variable parameter model with large Cd2 values at low quality agree fairly well with the present experimental results.

## ACKNOWLEDGEMENT

15. The author wishes to thank Mr. M. Nagatome and Mr. S. Aso of Mitsui Shipbuilding Co. for their assistance on the experimental works.

## REFERENCES

1. Retting, W.H. et al, 'RELAP 3 ----- A computer program for reactor blowdown analysis', IN-1321, Idaho Nuclear Corporation, 1970.

2. Moore, K.V. and Retting, W.H. 'RELAP 4 ----- A computer program for transient thermal-hydraulic analysis', ANCR-1127, Aerojet Nuclear Company, 1973.

3. Wilson, J.F. et al, 'The velocity of rising stream in a bubbling two-phase mixture', Trans. Am. Nucl. Soc. 1962, 5, p.151.

4. Sobajima, M., 'An analysis of transients in experiments of loss-of-coolant accidents', Nucl. Sci. Engr., 1976, 60, p.10.

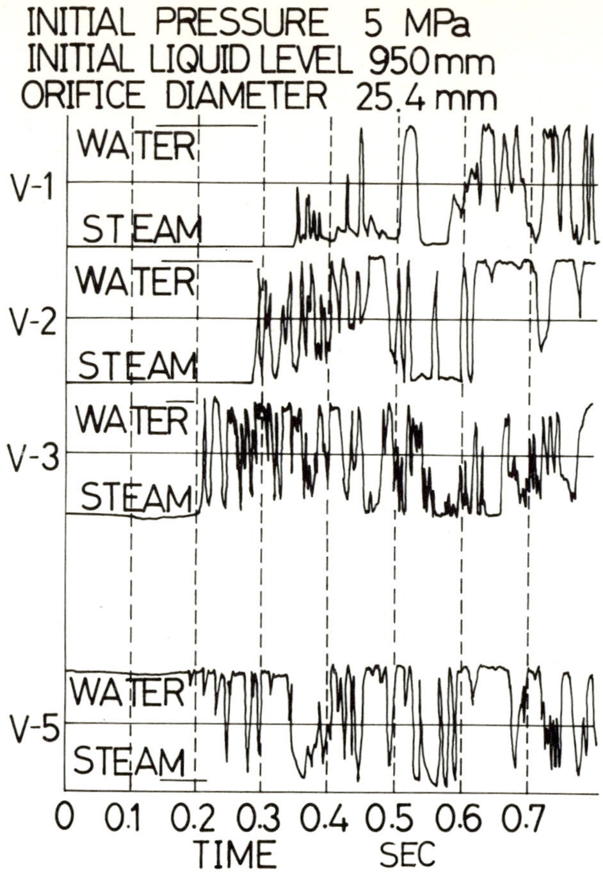


**Fig. 1: Oscillograph records of void meters just after the initiation of blowdown**

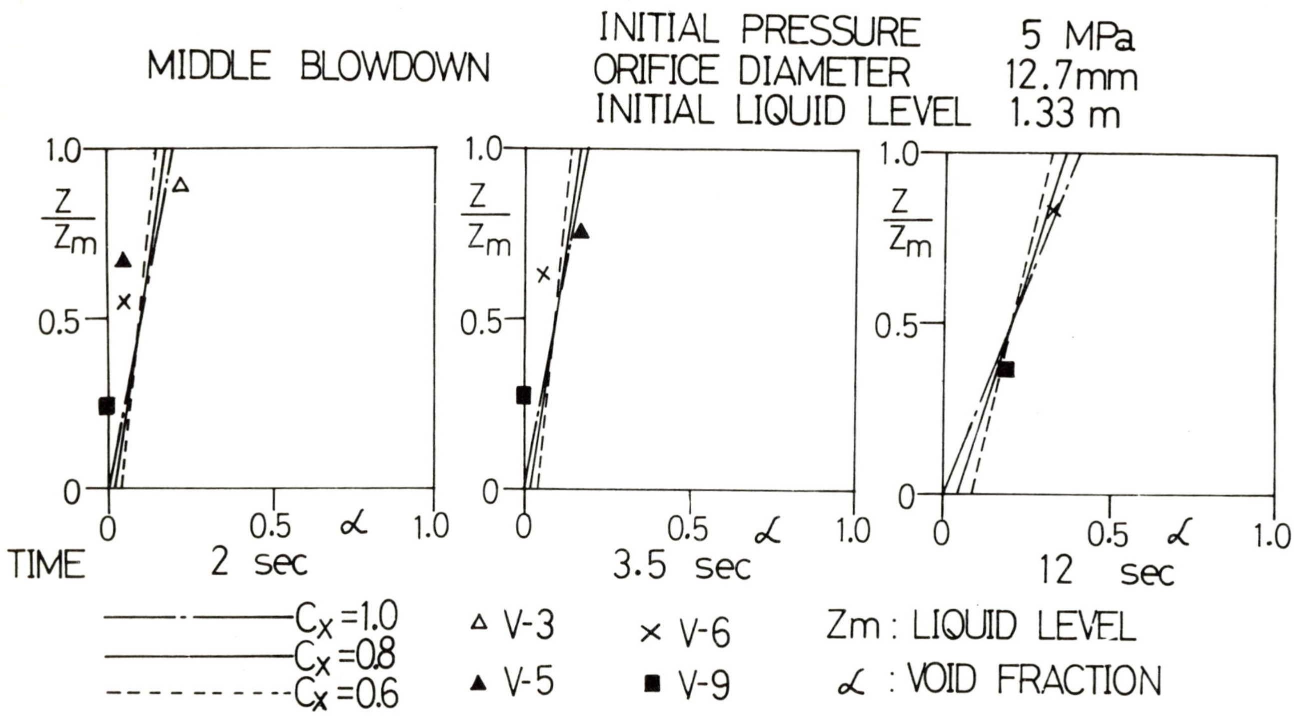


**Fig. 2: The change of vertical distributions of void fractions with time**

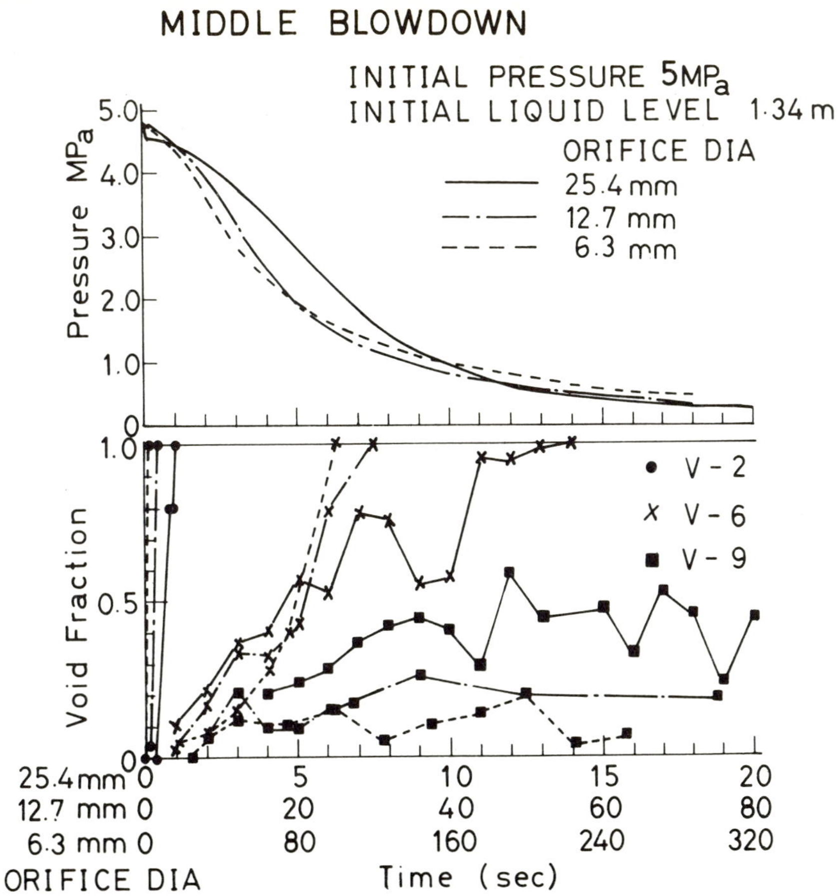


**Fig. 3: Comparisons of transient characteristics with the converted time scales for three orifice diameters**

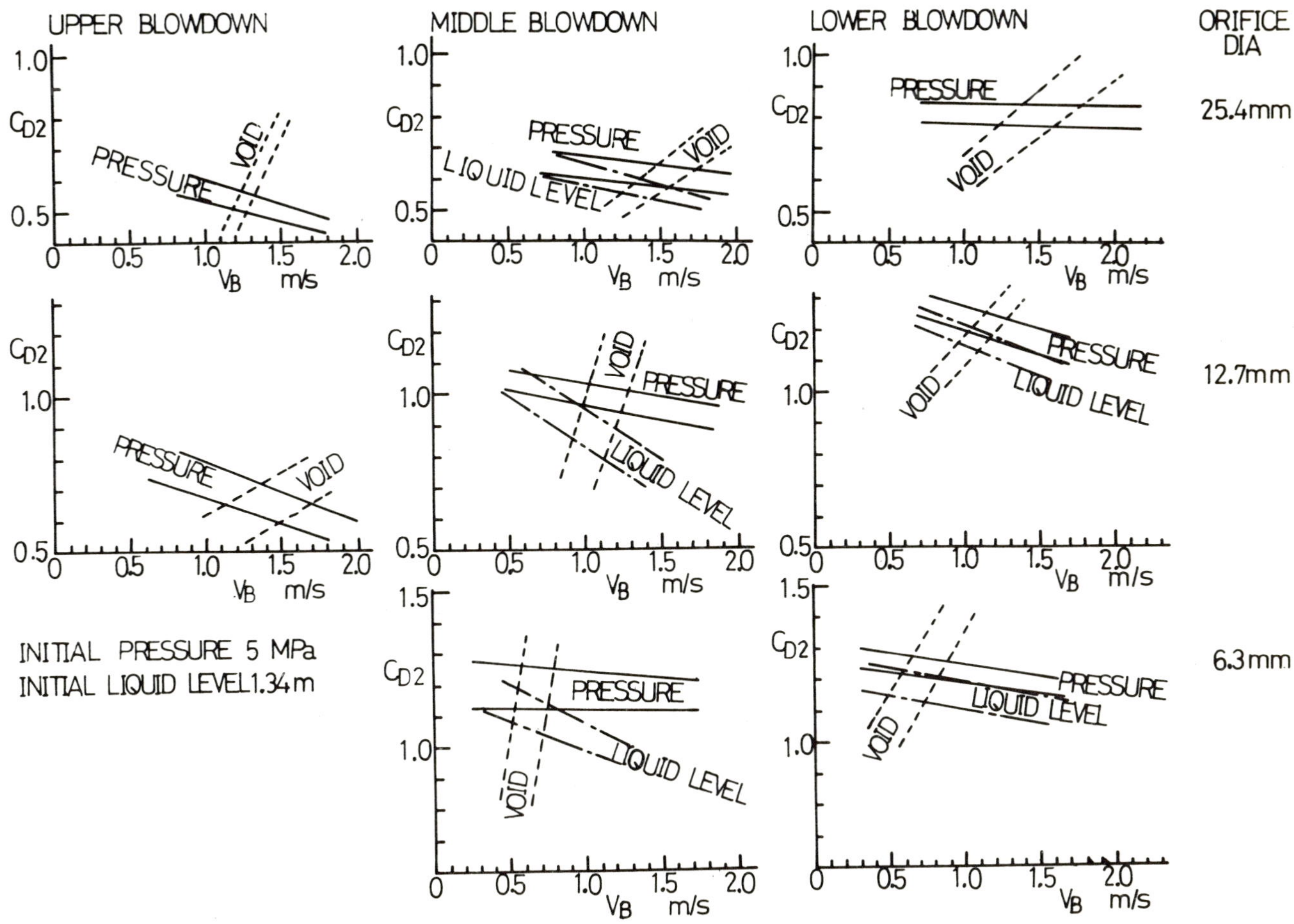


**Fig. 4: Cd2 and Vb values for the analytical results to fit the experimental data**

C209/77

# BLOWDOWN INTO FULL-PRESSURE CONTAINMENT

J. MARSHALL, BSc and P.G. HOLLAND, BSc, MSc
Australian Atomic Energy Commission, Reseach Establishment, Lucas Heights, NSW, Australia

SYNOPSIS Experimental results are presented of the blowdown of a 14 ℓ pressure vessel containing saturated steam/water at 10.4 and at 7 MPa into a 1.8 $m^3$ vessel. Various condensation heat transfer schemes are used in a simple theoretical model to estimate the containment pressure transient and these estimates are then compared with experimental results. The pressure rise is generally overestimated, sometimes by very large amounts, and none of the schemes normally used in reactor safety calculations take into account all the effects likely to arise from differing initial conditions. A simple modification to an existing scheme gives a better estimate under the experimental conditions so far investigated.

## INTRODUCTION

1. In estimating the pressure rise in a full-pressure containment during loss of coolant accident (LOCA) a parameter of most vital importance is the heat transfer between the containment atmosphere and the containment structure. Safety studies normally demand that during the initial transient, or blowdown stage, engineered safety devices such as sprays or fan-blown heat exchangers be not considered operative. The major heat transfer occurring then will be by natural condensation on containment walls or structures. In such circumstances the data used in the major computer codes employed in reactor safety studies, while conservative, may not be very realistic. For example, the condensation heat transfer coefficients offered in CONTEMPT [1], derived from Uchida et al. [2], gave considerable overestimation of containment pressure in experiments on the CVTR containment reported by Schmitt et al. [3]. Measurements indicated average condensation heat transfer coefficients of about five times the Uchida value during the steam-inflow period and use of these higher values in CONTEMPT resulted in more realistic pressure history calculation. A similar effect was found earlier in the present series of experiments [4]; a reasonable fit to measured pressure and temperature transients over a wide range of initial air content in the low pressure vessel (LPV) simulating a containment was obtained by taking the heat transfer coefficient at five times the Uchida values for the blowdown phase and reverting to the original Uchida values after the blowdown high pressure vessel (HPV) had emptied of liquid.

2. This situation has been investigated further and the paper gives some of the measurements of the HPV blowdown and outlet choking pipe conditions. These are governed by the outlet mass and energy flow which are other main parameters in containment pressure rise estimation.

## OPERATION OF THE RIG

3. The rig with the main instrumentation is outlined on Figure 1. Dimensions of vessels and pipes are given in Table 1. The HPV is filled to the required level, the water is boiled with the vent valve open to purge the air, and the amount of condensate is measured. The vent valve is then closed and the pressure raised to the required value by further heating. Time is allowed for the conditions to settle and to minimise temperature gradients within the HPV. The two bursting discs have a rupture pressure of about three-quarters of the final HPV pressure and the interspace is filled with nitrogen at half HPV pressure. Blowdown is initiated by releasing the nitrogen pressure. A short pipe section, the 'choking section', defines the flow area for the HPV outflow. It has a rounded inlet with parallel flow path up to a slow expansion section and was intended to give conditions in which choking would occur at the outlet of the parallel section and where the pressure measured in a wall tap at this plane (P3) would indicate choking pressure. Efflux from the HPV passes through the expansion pipe (EP) into the large LPV which represents a full pressure containment.

4. All pressure measurements are made by diaphragm/strain gauge transducers and the temperature measurements by 1 mm diameter stainless steel sheathed thermocouples. The HPV is lagged but the remainder is not. Long term temperature transients in the LPV indicate that the external heat loss rate is very small particularly if considered over the short initial blowdown period (less than 10 seconds). The water masses given in the experimental results are the heated mass in the HPV. In addition to this heated mass there is about 0.3 litres of water unheated in various dead volumes including the outlet pipe.

5. Initial experiments [4] were made with the inside walls of the LPV left as bare steel. These became slightly rusted and for later tests were cleaned by wire brushing and treated with two coats of a carbon-based boiler paint Apexior No.1. (British Paints) suitable for use with high temperatures and rapid temperature changes.

 

## SIMPLE LUMPED PARAMETER MODEL

6. To facilitate design of the experiments and understanding of the results the system has been analysed using a simple lumped parameter model as fully described in [4]. The fluid conditions within the HPV and LPV are governed by mass, volume and energy conservation equations with the steam/water content considered to be continuously in saturation equilibrium. In the LPV there is heat transfer from the steam/water to the air and to the containment structure. Heat transfer to air is considered to be complete and instantaneous but heat transfer to the vessel walls may be governed by the conditions of instantaneous atmospheric air/steam mass ratio etc. which are discussed later. LPV wall conditions are determined by a lumped model with heat flow from the surface through a half-thickness conductance into a thermal mass. The flow between the vessels is derived from Moody [5] with an area orifice coefficient of 0.74; the inlet quality is obtained from the extreme assumption that there is single phase water at the outlet until the HPV liquid content is near zero and single phase steam subsequently. The equations are written in terms of differentials with respect to pressure and of differential pressure with respect to time, and are then integrated numerically for solution.

7. This model is extremely simple but contains many features similar to more sophisticated analyses. The use of the Moody model with simple orifice coefficient for flow and the LPV internal wall lumped heat transfer model are perhaps the most drastic basic simplifications. However, the codes RELAP-3 and CONTEMPT LT022 have also been used to calculate conditions in the HPV and LPV respectively. These codes have given pressure transients very similar to the present model when similar assumptions, e.g. the Uchida condensation heat transfer data, are used. Large codes like these are difficult to modify and so the present model, which is flexible, was derived. It is used here to illustrate the experimental results and as a vehicle for testing the likely utility of various condensation heat transfer proposals. Results of calculations with this model are shown as dotted lines on the figures giving experimental results.

## EXPERIMENTS

### Blowdown Measurements

8. Figure 2 shows the HPV pressure (P1) and pressure at the choking section outlet (P3) for an initial condition of 10.4 MPa with initial fill of 7.1 kg. One feature of interest is the very sharp knee in the P3 transient which occurs just before the smoother knee in the P1 transient. These knees are closely described by the theoretical model in which they occur as a result of the assumed quality step change from 0 to 1 at the time when the HPV empties of liquid, giving some support to this simple assumption. Pressure ratios (choking/inlet stagnation) calculated by the method of Moody [5] are nearly independent of inlet stagnation pressure in the present range and are about 0.66 when the inlet quality is zero and about 0.56 when it is unity. These ratios are somewhat higher than the measured ratios (P3/P1), particularly for the 10.4 MPa case, but the general behaviour of sudden change just before the P1 knee is the same as obtained in the model. There is doubt as to whether P3 is always actually at the plane of choking. At this plane the pressure gradient will be high and a small displacement of the plane will give large measured pressure variation. The area coefficient used in the model was derived from earlier experiments at 7 MPa and was chosen to fit the time of occurrence of the knee. It has given reasonable fit to all subsequent tests with a wide range of initial mass and pressure but with the same rig configuration.

### Containment Measurements

9. Figures 3 and 4 show the LPV pressure transients resulting from two blowdown cases. The initial LPV air pressure is at one atmosphere with relative humidity expected to be approaching 100% because of a small pool of water near the drain outlet. The transients show a time delay of about 0.1 s after the bursting disc rupture and then a rapid rise to a peak at just after the knee in the HPV transients, followed by a slow fall.

10. Temperature measurements were also made and show substantial gradients along the vessel walls, particularly during the initial blowdown phase, indicating a considerable variation in the effectiveness of wall heat transfer. This appears to be largely due to the presence of air, since experiments in near vacuum conditions [4] have shown much more uniform wall temperatures.

## CONDENSATION HEAT TRANSFER

11. The heat transfer conditions between the atmospheric fluid (air/water/steam) and the containment structure and walls are most difficult to define. The situation differs considerably from that dealt with in normal situations where the rate of condensation is governed by the rate of diffusion of steam through air and the rate of heat transfer through surface liquid films. The main features of containment conditions are the large average air/steam mass ratio and the unknown and constantly varying three-dimensional flow pattern. Since detailed calculation of such conditions, if possible, would be prohibitively difficult and expensive, it is customary to try to find a simple coefficient suitable for use in lumped parameter models.

12. The condensation heat transfer coefficients derived by Richardson et al. [1] from the results of Uchida et al. [5] cover a wide range of air/steam mass ratio. They were apparently derived from experiments with a small vertical plate in a still atmosphere situation. They were correlated for the present work by the expression

$$H = 450\ r^{-0.8}\quad \mathrm{W\ m^{-2}\ K^{-1}}$$

up to a maximum of 1590 $\mathrm{W\ m^{-2}\ K^{-1}}$, where r is the average air/steam mass ratio. For the present experiments the expression considerably overestimates the pressure rise (Figures 3 & 4). As previously mentioned a similar effect was also found in the CVTR experiments [3] where a better estimate of peak pressure was given by using coefficients up to about five times the Uchida values during the period of incoming steam flow. In the subsequent period much lower values appeared to be appropriate. For the present experiments, taking heat transfer coefficients of five times the Uchida values during the blowdown phase and reverting to original values for subsequent periods give greatly improved fit as shown by the 'modified Uchida' curves in the Figures. This procedure may perhaps be justified on the general grounds that in the blowdown phase the incoming steam velocity is relatively high so that the air would be swept away from the containment surfaces and high heat transfer might be expected. In the subsequent phase the velocities would be low and the heat transfer then become similar to that in steady state quiescent conditions. In the calculational model the end of blowdown is taken as the point at which the HPV becomes empty of liquid.

13. A review by Slaughterbeck [6] of heat transfer coefficients for use in containment response analysis recommended that a peak heat transfer coefficient be determined from the expression

$$H_{max} = 0.603 \left(\frac{Q}{V t_p}\right)^{0.62} \quad W\ m^{-2}\ K^{-1}$$

where $Q$ is the total energy in the blowdown vessel (J)
$V$ is the volume of the containment ($m^3$)
$t_p$ is the time of containment pressure peak (s) .

14. The instantaneous value of H should then be varied linearly from zero to $H_{max}$ over the period up to $t_p$. After $t_p$ the Uchida values should be used. The peak time $t_p$ is interpreted here as the time of the peak in the experimental transient. Results using Slaughterbeck's method are also shown on Figures 3 and 4 and give a reasonable peak pressure in these cases, although the actual shape of the calculated transient is unsatisfactory. However, the method deals only with initial air pressure of one atmosphere and has no factors for other conditions. It was derived from unpublished Japanese work [6] presumably on small rigs similar to the present size and may be quite inappropriate for anything different. A warning to this effect apparently accompanied the original work and so it may be suitable only for a narrow range of system size and power.

15. A further method is used by Brosche [7] in the code ZOCO V. This is given in work by Henderson and Marchello [8] for condensation on horizontal pipes as

$$H = H_N/(1 + 0.51\ y) \quad W\ m^{-2}\ K^{-1}$$

where $H_N$ is the Nusselt heat transfer coefficient for the pure condensable
$y$ is the mole percentage fraction of non-condensable.

16. The original work covers only a limited range in air/steam mass ratio and the expression gives heat transfer values rather lower than the Uchida values for the high air/steam mass ratios in the present cases.

DISCUSSION

17. The simple theoretical model used here cannot be expected to give close fit to the experimental results, but it should enable a comparison to be made between the various heat transfer coefficients used to estimate the pressure rise in a full-pressure containment during a blowdown. Clearly none of those considered here is satisfactory. The simple Uchida values are far too low during the initial pressure rise phase, although it should be noted that they provide a very good fit for the subsequent period (for example in the Slaughterbeck curves on Figures 3 and 4).

18. One point of interest in comparing the results of the 10.4 MPa transient with the 7 MPa transient is that the LPV peak pressure for the high pressure blowdown does not increase as much as predicted by the Uchida based calculations. This may be because the initial mass and energy flow rates are much higher and so the initial heat transfer will also be higher. The Slaughterbeck recommendation has the advantage of increasing heat transfer with increased initial HPV energy and so tends to take this effect into account. However, the lack of factors for initial air pressure, the uncertainty as to its suitability for different system scale and also the requirement for predetermining the time of peak pressure, all make it unsatisfactory.

19. The modified Uchida method has given considerably improved performance, but it is rather arbitrary. It can deal adequately with low initial air pressure, at least to 0.5 atmosphere [4], but it does not take into account the apparent changes in heat transfer caused by the higher initial HPV pressure. Work is proceeding to obtain more data on which to base the development of a heat transfer scheme including such effects.

CONCLUSIONS

20. The present experiments tend to confirm that the condensation heat transfer coefficients generally used in estimating containment conditions during a LOCA are likely to be conservative. The Uchida data appear to provide a suitable basis for estimation of average heat transfer in the post blowdown phase but are not appropriate, without modificiation, to the period of pressure rise. Over the experimental conditions investigated to date the 'modified Uchida' scheme appears to provide improved estimation. More work is needed to determine whether a simple heat transfer scheme can be found and justified which, when used in lumped parameter models, would provide a satisfactory estimate of containment pressure transients over a wide range of conditions.

ACKNOWLEDGEMENT

21. Acknowledgement must be given for the contribution of C. Campbell in construction and operation of the rig.

REFERENCES

[1] RICHARDSON, C.C., FINNEGAN, L.J., WAGNER, R.J. and WAGGE, J.M. - 'CONTEMPT - a computer program for predicting the containment pressure-temperature response to a loss-of-coolant accident'. IDO-17220, 1967.

[2] UCHIDA, H., OYAMA, A. and TOGO, Y. - 'Evaluation of post-incident cooling systems of light water power reactors'. 3rd Int. Conf. on Peaceful Uses of Atomic Energy, Vol. 13 p.63, 1964.

[3] SCHMITT, R.C., BINGHAM, G.E., NORBERG, J.A. - 'Simulated design basis accident tests of the Carolina's Virginia Tube Reactor containment'. IN 1403, 1970.

[4] MARSHALL J. and HOLLAND, P.G. - 'Effect of air content and mass inflow on the pressure rise in a containment during blowdown'. Proc. Second Australasian Conf. on Heat and Mass Transfer, Sydney, February 1977, pp. 339-346.

[5] MOODY, F.J. - 'Maximum flow rate of a single-component two-phase mixture'. Trans. ASME, Feb. 1965, pp. 134-142.

[6] SLAUGHTERBECK, D.C. - 'Review of heat transfer coefficients for condensing steam in a containment building following a loss-of-coolant accident'. IN 1388, Sept. 1970.

[7] BROSCHE, D. - 'ZOCO V a computer code for the calculation of time and space-dependent pressure distributions in reactor containments'. Nucl. Eng. Design, 23 (1972) pp. 239/272.

[8] HENDERSON, C.L. and MARCHELLO, J.M. - 'Film condensation in the presence of a non-condensable gas'. J. Heat Trans., Aug. 1969, p.447/450.

TABLE 1
RIG DIMENSIONS (Reference 4)

| Component | Volume $m^3$ | Diameter mm | Length m | Wall thickness mm | Wall Surface area $m^2$ |
|---|---|---|---|---|---|
| HPV | 0.0136 | 97.2 | 1.86 | 8.6 | 0.571 |
| Outlet pipe | $2\ 10^{-4}$ | 15.9 | 0.87 | 1.6 | 0.046 |
| Choking Section | $8.6\ 10^{-6}$ | 7.94 | 0.173 | 5.6 | $4.31\ 10^{-3}$ |
| Discharge pipe | $8\ 10^{-5}$ | 15.9 | 0.4 | - | 0.02 |
| EP | 0.0142 | 92.0 | 2.13 | 5.0 | 0.65 |
| LPV | 1.84 | 914. | 3.0 | 6.4 | 8.57 |

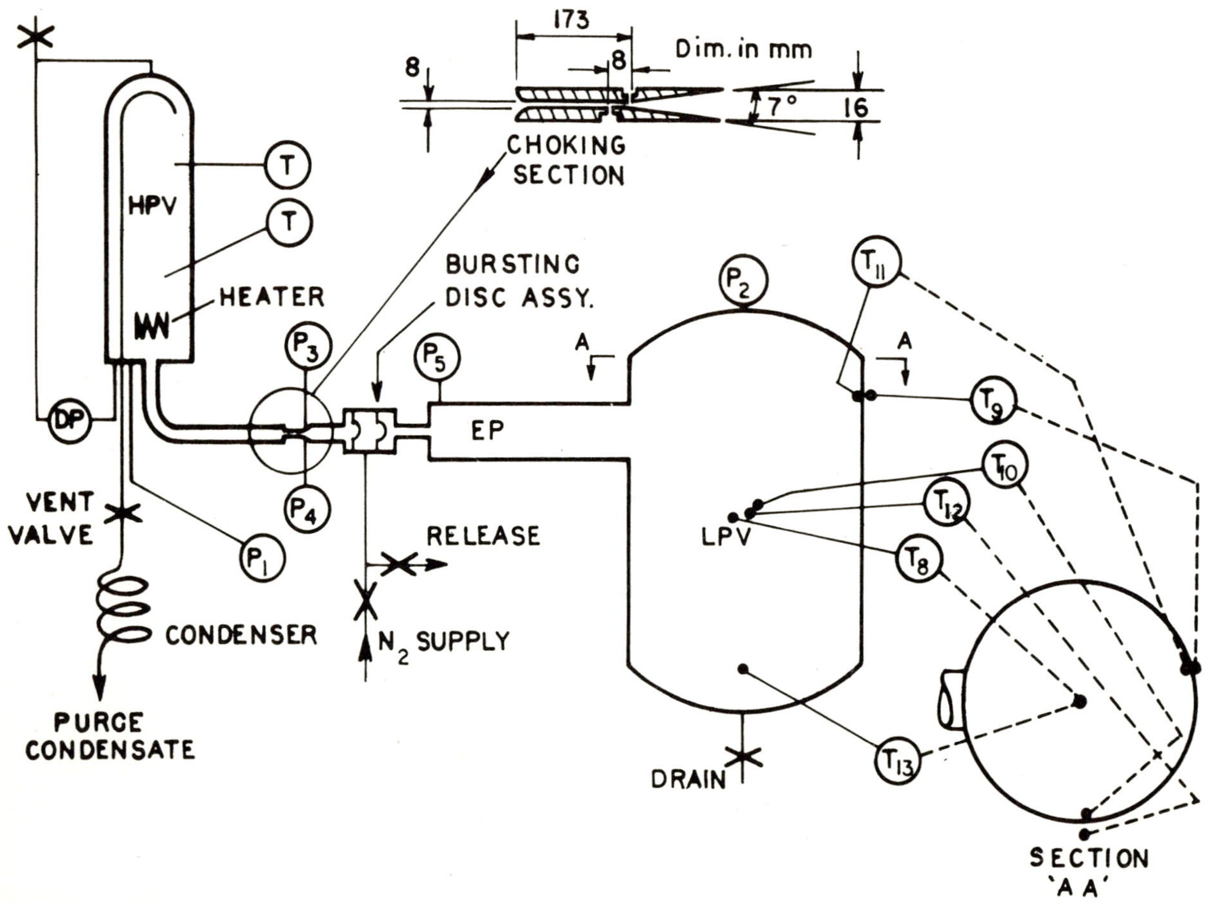


Fig. 1: Experimental rig

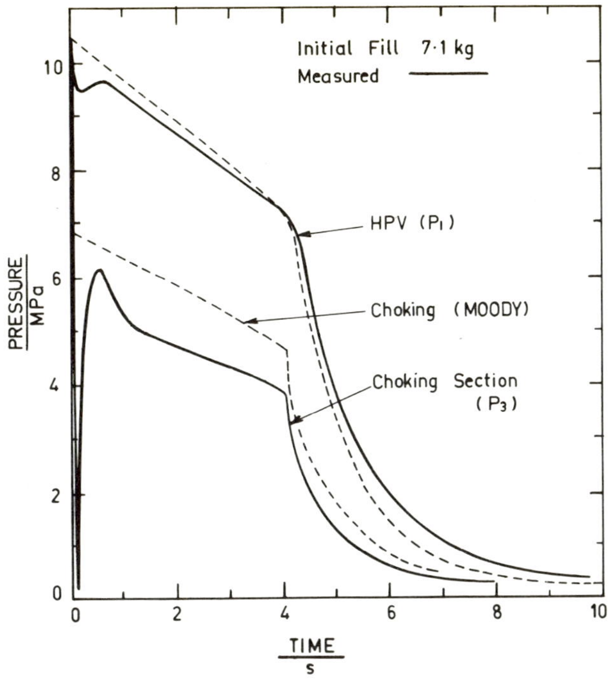


Fig. 2: Blowdown transient : 10.4 MPa

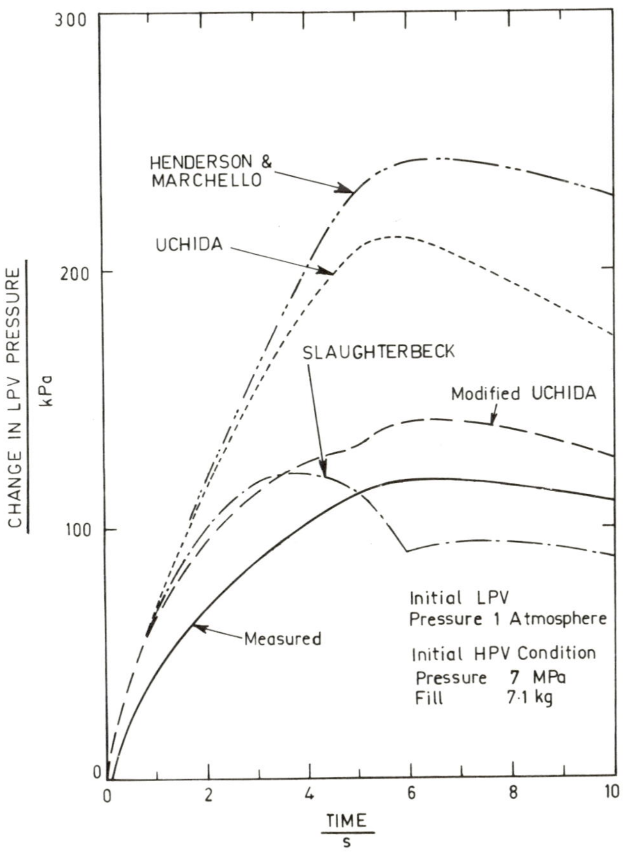


Fig. 3: LPV pressure transients : HPV 7 MPa

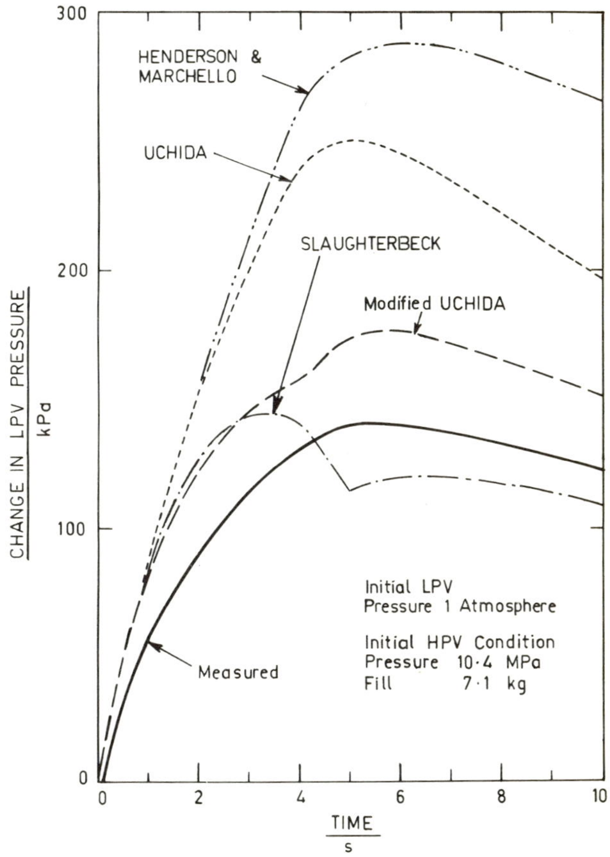


Fig. 4: LPV pressure transients : HPV 10.4 MPa

# C210/77

# SAFETY EVALUATION OF INLET-TUBE BREAK FOR A PRESSURE TUBE TYPE REACTOR

H. KATO, BSc,
Power Reactor & Nuclear Fuel Development Corporation, Tokyo, Japan
H. YAMANOUCHI, BSc, T. MANO, BSc,
Hitachi Ltd, Ibraraki-ken, Japan
and
T. KITAHARA, BSc
Power Reactor & Nuclear Fuel Development Corporation, Tokyo, Japan

SYNOPSIS Safety evaluation of inlet tube break for the pressure tube reactor, FUGEN is described. The investigation was conducted by using the computor code SENHOR and the experimental results obtained at the Ō-arai engineering center, PNC. It was found that, for a smaller rupture at the inlet tube with a small leakage (≤ 10 kg/s), sufficient core flow is expected even in the ruptured channel and the MCHFR is expected to remain above 1.0 under full power operating condition. A postulated inlet tube break accident with large leakage (> 10 kg/s) is also evaluated in comparison with the experiment.

## INTRODUCTION

1. The purpose of this study is to affirm the adequacy of ECCS design of a pressure tube type reactor, FUGEN (ref. 1), and to extend it to those of the similar large power reactors. For the above purposes, series of experiments including the blow-down, the fuel heat-up and the ECCS performance have been carried out at Ō-arai Engineering Center, PNC; namely, large pipe rupture in liquid phase (down-comer rupture) and that in steam phase (main steam pipe rupture) (ref. 2). On the other hand, analysis code, SENHOR (ref. 3), which analyzes the behavior of the loss of coolant accident for the pressure tube type reactors, has been developed to compare with the experimental results. For realizing the adequacy of ECCS of FUGEN, large pipe rupture in liquid phase (the down-comer rupture) (ref. 4) and also that in steam phase (the main steam pipe rupture) have been evaluated. For the small and intermediate rupture in liquid phase, some experiments have also been carried out; and, a capable function of analyzing such a small rupture accident has been added to the code (ref. 5).

2. In this report, for realizing the adequacy of ECCS for small and intermediate rupture in liquid phase, a postulated inlet tube rupture accident for FUGEN was evaluated, in comparison with the experiments.

## ANALYTICAL MODEL

3. For evaluating FUGEN inlet tube small rupture accident and full scale browdown tests, attention has been given to development of adaptable hydraulic and heat up calculation models. The outline of analytical models are explained in this section.

### Blowdown model

4. Hydraulic consequences during blowdown phenomena are obtained by solving simultaneous equations which describe mass, momentum and energy conservation. In the case of two-phase flow, these equations are expressed as follows:

$$\frac{\partial}{\partial t}\{\alpha\rho_g-(1-\alpha)\rho_\ell\}+\frac{\partial}{\partial z}\{\alpha\rho_g v_g+(1-\alpha)\rho_\ell v_\ell\}=0, \qquad (1)$$

$$\frac{\partial}{\partial t}\{\alpha\rho_g v_g+(1-\alpha)\rho_\ell v_\ell\}+\frac{\partial}{\partial z}\{\alpha\rho_g v_g^2+(1-\alpha)\rho_\ell v_\ell^2\}+\frac{\partial P}{\partial z}$$
$$=-F-\{\alpha\rho_g+(1-\alpha)\rho_\ell\}\sin\theta, \qquad (2)$$

$$\frac{\partial}{\partial t}\{\alpha\rho_g(h_g+\frac{v_g^2}{2})+(1-\alpha)\rho_\ell(h_\ell+\frac{v_\ell^2}{2})\}$$
$$+\frac{\partial}{\partial z}\{\alpha\rho_g v_g(h_g+\frac{v_g^2}{2})+(1-\alpha)\rho_\ell v_\ell(h_\ell+\frac{v_\ell^2}{2})-\frac{\partial P}{\partial t}$$
$$=U+\{\alpha\rho_g v_g+(1-\alpha)\rho_\ell v_\ell\}\sin\theta. \qquad (3)$$

where
$\alpha$ = void ratio, t = time, z = distance, $\rho$ = density, $v$ = velocity, $\Theta$ = incline of pipe, p = pressure, F = friction, h = specific enthalpy, U = heat addition rate
suffix: $\ell$ = liquid phase, $g$ = gas phase.

5. Calculations of coolant flow in pipes are based on finite differential scheme. Quasi-steady state assumption is used for pressure gradient and flow velocity calculation. Time derivatives of pressure and flow velocity are neglected and flow rate are determined by matching pressures at both ends of pipes with reservoir pressures. Interia effect of flow is evaluated with change of flow rate and considered by modifying friction term. Energy transport is calculated by using simplified method of characteristics. In these calculations the effects of momentum flux, inertia and gravity force are normally taken into consideration. Pressure propagation can not be described in this model, but the effect of pressure wave is negligible for our purpose.

6. For estimation of two-phase critical flow rate and slip ratio, following correlations are available.

Critical two-phase flow ---- Moody (ref. 6), Ogasawara (ref. 7), Fauske (ref. 8)

Slip ratio---$\sqrt{\rho_\ell/\rho_g}$, Smith(ref.9), Zivi(ref.10)

In analysis shown in this report, Moody and $\sqrt{\rho_\ell/\rho_g}$ correlations are used for critical flow and slip ratio, respectively.

7. For simulating rupture of part of inlet tube, flow model of pipe with branch tube was developed. Pressure at the connection is determined by keeping balance of flow-in and flow-out of coolant. Pressure gap at the connection is estimated with local pressure loss coefficients which are normally used in hydrodynamic calculation.

8. Steam drum and lower header are treated as water reservoirs. Effects of internal structures such as baffle plate and separator in steam drum are taken into consideration. It is obvious that elevation of water surface in reservoirs effects blowdown consequences so that the change of water level is calculated in detail. Bubble rise velocity in reservoirs of 30 cm/s is adopted, since the experimental pressure behavior curve of steam phase rupture has been well analyzed using this value (ref. 3).

9. When large amount of flow into downcomer from steam drum takes place, vortex carrys steam from steam phase over the water surface. Steam quality at inlet of downcomer is calculated from flow area of steam phase and steam velocity into downcomer. The flow area is expressed as a function of water surface level heights which simulates a curve of vortex. In the present analysis, the flow area ($A_s$) of the steam is calculated by the following equation for the simplicity,

$$A_s = \begin{cases} \pi \left( \frac{D}{2} - a H_\ell \right)^2 & \text{.... for } 0 \leqq H_\ell \leqq \frac{D}{2} \\ 0 & \text{.... for } H_\ell > \frac{D}{2} \end{cases} \tag{4}$$

where D = diameter of downcomer
$H_\ell$ = water level height
a = input constant ($0 \leq a \leq 1$), a=1 is used in the present analysis.

The steam velocity ($v_s$) is calculated by the following equation,

$$v_s = v_\ell \sqrt{\frac{(1-\chi)\rho_\ell + S^2 \chi \rho_\ell}{(1-\chi)\rho_g + \chi \rho_\ell}} \tag{5}$$

where, $v_\ell$ = velocity of liquid
$S = \sqrt{\rho_\ell/\rho_g}$
$\chi$ = quality $\left(= \frac{S\alpha\rho_g}{S\alpha\rho_g + (1-\alpha)\rho_\ell}\right)$
($\alpha$ = void fraction in liquid phase)
$\rho_g$ = density of vapour in liquid phase
$\rho_\ell$ = density of liquid.

This equation is derived from the consideration that the kinetic energy of steam is equal to that of liquid (two phase mixture), since they depend on the same pressure difference between water surface and downcomer inlet.

Heat up calculation

10. Temperature distribution in a fuel rod is analyzed in scheme of implicit finite differential of transient heat conduction equation. Only radial heat conduction is considered and temperature distribution is assumed to be axial symmetrical.

$$\rho c \frac{\partial T}{\partial t} = \frac{k}{r} \frac{\partial}{\partial r}\left(r \frac{\partial T}{\partial r}\right) + Q \tag{6}$$

where
$\rho$ = density, c = specific heat, k = heat conductivity, t = time, T = temperature, r = radial distance, Q = heat production rate per unit volume.

11. For estimating critical heat flux and post dry out heat transfer, following correlations are available.

Critical Heat Flux ----- Nagai (ref. 11), HTL empirical correlation (ref. 14)
Post Dryout Heat Transfer Coefficient --- Groeneveld (ref. 12), Kays (ref. 13)

We used HTL empirical correlation for critical heat flux that was obtained for the FUGEN type fuel assembly with 26 cm of spacer pitch. The HTL experiments show the distinctive dependency of critical heat flux on the spacer pitches. The critical heat flux ($\phi_{BO}$) obtained for the FUGEN type fuel assembly (28 rods cluster with 26 cm spacer pitch) is expressed as follows,

$$\phi_{BO} = \begin{cases} (2.2-3\chi)\times10^6 \text{ kcal/m}^2\text{h} & \text{... for } \phi_{BO} \leq 0.5 \\ 1.4(1-\chi)\times10^6 & \text{... for } 0.5 \leq \phi_{BO} \leq 0.85 \end{cases} \tag{7}$$

where $\chi$ = quality. This correlation was obtained under the steady state condition of 70 ata for upwards flow.

12. For predicting onset of dryout during blow-down, we adopted a simple method based on the blow-down experiment performed at the full scale experimental facility that simulates FUGEN primary cooling system (ref. 2).

The method is as follows:
- Dryout in a ruptured channel starts when steam drum becomes empty.
- Dryout in the not-ruptured channels starts when the water level reaches 1/2 core height.

This method is applicable to the onset of dryout during blow-down without stagnant flow.

EVALUATION FOR FUGEN INLET TUBE BREAK

13. Emergency Core Cooling System (ECCS) for FUGEN shown in Fig. 1, consists of Accumulated Pressure Core Injection (APCI), High Pressure Core Injection (HPCI) and Low Pressure Core Injection (LPCI), in order to keep a fuel cladding integrity for all loss of coolant accidents (LOCA). At large rupture accident, APCI injects accumulated coolant stored in accumulater to lower header. At small rupture accident, HPCI sprays the coolant stored in condensate storage tank into steam drum. HPCI operation causes the core to depressurize and promote core reflooding by LPCI. At both of small and large rupture accidents, LPCI injects the coolant stored in condensate tank into the lower header after reactor pressure was depressurized below 40 $Kg/cm^2g$. LPCI cools the core for a long period after accident. Up to now, reactor safety and ECCS efficiency for down-comer break accident

and main steam pipe break accident have been confirmed satisfactorily by comparison with experiments carried out at Ō-arai test loop.

14. For inlet tube leak accident during full power operation, the relation of the minimum critical heat flux ratio (MCHFR) and leak flow rate was calculated as shown in Table 1. The core flow rate in the hottest channel is approximately 8.9 kg/s and the MCHFR is 1.9 under the full power operating condition. When an inlet tube leak accident with leak rate of 5 kg/s takes place, the core flow rate in the ruptured channel decreases to 7.0 kg/s and the MCHFR decreases to 1.8. Moreover, when the leak rate increases to 10 kg/s, the core flow rate decreases to 4.5 kg/s and the MCHFR decreases to 1.2. The third column shows the total flow rate (leak flow rate + core flow rate). From this calculation, it is found that the MCHFR is still kept above 1.0 even if the leak rate reaches 10 kg/s. This rupture size corresponds to 25% (by area) of the inlet tube.

Table 1

Hottest channel core flow rate versus inlet-tube leak rate

| Leak Flow Rate | Hottest Channel Core Flow Rate | Channel Total Flow Rate | MCHFR |
|---|---|---|---|
| 0 Kg/s | 8.9 Kg/s | 8.9 Kg/s | 1.9 |
| 5 Kg/s | 7.0 Kg/s | 12.0 Kg/s | 1.8 |
| 10 Kg/s | 4.5 Kg/s | 14.5 Kg/s | 1.2 |

15. It is possible to detect this amount of leakage easily by either inlet humidity of primary containment vessel (PCV) air recirculation system, PCV temperature or PCV drain sampling rate. Therefore, according to detect leakage before rupture size reaches severe dimension, the reactor can be scramed safely without fuel dryout.

16. In the case of postulated inlet-tube rupture accident, blow-down and fuel heat up characteristics are calculated as shown in Fig. 2. It is assumed that loss of electric power accident occurres at the same time of LOCA initiation. F.J. Moody's correlation is used for critical discharge flow model. Discharge coefficient of 1.0 is used. On-set of fuel cladding dryout for a ruptured channel and for not-ruptured channels are postulated to be initiated when the steam drum becomes empty and when the coolant level decrease to 1/2 core height, respectively. After accident, water level decreases and HPCI sprays the coolant in the steam drum. Then, the reactor is depressurized and LPCI can operate below 40 $Kg/cm^2g$. After LPCI starts, the not-ruptured channels are reflooded. Then, coolant level rises up to the steam drum through the not-ruptured channels and the coolant flows to the ruptured channel.

17. Fuel cladding temperature at the ruptured channel starts to heat up when steam drum becomes empty. Steam cooling effect by HPCI is expected for post dryout. Assuming adiabatic after 250 sec, fuel cladding temperature increases rapidly. After 420 sec, when LPCI coolant flows to the ruptured channel, the fuel cladding temperature quenched. Peak cladding temperature is approximately 600°C. Fuel cladding temperature at the not-ruptured channels doesn't heat up because water level is kept above 1/2 core level by ECCS coolant injection.

COMPARISON OF CALCULATED AND EXPERIMENTAL RESULTS

18. In order to verify the computer code adopted to estimate "FUGEN" safety in previous section, calculated results have been compared with full scale blowdown experiment. The comparison between calculated and experimental results is described in this section.

Full scale blowdown experiment

19. Full scale blowdown experiments have been carried out at the test facility of O-arai Engineering Center.

The flow sheet of the facility is shown in Fig. 3. This facility simulates full scale 1/8 section of "FUGEN" and integrated blowdown tests are able to be carried out.

Features of the test facility are as follows (ref. 15).

| | | |
|---|---|---|
| Channel | Number | 25 (5 heated) |
| | Heated Length | 3.7 m |
| Operating Condition | Pressure | 74 $kg/cm^2$ |
| | Temperature | 300°C |
| | Flow Rate | 36.6 kg/s |
| | Head | 8 $kg/cm^2$ |
| Power | | 1.4 MW |
| ECCS | | APCI,HPCI,LPCI |

Blowdown experiments of inlet tube rupture with the facility are called LIP series and 25 experiments have been performed (ref. 2).

Range of experimental conditions are,
Initial Pressure .... 15 ∿ 73 $kg/cm^2$
Rupture Size ........ 9 ∿ 62.3 mmϕ (2.1∿100%)

20. Analytical prediction has been compared with LIP-11 experiment, the initial conditions of which are as follows;
Initial Pressure .......... 72.8 $kg/cm^2$
Rupture Size .............. 62.3 mmϕ
Initial Water Inventory ... 11.6 ton

Prediction of experiment

21. In this calculation discharge coefficient is assumed to be 0.7. The same value has been used for calculation of large size pipe rupture experiments at downcomer or main steam line. For predicting onset of dryout, we adopted a convenient empirical method obtained from experimental and analytical results (ref. 2).

- Dryout in ruptured channel starts when steam drum becomes empty.
- Dryout in not-ruptured channels starts when reactor core water level reaches 1/2 core height.

It has been proved from experiments that onset of dryout prediction in ruptured channel is 25∿50 sec. earlier than experimental data for 36∿62.3 mmϕ inlet tube rupture. For predicting onset of dryout in not-ruptured channels, this method appeards to give reasonable results. Steam cooling is expected in ruptured channel after onset of dryout so that heat transfer coefficient of steam forced convection is calculated by Kays' correlation (ref. 13). For not-ruptured channels, adiabatic condition is assumed in order to obtain conservative estimation.

Comparison between calculated and experimental results

22. Figure 4 shows the comparison between experimental and calculated results obtained on the basis of above-mentioned assumptions. Prediction of blowdown phenomena gives close agreement with experimental data. Calculated depressurization rate was somewhat different from experiment at early stage of blowdown. It seems that calculated steam carry under from steam drum into ruptured channel is less than experiment. There is no problem for safety estimation since predicted pressure agrees with experiment after 150 seconds. Prediction of onset of dryout in ruptured channel is earlier than experiment with 50 seconds. This results is consistent with former comparisons between calculations and experiments (ref. 2). For onset of dryout in not-ruptured channels the prediction gives somewhat conservative results. Dryout in ruptured channel occurres earlier but the temperature rise rate is smaller than in not-ruptured channels. This result shows that steam cooling is effective in ruptured channel. As a result of comparison between calculated and experimental data, it is confirmed that the computer prediction gives reasonable description of blowdown phenomean during inlet tube rupture accident and the prediction of fuel rod heat-up gives conservative results.

DISCUSSION

23. The result of inlet tube rupture analysis for FUGEN is very conservative because of adopting the following two points; (1) Discharge coefficient of 1.0 is used (C=1.0). (2) Dryout in ruptured channel is assumed to be equal to steam drum empty time. If we use C=0.7 in the analysis for FUGEN, the reduction rate of core pressure is smaller, and the PCT becomes lower. Experiments shows that fuel dryout of ruptured channel starts 25∿50 sec later after steam drum becomes empty (ref. 2). If 25 sec is used, the net heat-up time become shorter, and the PCT becomes lower. Taking these two effects into consideration, the PCT is approximately 100°C lower than the calculated temperature shown in Fig. 2.

24. As mentioned above, for a postulated inlet tube rupture accident for FUGEN, the evaluation is performed very conservatively. The estimated PCT is approximately 600°C for a fuel assembly in a ruptured channel; and, for other channels (not-ruptured), water level does not decrease below the top of the fuels, and there is no temperature rise of the fuels, with the effect of ECCS. For a smaller rupture with small leakage (< 10 kg/s), sufficient coolant flow is expected even in the ruptured channel and the MCHER is expected above 1.0 under the full power operating condition. The PCT evaluated for small ruptures is considerably low compared with the PCT (990°C) for the design basis accident (down-comer break) previously reported (ref. 4).

25. Summarizing the study including the ones already done, it is concluded that the effect of the ECCS is favorable both for posturated small rupture and large rupture accidents.

26. From comparison with experiments, the followings are noticeable: Adoption of the value of 0.7 on the discharge coefficient is very favorable to analyze experimental results, which are large rupture in liquid phase (down-comer break), that in steam phase (steam pipe rupture), and small rupture in liquid phase (inlet-tube rupture). Concerning to the onset-time of dryout in the not-ruptured channels, adoption of 1/2 water level is adequate in decay heat region after reactor shut-down (heat flux; 1∿4 W/cm$^2$). For the initiation of heat-up in the ruptured channel, adoption of steam drum empty time is also adequate with conservatism, which gives earlier time of 25∿50 sec than experiments.

REFERENCES


(1) SHIMA S. and SAWAI S. The FUGEN project. CNA Annual Conf. 1973, 204.

(2) TOKUMITSU M. Full scale safety experiments of FUGEN. CREST specialist meeting on emergency core cooling for light watre reactors, Munich, 1972 Oct. ; PNC report 1972, N341 72-12.

(3) KAWABE R., MINATO A., OGASAWARA H., MASUOKA R. and KATO H. SENHOR-II, a computer code for loss-of-coolant accidents of pressure tube type reactors. Nucl. Eng. and Design, 1975, 35, No.1, P.13.

(4) KATO H., YAMANOUCHI H. and MASUOKA R. Safety evaluation of LOCA for a pressure-tube-type reactor. Trans. Amer. Nucl. Soc., 1974, 19, P. 291; PNC report, 1974, N341 74-13.

(5) MINATO A., KAWABE R., KATO H. and YAMANOUCHI H. A small-scale pipe-rupture analysis for pressure-tube-type reactors. Trans. Amer. Nucl. Soc., 1976, 24, P.292; PNC report 1976, N341 76-22.

(6) MOODY F.J. Maximum flow rate of a single-component two-phase mixture. Trans. ASME, 1965, Ser C.

(7) OGASAWARA H. A theoretical prediction of two-phase critical flow. Bulletin of the JSME, 1967, 10, No.38.

(8) FAUSKE H.K. Contribution to the theory of two-phase one-component critical flow. ANL-6633, 1962.

(9) SMITH S.L. Void fraction in two-phase flow: Correlation based upon an equal velocity head model. Proc. IME, 1970, 184, Pt 1.

(10) ZIVI S.M. Estimation of steady state steam void fraction by means of the principle of maximum entropy production. Tran. ASME. 1963.

(11) NAGAI N. Burn-out heat flux of fuel cluster, Hitachi Hyoron, 1970, 52, No.5.

(12) GROENEVELD D.C. An investigation of heat transfer in the liquid deficient regime, AECL-3281 1969.

(13) KAYS W.M. et al. Heat transfer in annular passages. Int. Journal of Mass and Heat Transfer. 1963, 6 P.537.

(14) KOBORI T. Critical heat flux measurements in a full scale rod cluster, Bulletin of the JSME, 1976, 19 No.131 P.540.

(15) SAWAI S. Present status of reactor safety R&D pressure tube type reactor. PNC report 1976, SN 341 76-21.

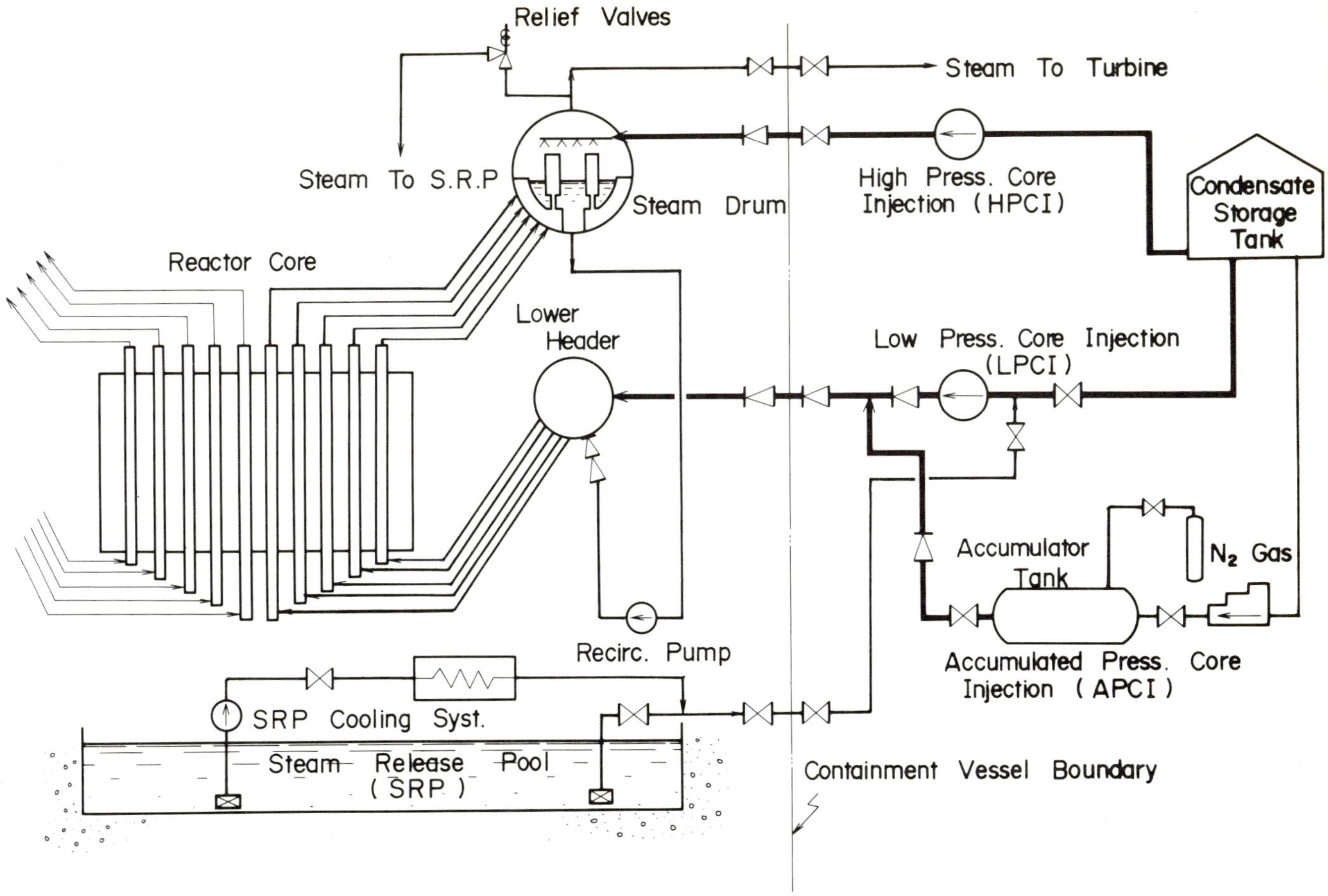


Fig. 1: Schematic flow diagram of FUGEN ECCS

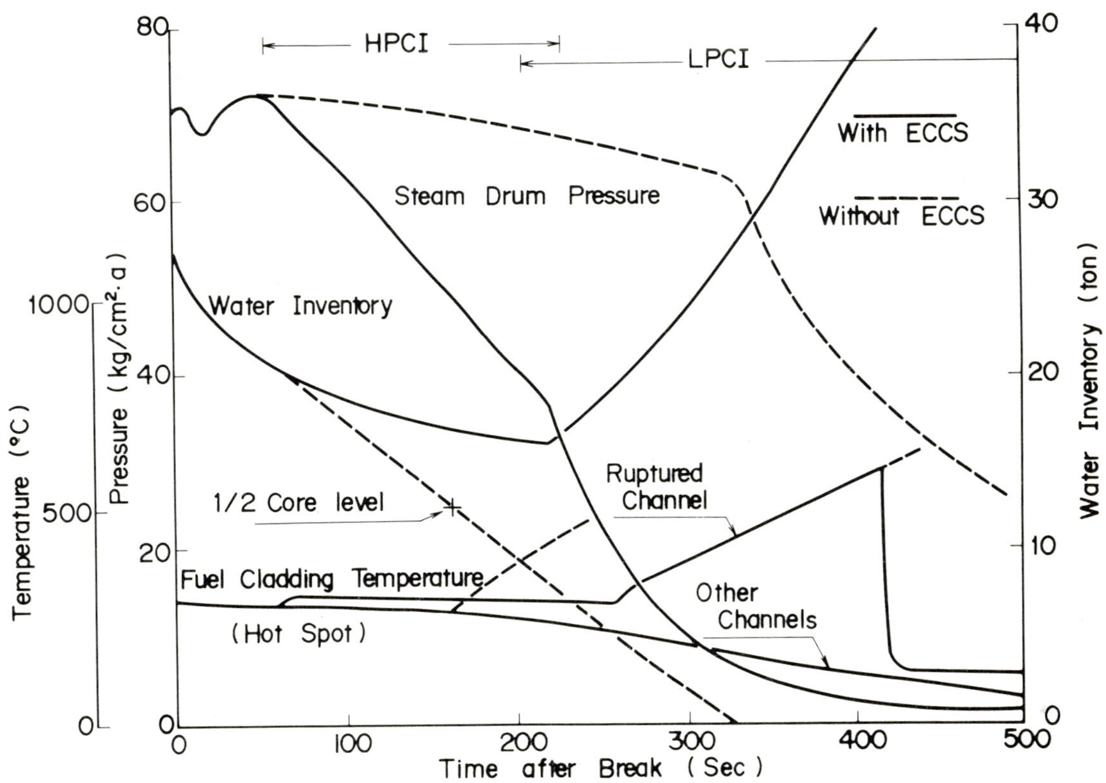


Fig. 2: Analysis of postulated inlet tube rupture accident for FUGEN

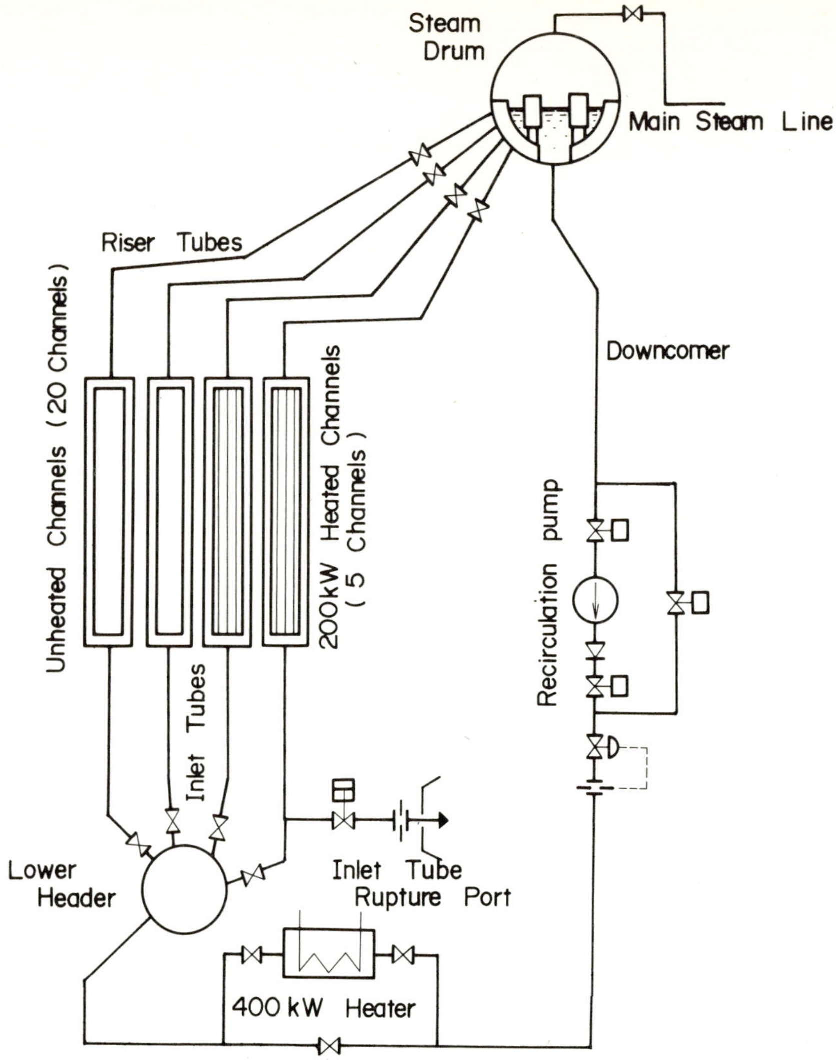


Fig. 3: Ō—arai inlet tube rupture test facility

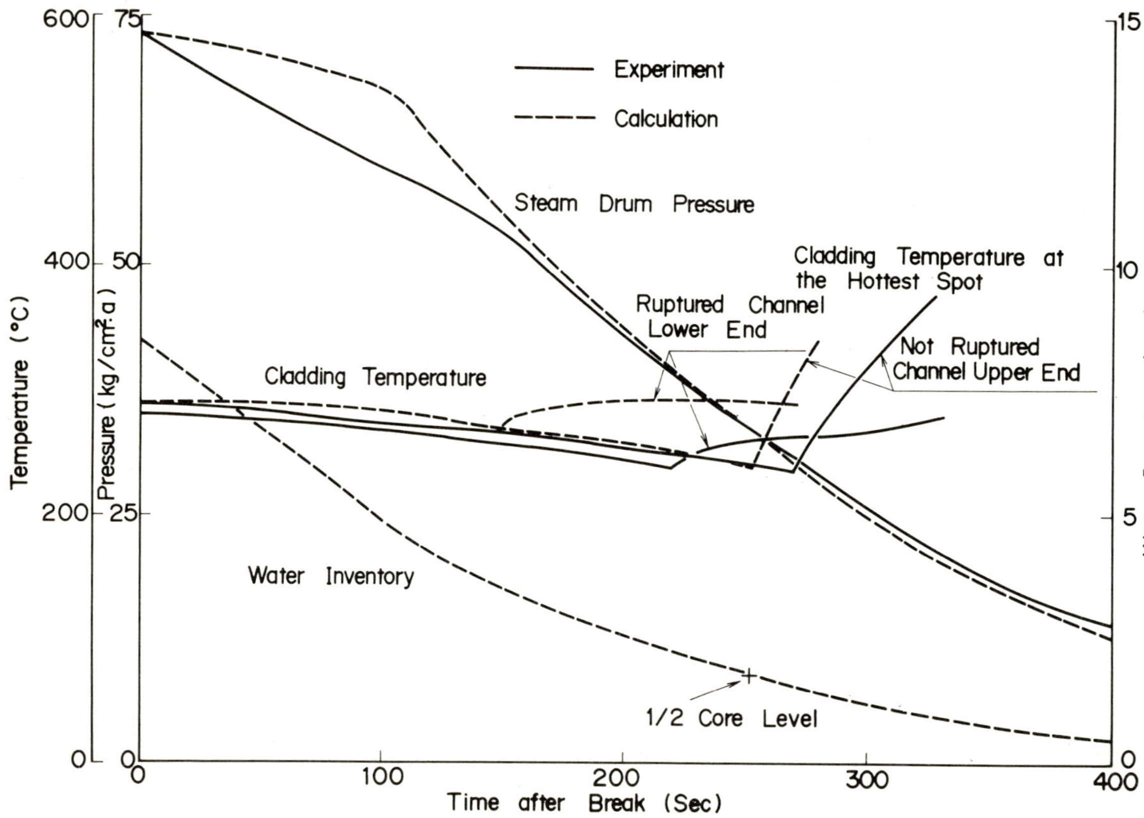


Fig. 4: Comparison with experiment (LIP—11) (62.3 mm dia. inlet tube break)

C211/77

# MODELLING ASPECTS OF TWO PHASE FLOW

F. MAYINGER, Dr.-Ing.
Institut für Verfahrenstechnik, T.U. Hannover, 6FR

SYNOPSIS In two phase flow scaling is much more limited to very narrowly defined physical phenomena than in single phase fluids. For complex and combined phenomena it can be achieved not by using dimensionless numbers alone but in addition a detailed mathematical description of the physical problem - usually in the form of a computer program - must be available. An important role plays the scaling of the thermodynamic data of the modelling fluid. From a literature survey and from own scaling experiments the conclusion can be drawn that Freon is a quite suitable modelling fluid for scaling steam - water mixtures. However, without a theoretical description of the phenomena nondimensional numbers for scaling two phase flow must be handled very carefully.

## INTRODUCTION

The transfer of experimental results, obtained under scaled conditions, to the reality in an industrial plant is an old problem of engineering. In single phase hydrodynamics and heat transfer we have been accustomed to use scaling and modelling laws for many years. Thermohydraulic occurences are equivalent, if the velocity field, the temperature field and the pressure field are similar. These fields are described by the well-known conservation laws for mass, energy and momentum from which can be derived dimensionless numbers like the Reynolds-, the Froude-, the Euler- or the Nusselt-number and which have to be identical in the test and in the original conditions.

Even for the simpler case of single phase flow it is not possible to satisfy scaling laws completely, e.g. in mixed convection either the forced flow can be scaled with the Reynolds-number or the buoyant induced one can be scaled. Such restrictions become more severe - as we will see - in two phase flows.

Dimensionless numbers for scaling can be derived from:

- differential equations completely describing the physical phenomenon,
- by arranging a matrix from the exponents of the influencing parameters and their dimensions
- and finally simply by trial and error putting together the variables to dimensionless groups.

By these methods or sometimes just by intuitive experience several dimensionless numbers were found for two phase flow and for thermohydraulic processes with phase change already several years ago. Table 1 gives some examples of them. They are usually restricted to a single effect or a simple phenomenon, e.g. the Weber-number to droplet formation and the Buoyancy number to free convection. An exception is the Martinelli-number X, which was found by Martinelli studying two phase pressure drop, but which seems to have a much more general validity.

Whereas in single phase flow only the velocity and the temperature field have to be considered, in two phase the density distribution - i.e. the void fraction $\varepsilon$ or the quality $\dot{x}$ - also has to be taken into account.

In a discussion of modelling aspects one of the first questions is, what has to be scaled in two phase flow. In single phase systems usually the size of an apparatus must be scaled, i.e. the experiments are done with small scaled down test objects. In two phase flow certainly there is the question of the proportion too, but a much more severe problem arises from the large heat input, which is needed for two phase flow and boiling experiments. A geometrical scaling down is usually not possible, because the characteristic, fluiddynamic, volumetric units - e.g. bubbles - are almost independent of the dimensions of the apparatus. Therefore the main emphasis in modelling two phase flow is to scale the original fluid - usually water - with a modelling fluid, which eases the experimental problems because of lower and more convenient parameters of pressure and temperature and needing less energy input.

## THERMODYNAMIC SCALING OF THE FLUID

Compared to single phase flow the second phase brings with it much more complication for modelling aspects, especially with respect to the thermodynamic properties of both phases, which have to be similar in the original and in the modelling fluid. Also the phase change behaviour must be comparable. The very first criterion for selecting a modelling fluid, however, is that its thermodynamic properties should be known for the modelling fluid as well as for the original fluid. Besides water the thermodynamic properties of the refrigerants (R11, R12, R113) are well researched and precisely known. In addition these fluids offer convenient experimental conditions. Therefore they are usually used as modelling fluids, especially to scale water.

The first aim for achieving similarity of the thermodynamic conditions is to ensure that the important influencing properties are comparable. This usually can be completely done only for one property and very often the thermodynamic state is selected in such a way that the density ratio $\varrho_l/\varrho_g$ between liquid and vapour is the same in the modelling and in the original fluid. This can be a very serious restriction when other properties are as important as the density.

If there was a common equation of state for both fluids, one could make use of the thermodynamic theorem of corresponding states. One of the simplest equations of state is the Van der Waals-equation, which reduced with the critical data, has the form

$$\left(\frac{p}{p_c} + \frac{3}{v/v_c}\right)\left(3\frac{v}{v_c} - 1\right) = 8\frac{T}{T_c} \tag{1}$$

and contains only universal constants not restricted to a certain vapour. Certainly the Van der Waals equation allows only a very rough and unprecise prediction of the thermodynamic properties but it gives the hint, that the critical data $p_c$, $T_c$ and $v_c$ may be useful to get a corresponding state for comparable thermal, caloric and even transport properties (ref. 1).

In a single component system the two phases can coexist only along the saturation line, which means, that one thermodynamic property fixes the whole state. So using the critical properties one can either scale with the critical pressure ratio $p/p_c$

$$\left(\frac{p}{p_c}\right)_O = \left(\frac{p}{p_c}\right)_M \tag{2}$$

or with any other reduced value of temperature, density, viscosity and so on. Comparisons have proved that the pressure ratio is a useful parameter for scaling between water and refrigerants. As shown in fig. 1, there is only a deviation of a few percent in the density ratio $\varrho_l/\varrho_g$ of liquid and vapor for water and R12 if scaled with the critical pressure ratio, which means that in this case scaling with $p/p_c$ and with $\varrho_l/\varrho_g$ gives almost the same results.

In many hydrodynamic and heat transfer problems the density ratio may play the most important role but there are other problems like bubble formation, flow pattern, flashing and so on, where other properties like surface tension, viscosity, thermal conductivity are the governing thermodynamic values. A comparison of some properties scaled with the critical pressure ratio is given in fig. 2, where data for viscosity, thermal conductivity and surface tension for water and R12 are compared.

The absolute values of these data differ considerably but the R12 data can easily be transduced to water conditions by simply multiplying with an individual factor as demonstrated in the figure with the dotted lines. This factor then has to be taken into account in the fluid dynamic scaling, e.g. the viscosity factor in the Reynolds-number

$$Re = \frac{\dot{m}d}{\eta} \; ; \; Re_O = Re_M \; ; \dot{m}_M = \dot{m}_O \frac{\eta_M}{\eta_O} = k \cdot \dot{m}_O \cdot \eta_M \tag{3}$$

where according to fig. 2 one gets $k_l = 2{,}15$ and $k_g = 0{,}8$. In two phase flow even for this simple problem one therefore has two correction factors, one for the liquid and another one for the gaseous phase and therefore one has to decide which phase is the more important one governing the problem. For example scaling mass flow rate in two phase flow one can primarly only use the viscosity either of the liquid or of the gas. For the ratio of liquid and vapor mass flow rate an additional condition exists, namely the quality $\dot{x}$ or the void fraction. That means, one has to neglect the viscosity influence of one of the phases. From this example already it can be seen that for two phase scaling one has to subdivide into two groups of parameters, namely primary and secondary ones.

In transient conditions, e.g. with flashing in a sea desalination plant or during a loss of coolant accident in water cooled nuclear reactors one must also take into account the first derivatives with respect to pressure of some of the thermodynamic properties. As demonstrated by Belda (ref. 2) and Baker (ref. 3) these derivatives also can be made coincident by simple multiplying factors. Fig. 2 contains examples of two of the derived properties which play an important role in fluid dynamic behaviour during a loss of coolant accident.

## MODELLING OF SIMPLE AND SEPARATE FLUID DYNAMIC EFFECTS

Under simple circumstances modelling can be successfully achieved using simple dimensionless numbers as for the well known applications of pressure drop and heat transfer in single phase flow. For more complicated cases the dimensionless numbers will give only approximate hydrodynamic and thermodynamic conditions for running the tests and the transfer to the original conditions has to me made by a detailed mathematical description of the physical phenomena.

A simple and separate phenomenon in two phase flow, e.g. is the formation of flow pattern, i.e. the distribution of the two phases in the cross section of a channel. Quandt (ref. 4) deduced predictions for flow pattern from the assumption that the forces - i.e. surface tension, buoyancy and pressure drop - acting on the fluid must be responsible for the phase distribution. From these considerations one can derive dimensionless numbers as e.g. the Weber-number and a dimensionless group, which is the product of the Laplace-constant and a characteristic length, e.g. the bubble diameter.

$$N_{annular\,flow} = \frac{\rho\, w^2 d}{\sigma} \qquad N_{bubble\,flow} = \frac{\sigma}{\rho\, g\, d^2} \tag{4}$$

However, the method of Quandt was not very successful in predicting the regimes of flow pattern. For vertical flow the flow regime map by Bennett (ref. 5) is often used. Zetzmann (ref. 6) demonstrated that simply by modifying the abscissa of the Bennett-diagram with the critical pressure ratio, the flow regime map can be generalized for different fluids and pressures as shown in fig. 3.

The picture is less clear in the case of the void fraction $\varepsilon$ and the slip ratio $s = w_g/w_l$ which are closely connected with the quality $\dot{x}$ by

$$s = \frac{1-\varepsilon}{\varepsilon} \frac{\dot{x}}{1-\dot{x}} \frac{\rho_l}{\rho_g} \quad (5)$$

The slip ratio and through it the void fraction are influenced by the buoyancy forces and by the pressure drop in the channel, so empirical correlations for predicting void fraction usually contain the Froude-number or the Martinelli-parameter X. But correlations with these dimensionless numbers do not claim general validity for a variety of fluids. A promising step for generalizing the prediction of void fraction was done by Nabizadeh (ref. 7). Starting from the equation of Zuber and Findlay (ref. 8)

$$\varepsilon = \frac{\dot{x}}{\rho_g}\left[C_0\left(\frac{\dot{x}}{\rho_g} + \frac{1-\dot{x}}{\rho_l}\right) + \frac{1{,}18}{\dot{m}}\left(\frac{\sigma \cdot g(\rho_l - \rho_g)}{\rho_l^2}\right)^{0{,}25}\right]^{-1} \quad (6)$$

and supported by a large number of reliable measurements he developed an empirical correlation for the $C_0$-factor in Zuber-Findlay's equation

$$C_0 = \left(1 + \frac{1-\dot{x}}{\dot{x}} \cdot \frac{\rho_g}{\rho_l}\right)^{-1}\left[1 + \frac{1}{n}\, Fr^{-0{,}1} \cdot \left(\frac{\rho_g}{\rho_l}\right)^n \cdot \left(\frac{1-\dot{x}}{\dot{x}}\right)^{\frac{11n}{9}}\right] \quad (7)$$

$$n = \sqrt{0{,}6\,\frac{\rho_l - \rho_g}{\rho_l}} \quad (8)$$

Applying his correction he gets quite good agreement between correlated and measured data for $H_2O$, $D_2O$, R12 and R113 as demonstrated in fig. 4. This is already an example, in which dimensionless numbers alone are not sufficient for modelling two phase flow.

A very early first step for modelling pressure drop was done by Martinelli (ref. 9). He gave a simple correlation between the two phase multiplier $\Phi^2$ and the parameter X (see Table 1).

$$\left(\frac{\Delta p}{\Delta l}\right)_{2ph} = \Phi_l^2 \left(\frac{\Delta p}{\Delta l}\right)_{l\,only} \qquad \Phi_l^2 = f_1(X)$$
$$\left(\frac{\Delta p}{\Delta l}\right)_{2ph} = \Phi_g^2 \left(\frac{\Delta p}{\Delta l}\right)_{g\,only} \qquad \Phi_g^2 = f_2(X) \quad (9)$$

In a detailed experimental and theoretical analysis Friedel (ref. 10) starting from Baroczy's (ref. 11) deliberations elaborated scaling rules for two phase pressure drop. He demonstrated that the two phase multiplier (equation 9) is identical in water and in R12, if the following dimensionless groups are equal under test and under original conditions.

The group with the ratio of the Froude- and the Reynolds-number contains a scaling factor k which is a slight function of the pressure and varies between 1,68 and 1,75. With good approximation a mean value of 1,7 can be taken for k in a pressure range of $50 < p_{H_2O} < 90$ bar and $6 < p_{R12} < 10{,}8$ bar.

$$\left.\frac{(\eta_l/\eta_g)^{0{,}2}}{(\rho_l/\rho_g)^{0{,}9}}\right|_O = \left.\frac{(\eta_l/\eta_g)^{0{,}2}}{(\rho_l/\rho_g)^{0{,}9}}\right|_M \quad (10)$$

$$\left.\frac{Fr_l}{Re_l^{0{,}25}}\right|_O = k\left.\frac{Fr_l}{Re_l^{0{,}25}}\right|_M \qquad \dot{x}|_O = \dot{x}|_M \quad (11)$$

Friedel adapts the thermodynamic state by choosing the same viscosity/density-ratio in the model and in the original. Then the mass flow rate in the test is fixed by the scaling factor k via the Reynolds-number of the liquid. Proceeding so, measured data in water and in R12 fall along the same line as demonstrated in fig. 5, where the two phase multiplier is plotted versus the quality $\dot{x}$.

MODELLING COMBINED EFFECTS

The fluid dynamic behaviour in an engineering apparatus like a boiler, a nuclear reactor or distillation and extraction columns is affected by a number of phenomena very strongly combined together and usually fluid flow and heat transfer are linked together. There is no general rule how to treat these problems and therefore a few exemples - namely burnout behaviour and the blowdown conditions during a loss of coolant accident in a nuclear reactor - shall be discussed.

There are several proposals in the literature for scaling critical heat flux from measurements with refrigerants to water. All start from the simple thermodynamic condition that the density ratio of liquid and vapor has to be the same in the modelling and in the original fluid. Stevens (ref. 12, 13) prefered a more empirical procedure and made a direct comparison of the burnout data in R12 and water with simple scaling factors. He postulated that in addition to the same density ratio and to the identical geometry the quality, or the subcooling at the inlet and the outlet of the heated channel have to be the same in the model and in the original. Therefore only the mass flow rate remains as a free variable to be adjusted in the modelled conditions.

Barnett (ref. 14) used dimensionless analysis for deriving burnout scaling laws. He listed separately all those parameters, thought to mainly govern the physical behaviour of the phenomenon and those, which were thought to have only minor influence to provide primary and secondary groups.

$$\frac{\dot{q}}{h_{lg}^{1{,}5}\rho_g} = f\left(\frac{l}{d}, \frac{d \cdot \rho_g \cdot h_{lg}^{0{,}5} c_l}{\lambda_l}, \frac{\dot{m}}{\rho_g h_{lg}^{0{,}5}}, \frac{\Delta h}{h_{lg}}, \frac{\rho_l}{\rho_g}, \frac{\nu_l \cdot \rho_g \cdot c_l}{\lambda_l}, \frac{\sigma \cdot c_l}{h_{lg}^{0{,}5} \cdot \lambda_l}, h_{lg} \cdot \rho_g \cdot \frac{d(\rho_l/\rho_g)}{dp_s}\right) \quad (12)$$

From these groups Barnett developed several scaling factors F depending which parameters one chooses as primary ones. Examples of scaling factors, elaborated and derived by Barnett, are shown in table 2.

Bouré (ref. 15) clearly stated that a comprehensive scaling of the critical heat flux is not possible and that only a restricted similarity of the boiling crisis phenomenon can exist. He distinguishes - similar to Barnett - two groups of parameters, primary parameters and secondary ones. The latter are accomodated by correction factors determined experimentally and applied to the primary parameters. Doing so, he obtained the following four dimensionless groups

$$N_1 = K_1 \frac{\rho_l}{\rho_g} \quad ; \quad N_2 = K_2 \frac{\Delta h_{inl}}{h_{lg}}$$
$$N_3 = K_3 \frac{\dot{q} \cdot l}{h_{lg} \cdot \dot{m} \cdot d} \qquad N_4 = K_4 \frac{\dot{m}}{\rho_l \sqrt{g l}} \tag{13}$$

Ahmad (ref. 16) selected three dimensionless groups and arranged them in the modelling parameter $\psi$:

$$\psi = \underbrace{\left(\frac{\dot{m}^2 \cdot d}{\rho_l \cdot \sigma}\right)^{0,67}}_{We} \cdot \underbrace{\left(\frac{\dot{m} \cdot d}{\eta_l}\right)^{-0,133}}_{Re_l^*} \cdot \underbrace{\left(\frac{\dot{m} \cdot d}{\eta_g}\right)^{-0,2}}_{Re_g^*} \tag{14}$$

He expressed the critical heat flux $\dot{q}$ in the form of the modified boiling number as a function of the modelling parameter

$$N_{BO}^* = \frac{\dot{q}}{\dot{m} \cdot h_{lg}} = f\left(\psi, \frac{\Delta h_{sub}}{h_{lg}}, \frac{\rho_l}{\rho_g}, \frac{l}{d}\right) \tag{15}$$

using identical geometrical, density and quality conditions in the test and in the original.

Hein and Kastner (ref. 17) compared measurements in water and in R12 and scaled them according to Bouré as shown in fig. 6. They found that there is an influence of the mass flow rate on the factor $K_3$ (see eq. 13), which can be taken into account by a simple mass flow dependent correcting function as demonstrated in fig. 6.

The influence of the mass flow rate becomes greater with non uniform heat flux distribution in the heated channel, mainly, for rod clusters with different power input to the individual rods. This is due to a mixing interaction between the subchannels which are formed by the different rod arrays. This exchange of mass and energy is not taken into account in the scaling and is another example of the limitation to achieve complete similarity in two phase flow.

The thermohydraulic behaviour becomes even more complicated if transient conditions are considered, e.g. a blowdown during a loss of coolant accident in a reactor. However, blowdown tests done with R12 (ref. 18) and compared with similar measurements performed with water (ref. 19) gave the promising result that scaling is possible if the thermohydraulic conditions are carefully adapted. The blowdown occurence is too complex to overcome the scaling problems simply by using dimensionless numbers. One has to develop a physical conception of the thermohydraulic events and formulate it mathematically in the form of differential equations. Emphasis has to be given to imitate the very severe flashing conditions and the resulting acceleration of the fluid, which can be guaranteed by adapting the same local and temporal pressure ratio. In addition the critical (sonic) mass flow rate and different heat storage conditions have to be scaled. All these problems only can be overcome by a detailed theoretical analysis used as a basis for scaling. This analysis has to predict the results of R12-tests and the water tests as well.

There are two thermohydraulic aspects of interest during blowdown, one is the dryout-delay-time, i.e. that time until dryout occurs, and the other is the heat transfer in the post dryout condition. Starting from the constitution equations for mass, energy and momentum, using equations for the phase velocities derived from pressure drop deliberations and taking into account Hewitt's investigations (ref. 20) into the entrainment behaviour, Belda (ref. 2) developed an analysis for modelling dryout-delay-time which results in a differential integral equation, describing the film thickness as a function of time and place.

$$\frac{1}{\rho_l \cdot l}\left(\dot{M}_l(l,t) - \dot{M}_l(0,t) + \dot{M}_E(l,t)\right) + \frac{\partial A_l}{\partial t} + A_l \cdot \left(\frac{1}{\rho_g} \cdot \frac{\partial \rho_l}{\partial t} + \frac{1}{h_{lg}}\left(\frac{\rho_g}{\rho_l} \cdot \frac{\partial h_g}{\partial t} + \frac{\partial h_l}{\partial t}\right)\right) + A_{ges}\left(\frac{\int \dot{q}(z) \cdot dz \cdot U_b}{\rho_l \cdot h_{lg} \cdot l \cdot A_{ges}} - \frac{\rho_g}{\rho_l} \cdot \frac{1}{h_{lg}} \cdot \frac{\partial h_g}{\partial t}\right) = 0 \tag{16}$$

In equation 16 the local and temporal values of mass flow rate and quality in the rod cluster have to be known from blowdown calculations, which can be performed e.g. with the well-known computer codes like RELAP-4 or BRUCH. For modelling use one can derive from equation 16, the conservation laws and the phase velocities, dimensionless numbers as listed in table 3.

By scaling the hydrodynamic and thermodynamic conditions in such a way quite good agreement between water and R12 behaviour was found as can be seen from fig. 7, where the temporal course of system pressure, mass flow rate, void fraction and pressure drop in the heated section - fuel element or test channel - are plotted for water and for R12 versus the blowdown time.

This conformity, however, does not mean, that the dryout-delay-time is the same in R12 and in water, because only the hydrodynamic aspect of the process is analogous and all the thermodynamic details could not be scaled completely. However, dryout-delay-time can be obtained by calculation from equation 16, which as figure 8 demonstrates, gives good agreement between measured and predicted values in R12. The comparison was done for different locations of the break - in the hot and in the cold leg of the primary system - for a variety of break areas and for different heat flux densities at the moment before the accident occurs.

Also comparisons with water experimental data at

hand show good agreement with the theory as can be seen from fig. 9. This encourages to draw the conclusion that modelling on the detour of a theoretical analysis is a promising procedure for combined and complicated two phase flow phenomena.

To support this a comparison between heat transfer values measured in water and FREON during blowdown are shown in fig. 10, where the scaling was done via the well-known Nusselt-number and a Reynolds-number modified with the quality $\dot{x}$

$$Nu = 0{,}023\ Re^{0{,}8}\ Pr_g^{0{,}4}$$

$$Re = \frac{\dot{m} d}{\eta_g} \left[ \dot{x} + \frac{\rho_g}{\rho_l} (1 - \dot{x}) \right] \qquad (17)$$

The basic assumption idea in scaling this post-dryout heat transfer phenomenon is that only the vapour is cooling the heated walls directly. The liquid in form of droplets is lowering the temperature in the boundary layer and so having only an indirect effect on the heat transfer. In fig. 10 measurements done with water (ref. 19) and R12 (ref. 21) are compared with correlations from the literature. Keeping in mind that there are many uncertainties arising from the measuring technique the agreement is satisfactory.

## CONCLUSIONS

There are certainly a lot of other complicated two phase flow problems, which are of great technical interest. One has to mention for example the forced and free convection behaviour in a boiler, the flow and flashing in sea water desalination chambers as well as the mixing and demixing effects in fluid machinery components. Using dimensionless numbers for scaling one very carefully has to separate the primary ones determining the system behaviour and the secondary ones which are of only minor influence.

In not too complicated cases scaling can be achieved by empirical correlations, determined from the most important physical laws, governing the process. For many technical applications however, one is forced to very detailed representation of the thermohydraulic phenomena using computer codes.

This makes it necessary to get a better understanding of the thermohydraulic behaviour in two phase flow systems. A better theoretical approach would also help to generalize experimental data and make fuller use of the numerous experimental results in two phase flow, which up to now have not been correlated. So scaling is not only an expedient to save the expense of experimental research work it also helps in the comparison and the generalization of measured data. For the elaboration of modelling laws for two phase flow more effort needs to be put into the theoretical analysis of the processes.

## NOMENCLATURE

| | |
|---|---|
| $A$ | area, cross section |
| $c$ | specific heat |
| $C_o$ | distribution parameter |
| $d$ | diameter |
| $F$ | scaling factor |
| $F_{max}$ | maximum deviation |
| $g$ | acceleration due to gravity |
| $h$ | enthalpy |
| $h_{lg}$ | heat of vaporization |
| $k$ | scaling factor |
| $l$ | length |
| $\dot{M}$ | mass flow rate |
| $\dot{m}$ | mass velocity |
| $n$ | exponent |
| $P$ | primeter |
| $p$ | pressure |
| $\Delta p$ | pressure drop |
| $\dot{q}$ | heat flux |
| $S$ | Slip ratio |
| $t$ | time |
| $T$ | absolute temperature |
| $v$ | specific volume = $1/\varrho$ |
| $w$ | velocity |
| $x, y, z$ | length coordinates |
| $X$ | Martinelli parameter |
| $Z$ | dimensionless parameter, see fig.3 |

## GREEK SYMBOLS

| | |
|---|---|
| $\alpha$ | heat transfer coefficient |
| $\beta$ | see table 2 |
| $\gamma$ | see table 2 |
| $\varepsilon$ | void fraction |
| $\zeta$ | friction coefficient |
| $\eta$ | dynamic viscosity |
| $\vartheta$ | temperature |
| $\lambda$ | heat conductivity |
| $\nu$ | viscosity = $\eta/\varrho$ |
| $\varrho$ | density |
| $\sigma$ | surface tension |
| $\Phi$ | two phase multiplier |
| $\Psi$ | modelling parameter |

## DIMENSIONLESS GROUPS

| | |
|---|---|
| Fr | Froude-number |
| Gr | Grashof-number |
| N | bubble-number |
| $N_{BO}$ | boiling-number |
| Nu | Nusselt-number |
| Pr | Prandtl-number |
| Re | Reynolds-number |

## SUBSCRIPTS

| | |
|---|---|
| B | break |
| b | bubble |
| c | critical |
| CHF | critical heat flux |
| d | detachment |
| ex | exit |
| g | gas |
| h | heated |
| inl | inlet |
| l | liquid |
| lg | evaporation |
| max | maximum |
| M | model |
| O | original |
| o | at steady state condition |
| s | saturation |
| sub | subcooling |
| tot | total |

REFERENCES


1. U. Grigull u.a. Die Eigenschaften von Wasser und Wasserdampf nach der "1968 IFC-Formulation", Wärme- und Stoffübertragung 1, 1968, S. 202/213
2. W. Belda Dryoutverzug bei Kühlmittelverlust in Kernreaktoren, Dissertation, T.U. Hannover (1975)
3. J.L. Baker Flow regime transitions in vertical two phase flow, ANL 7093 (1965)
4. E. Quandt Analysis of Gas-Liquid Flow Patterns, A.J.C.E., Preprint 47, 6th National Heat Transfer Conference, Boston 1963
5. A. Bannett et al. Flow Visualiation Studies of Boiling at High Pressure, AERE-R4874 (1965)
6. K. Zetzmann Flow Pattern of Two phase Flow Inside Cooled Tubes, Nato Adanced Study Institute on Two Phase Flows and Heat Transfer, Istanbul 1976
7. H. Nabizadeh Übertragungsgesetze für den Dampfvolumenanteil zwischen Freon und Wasser, Kolloquium über Ähnlichkeitsgesetze in Zweiphasenströmungen, Hannover, 25.2.1976 and Dissertation Nabizadeh 1977
8. N. Zuber, J. Findlay Journal of Heat Transfer, Nov. 1965
9. R. Martinelli, R. Lockhart Proposed Correlation of Data for Isothermal, Two Phase Two Component Flow in Pipes, Chem. Eng. Progr. 45
10. L. Friedel Modellgesetz für den Reibungsdruckverlust in der Zweiphasenströmung, VDI Forschungsheft 572, 1975
11. C.J. Baroczy A systematic correlation for two phase pressure drop, Chem. Engng. Prog. Symp. Ser. 62 (1966)
12. G. Stevens, R. Staniforth Experimental Studies of Burnout Using Freon 12 at Low Pressures with Reference to Burnout in Water at High Pressure, Proc. of the Inst. of Mech. Eng. Vol. 180, Part 3 C, Paper 10, 1965/66
13. G. Stevens, R. Macbeth The use of Freon 12 to model forced convection burnout in water, ASME paper 70 HT 20, 1970
14. P. Barnett The scaling of forced convection boiling heat transfer, UKAEA-AEEW-R R134 (1963)
15. J. Bouré A method to develop similarity laws for two phase flows, ASME-paper 70-HT-25, 1970
16. S. Ahmad Fluid to fluid modelling of critical heat flux: a compensated distortion model, AECL-3663, 1971
17. D. Hein, W. Kastner Versuche zur kritischen Heizflächenbelastung unter stationären und instationären Bedingungen in Freon 12 und Wasser M.A.N.-Bericht 45.03.02, 1972
18. W. Belda, F. Mayinger, K.P. Viert Blowdownuntersuchungen mit dem Kältemittel R12, Proceedings der Reaktortagung des DAF, Berlin 1974, S. 7
19. KWU/R5-2885 Abschlußbericht über die Abblaseversuche mit 4-Stabbündeln, Januar 1974
20. G.F. Hewitt Annular Two Phase Flow, Oxford (1970)
21. K.P. Viert, F. Mayinger Vergleich verschiedener Wärmeübergangsbeziehungen im Filmsiedebereich bei Blowdownuntersuchungen, Paper 117, Reaktortagung 1975, Nürnberg


Weber number:

$$We = \frac{\rho \cdot d_d \cdot w^2}{\sigma}$$

Laplace constant:

$$La = \frac{\sigma}{g(\rho_l - \rho_g)} \quad [m^2]$$

Jakob number:

$$Ja = \frac{c \cdot \rho_l (\vartheta_w - \vartheta_{bulk})}{h_{lg} \cdot \rho_g}$$

Bubble Reynolds number:

$$Re_b = \frac{\rho_g \cdot w_b \cdot d_b}{\eta_g}$$

Boiling number:

$$N_{Bo} = \frac{\dot{q}}{h_{lg} \cdot \rho_g \cdot w_g}$$

Buoyancy number:

$$N_{Bu} = \frac{\rho_l - \rho_g}{\rho_l}$$

Subcooling number:

$$N_{sub} = \frac{\Delta h_{sub} \cdot (\rho_f - \rho_g)}{h_{lg} \cdot \rho_g}$$

Phase change number:

$$N_{ph} = \frac{\dot{m}_g \cdot (\rho_l - \rho_g) \cdot l}{\rho_g \cdot \rho_l \cdot w_l}$$

Slip ratio

$$S = \frac{w_g}{w_l}$$

Drift number

$$N_D = \frac{w_{g,surface}}{w_{f,surface}}$$

Martinelli parameter:

$$X = \left(\frac{\rho_g}{\rho_l}\right)^{0,5} \cdot \left(\frac{\eta_l}{\eta_g}\right)^{0,1} \cdot \left(\frac{1-\dot{x}}{\dot{x}}\right)^{0,9}$$

Table 1: Examples for dimensionless numbers in two-phase systems

$$\dot{q} = f\left(l, d, \dot{m}, p, \Delta h, h_{lg}, \rho_l, \rho_g, c_l, \lambda_l, \nu_l, \sigma, \frac{d(\rho_l/\rho_g)}{dp_s}\right)$$

$$\gamma = \frac{d(\rho_l/\rho_g)}{dp_s} \qquad \beta = \frac{d\vartheta_s}{dp_s}$$

| No. | properties | equation $\dot{q} = f(l, d, \dot{m}, p, \Delta h)$ | $F_q$ | $F_{d,l}$ | $F_{\dot{m}}$ | $F_{\Delta h}$ |
|---|---|---|---|---|---|---|
| 4 | $h_{lg}, \rho_l, \rho_g, c_l, \lambda_l, \gamma$ | $\frac{\dot{q}\cdot\gamma^{0.5}}{\lambda\cdot\sqrt{\rho_l}} = f\left(\frac{l}{d}, \frac{d\cdot c_l\cdot\sqrt{\rho_l}}{\lambda_l\cdot\gamma^{0.5}}, \frac{\dot{m}\cdot\gamma^{0.5}}{\sqrt{\rho_l}}, \frac{\rho_l}{\rho_g}, \frac{\Delta h}{h_{lg}}\right)$ | 22,43 | 0,717 | 1,914 | 11,72 |
| 5 | $h_{lg}, \rho_l, \rho_g, \beta, \lambda_l, \gamma$ | $\frac{\dot{q}\cdot\gamma^{0.5}}{h_{lg}\cdot\sqrt{\rho_l}} = f\left(\frac{l}{d}, \frac{d\cdot h_{lg}\cdot\gamma^{0.5}\cdot\sqrt{\rho_l}}{\beta\cdot\lambda_l}, \frac{\dot{m}\cdot\gamma^{0.5}}{\sqrt{\rho_l}}, \frac{\rho_l}{\rho_g}, \frac{\Delta h}{h_{lg}}\right)$ | 22,43 | 0,482 | 1,914 | 11,72 |
| 10 | $h_{lg}, \rho_l, \rho_g, \eta_g, \gamma$ | $\frac{\dot{q}\cdot\gamma^{0.5}}{h_{lg}\cdot\sqrt{\rho_l}} = f\left(\frac{l}{d}, \frac{d\cdot\sqrt{\rho_l}}{\eta_g\cdot\gamma^{0.5}}, \frac{\dot{m}\cdot\gamma^{0.5}}{\sqrt{\rho_l}}, \frac{\rho_l}{\rho_g}, \frac{\Delta h}{h_{lg}}\right)$ | 22,43 | 0,814 | 1,914 | 11,72 |
| 12 | $h_{lg}, \rho_l, \rho_g, c_g, \lambda_g, \beta$ | $\frac{\dot{q}\cdot\sqrt{\beta\cdot c_g}}{h_{lg}^{1.5}\cdot\rho_l} = f\left(\frac{l}{d}, \frac{d\cdot\sqrt{c_g\cdot h_{lg}\cdot\rho_l}}{\lambda_g\cdot\sqrt{\beta}}, \frac{\dot{m}\cdot\sqrt{\beta\cdot c_g}}{\sqrt{h_{lg}\cdot\rho_l}}, \frac{\rho_l}{\rho_g}, \frac{\Delta h}{h_{lg}}\right)$ | 23,06 | 0,317 | 1,968 | 11,72 |

Table 2: Scaling factors for critical heat flux according to Barnett (14)

| | |
|---|---|
| **heat flux** | $K_1 = \frac{\dot{q}(z)\cdot l\cdot P_h}{\dot{m}_{tot}\ A_{tot}\cdot h_{lg}}$ |
| **pressure** | $K_{21} = \frac{p_s}{\rho_l}\cdot\frac{\partial\rho_l}{\partial p}$ $\quad K_{23} = \frac{p_s}{h_{lg}}\cdot\frac{\rho_g}{\rho_l}\cdot\frac{\partial h_l}{\partial p}$ <br> $K_{22} = \frac{p_s}{h_{lg}}\cdot\frac{\partial h_l}{\partial p}$ $\quad K_{24} = \frac{p_s}{h_{lg}}\cdot\frac{\partial h_{lg}}{\partial p}$ |
| **acceleration** | $K_3 = \frac{\Delta p\cdot\rho_l}{\dot{m}^2_{tot}} = Eu_l$ <br> $K_4 = \frac{\Delta p\cdot\rho_g}{\dot{m}^2_{tot}}\cdot\frac{\rho_g}{\rho_l} = Eu_g\cdot\frac{\rho_g}{\rho_l}$ |
| **mass flow rate** | $K_5 = \xi_l \triangleq Re = \frac{\dot{m}_{tot}\cdot d}{\eta_l}$ |
| **geometry** | $K_6 = \frac{\sqrt{A_{tot}}}{l} \triangleq \frac{d}{l}$ |

Table 3: Dimensionless groups for dryout delay scaling during blowdown (Belda)

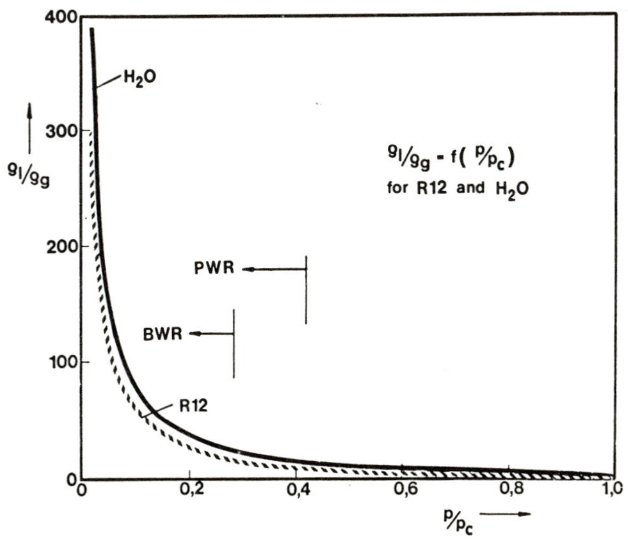


Fig. 1: Density ratio $\rho_l/\rho_g$ of water and R12 as function of reduced pressure

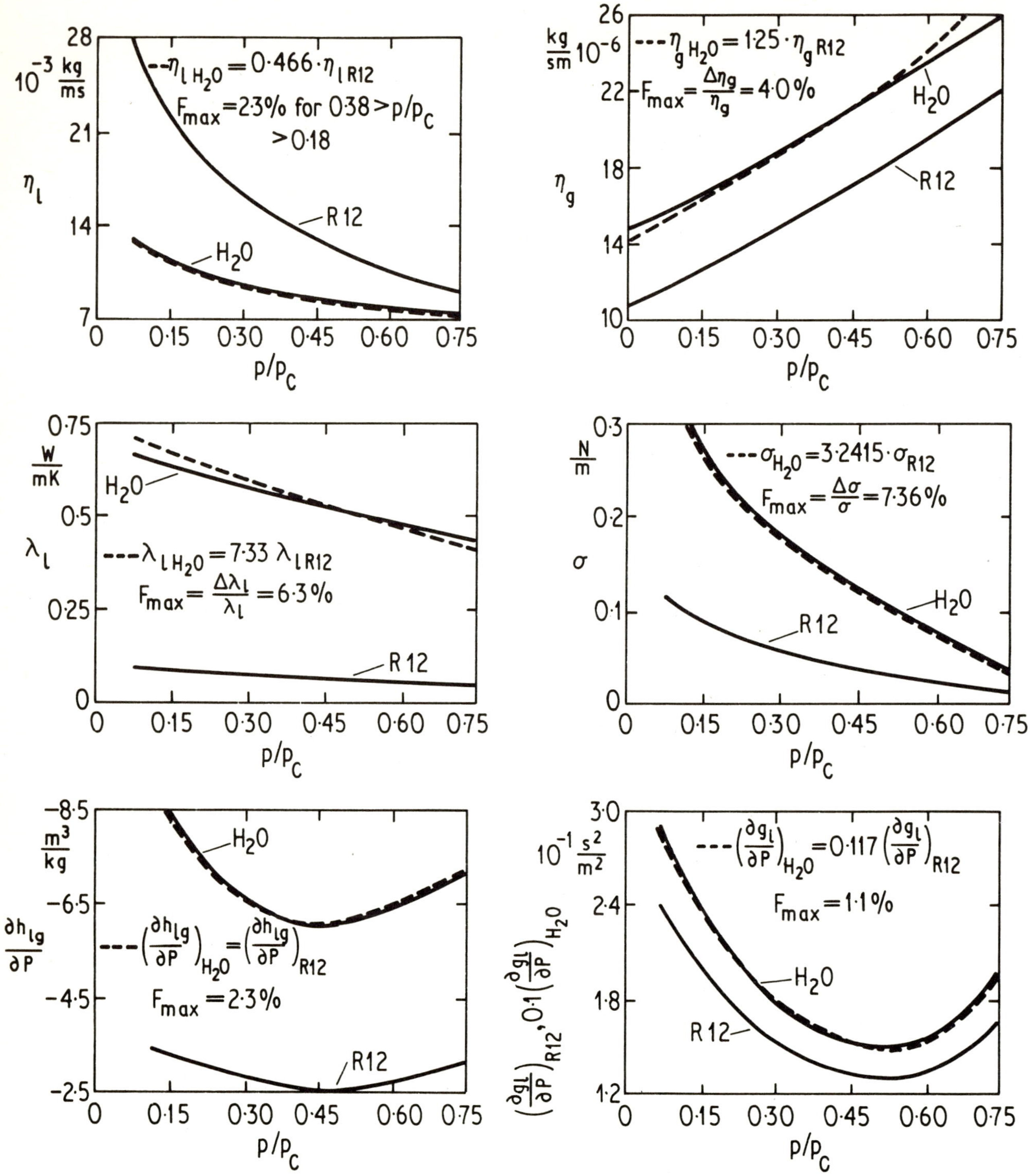


**Fig. 2: Comparison of thermodynamic, transport and derived properties (water and R12)**

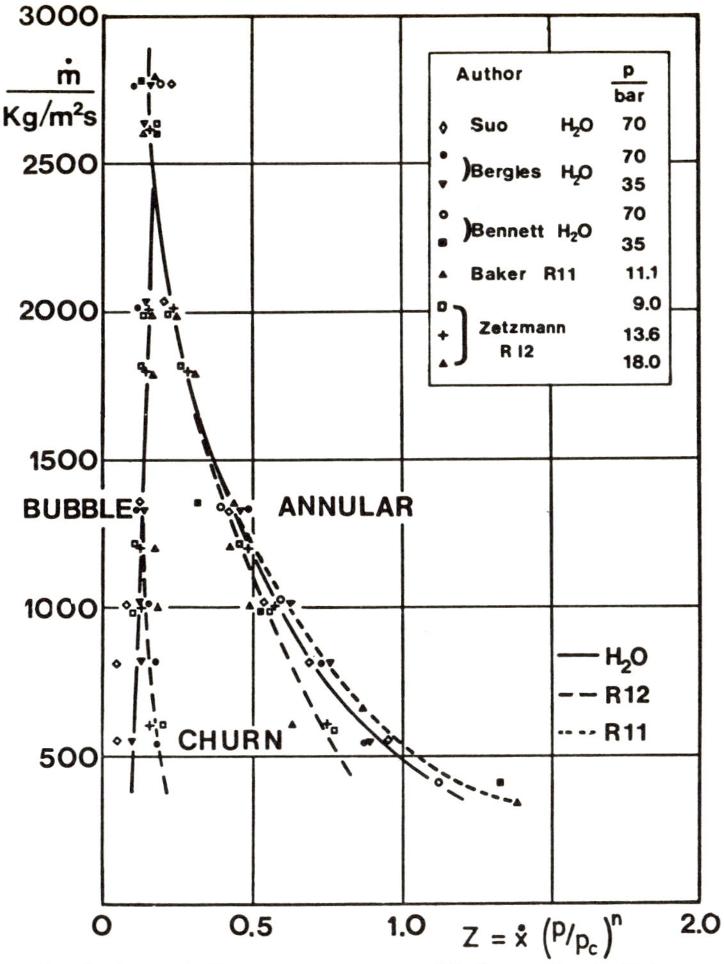


Fig. 3: Bennett's flow pattern map, modified according to Zetzmann (6)

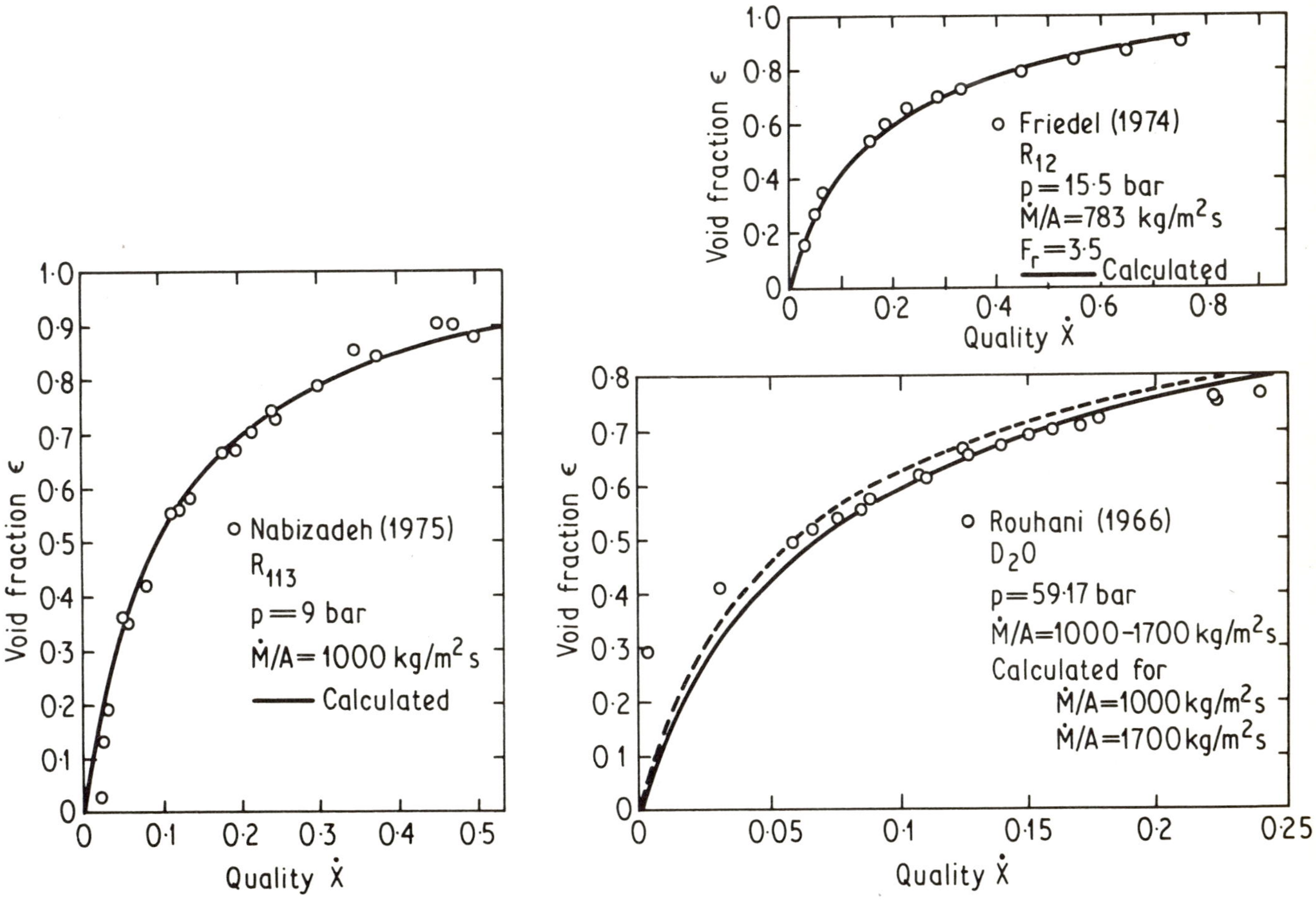


Fig. 4: Void fraction, comparison of water and Freon data, scaled with Zuber-Findlay's equation 8 modified by Nabizadeh (7)

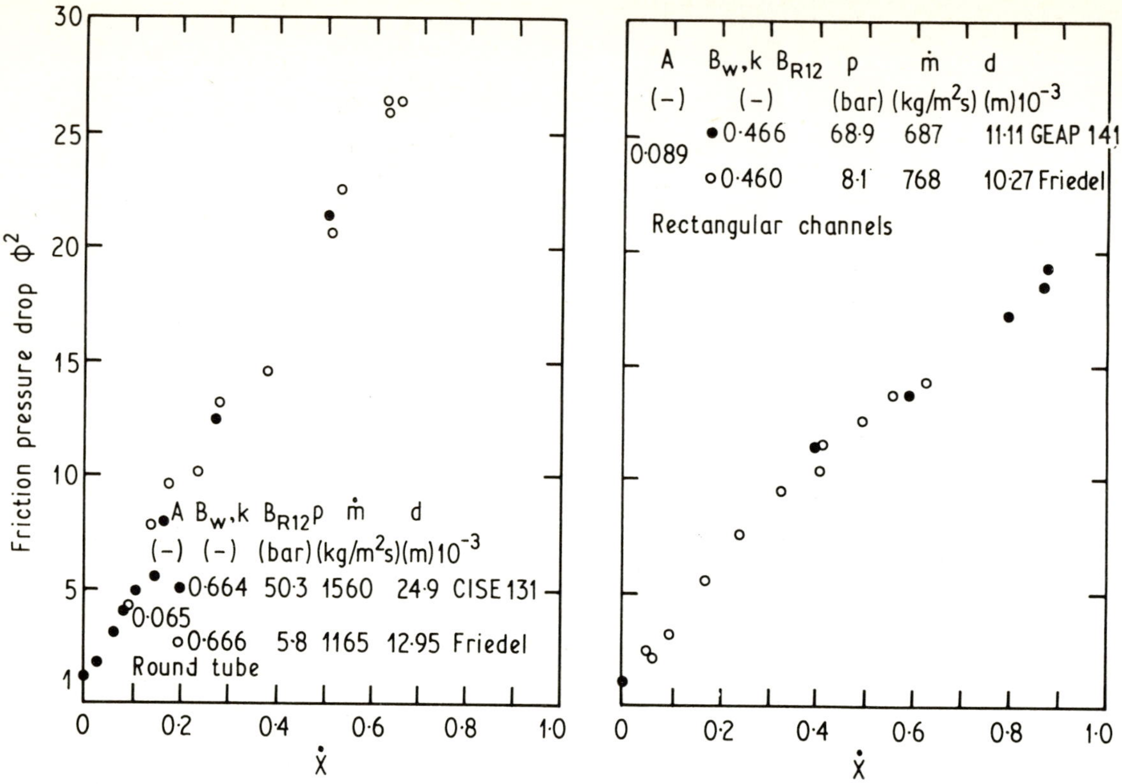


Fig. 5: Two-phase flow pressure drop multiplier scaled according to Friedel (10)

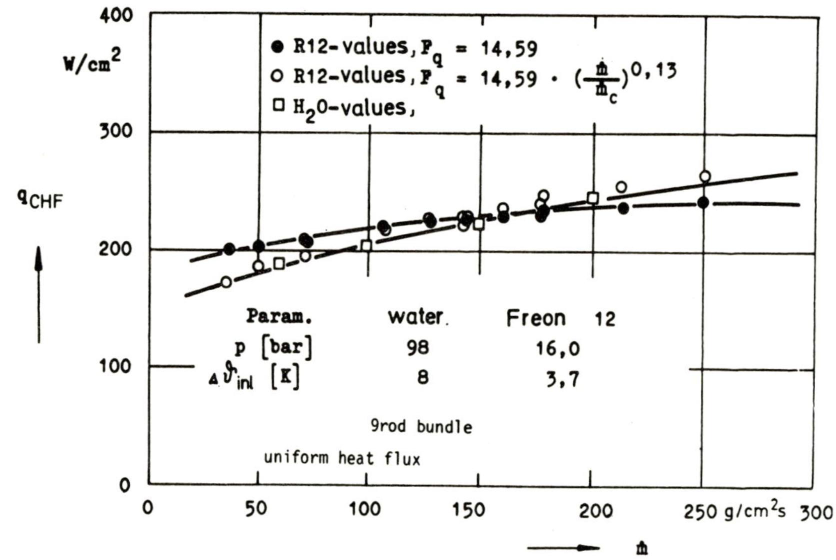


Fig. 6: Mass flow correction for critical heat flux scaling (Bouré, Hein)

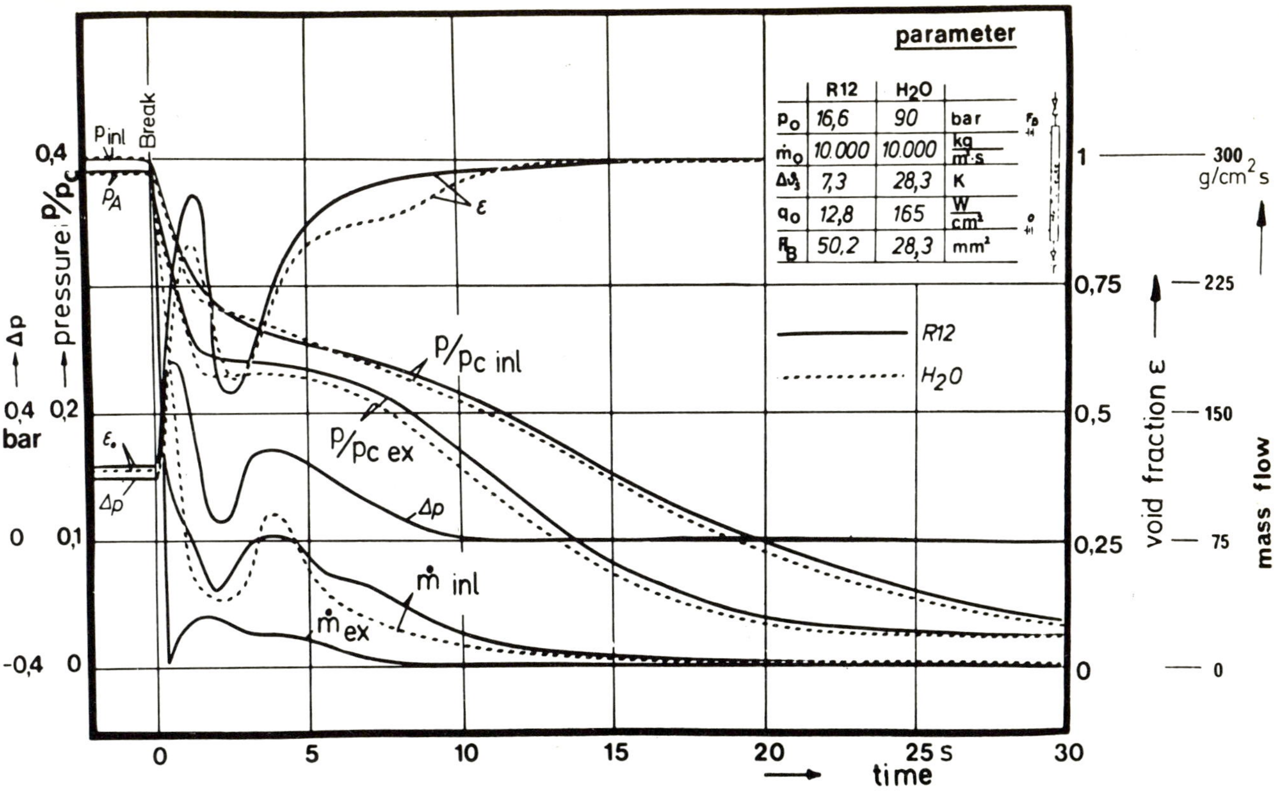


Fig. 7: Blowdown during loss of coolant accident, comparison between water tests and scaled R12 data

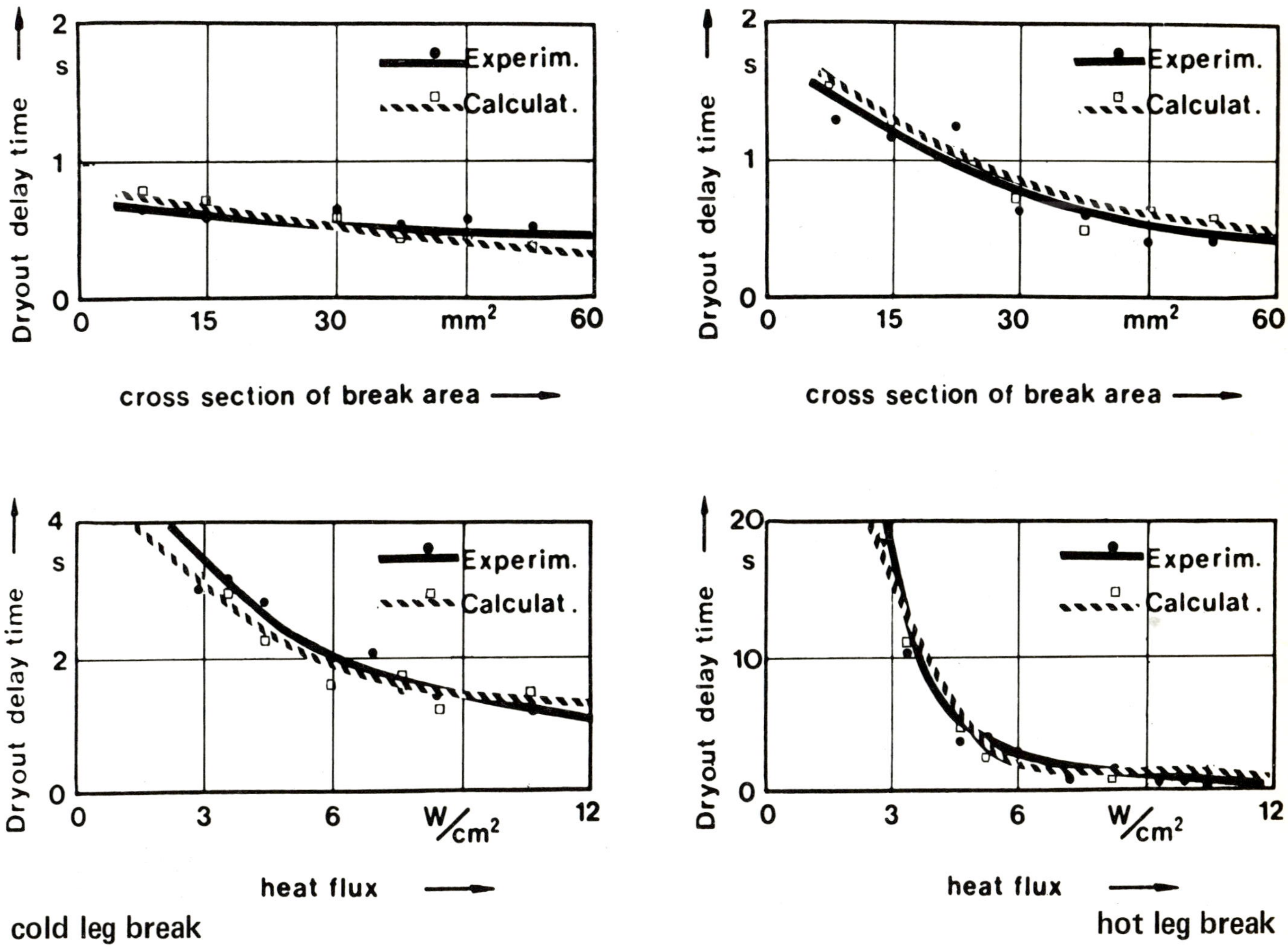


Fig. 8: Dryout-delay-time, measured (in R12) and calculated data

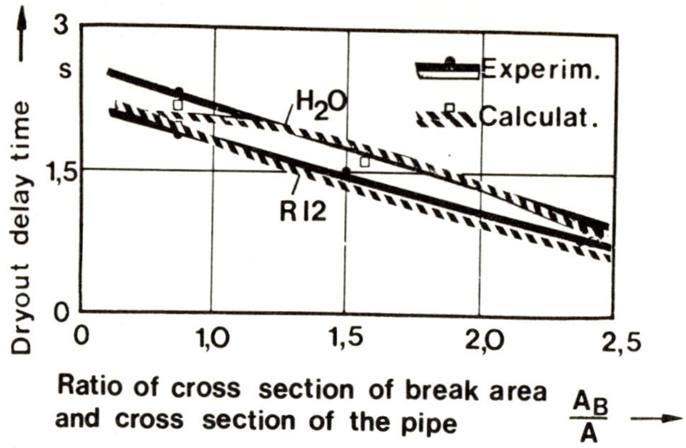


Fig. 9: Dryout-delay-time, comparison of experimental and theoretical data for water and for R12

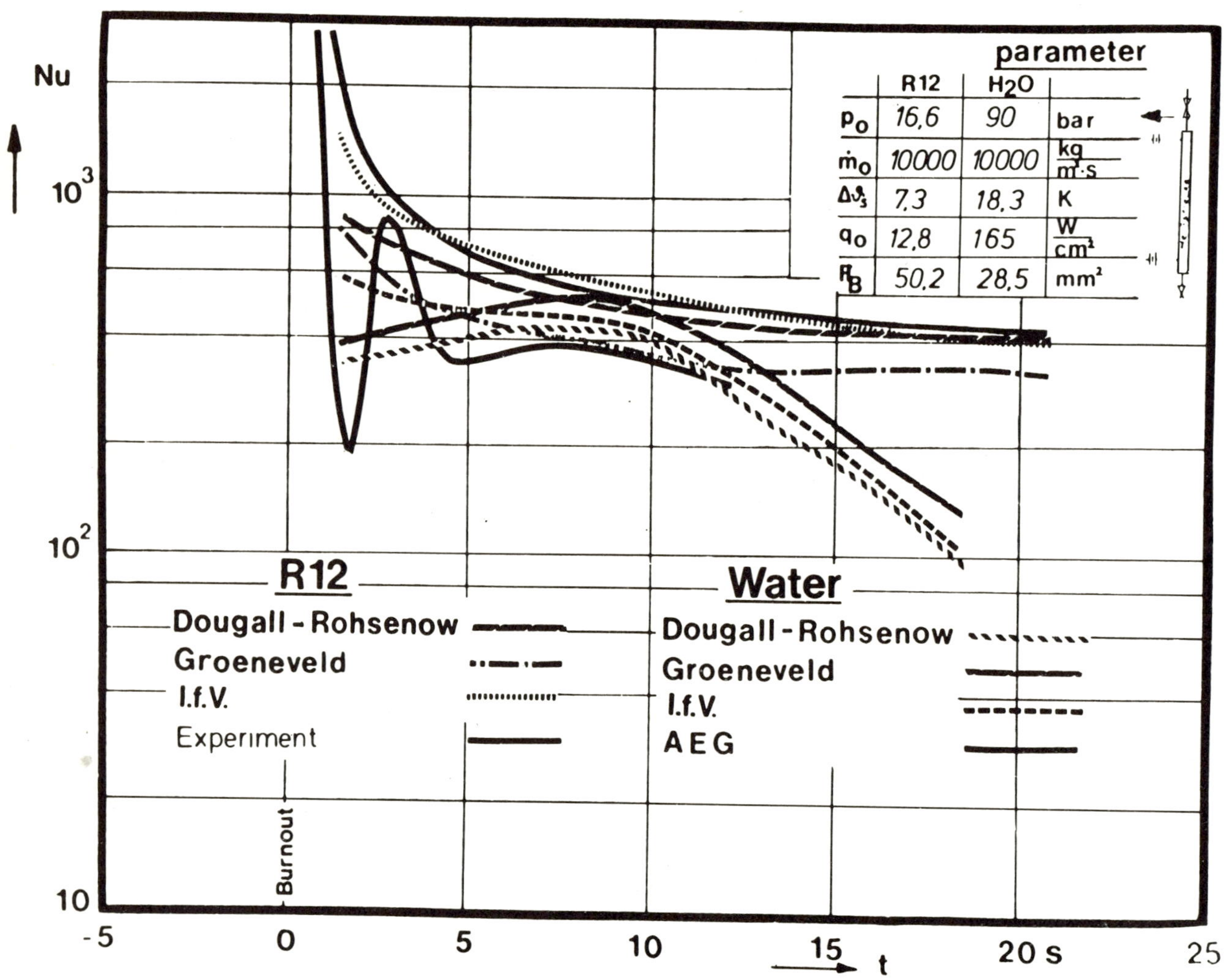


Fig. 10: Heat transfer in post dryout conditions during blowdown, comparison of water data and scaled Freon data

C212/77

# HEAT TRANSFER IN THE VICINITY OF A PARTIAL FLOW BLOCKAGE IN A REACTOR CORE

W. B. HALL, G. P. IOANNOU and J. T. TURNER
Simon Engineering Laboratories, University of Manchester

SYNOPSIS Experiments on an electrically heated model of a group of "ballooned" fuel elements have yielded data on pressure drop and heat transfer which can be used to predict temperatures in a reactor core under conditions of single phase steam cooling. In contrast to cladding deformations that are short in the axial direction, and which sometimes reduce cladding temperatures, the deformations tested show significant increases in temperature. A tentative method of generalising the data to deal with different configurations of blockage is proposed.

## 1. INTRODUCTION

In certain depressurisation accidents to water-cooled reactors the fuel pins may depend for their cooling on a flow of steam; they may then overheat, whereupon the cladding will deform under internal gas pressure. If localised rupture of the cladding occurs it is unlikely that cooling of the fuel will be further impaired. If, on the other hand, deformation takes the form of "ballooning" the consequence may be excessive fuel and cladding temperatures.

The paper examines one aspect of the problem: the heat transfer in a region of core in which a group of adjacent fuel rods have ballooned over a length of ten or twenty fuel rod diameters, so that the cladding on each rod touches that of its neighbours. Whether or not such conditions can arise depends upon the nature of the depressurisation fault and in particular on the subsequent temperature-time history of the cladding.

The consequences of partial blockage in a group of fuel rods clearly depend on whether the flow is able to by-pass the blocked region. Indeed, many experiments with partial blockages on all the sub-channels in a cluster, (so that by-passing is eliminated) have shown an improvement in heat transfer in the vicinity of the blockage. In the present tests the partially blocked region was embedded within a fairly large region of unblocked fuel rods, so that the flow could to some extent be diverted around the blockage. Pressure and temperature distributions were measured and a method of calculating the temperature distribution is tested against the measurements.

## 2. NOMENCLATURE

| | |
|---|---|
| $A$ | Flow area associated with one fuel rod |
| $d$ | effective diameter of coolant passage |
| $f$ | friction factor $\left(\frac{\Delta p.2d}{4\rho u^2 \ell}\right)$ |
| $\ell$ | length of blocked region (in the flow direction) |
| $\ell_1$ | distance from inlet to start of blocked region |
| $L$ | overall length of rod cluster (in the flow direction) |
| $\Delta p_o$ | pressure drop (see Fig.3 and equation 3) |
| $\Delta p_1$ | pressure drop (see Fig.3 and equation 2) |
| $\Delta p_2$ | pressure drop (see Fig.3 and equation 4) |
| $\Delta p_r$ | pressure drop (see equation 5) |
| $T_a, T_b$ | coolant temperatures at inlet and outlet of unblocked cluster (See Fig.4) |
| $T_1, T_2$ | coolant temperatures at upstream and downstream ends of blockage (see Fig.4) |
| $\Delta T$ | temperature difference between fuel rod and coolant (see Fig.4) |
| $\Delta T_c$ | increase in coolant temperature at the downstream end of blockage over that with no blockage |
| $\Delta T_s$ | increase in fuel rod surface temperature at the downstream end of blockage, over that with no blockage |
| $u$ | coolant velocity |
| $\mu$ | coolant viscosity |
| $\rho$ | coolant density |
| $\theta$ | $= \dfrac{T - T_a}{T_b - T_a}$ |
| $Re$ | $= \dfrac{\rho u d}{\mu}$, Reynolds Number |

Subscripts

| | |
|---|---|
| o | refers to unblocked passages |
| r | refers to blocked passages |
| c | refers to coolant |
| s | refers to fuel rod surface |

## 3. DESCRIPTION OF APPARATUS

The assembly consisted of an 18x18 array of rods on a 17mm square lattice, the overall length between the supporting grids being 540mm (Fig 1). Most rods had a uniform diameter of 12.7mm, but a group of 7x7 (arranged in the centre of the 18x18 array) had the shape shown in Fig.1. Within the 7x7 group of deformed rods were placed centrally either 1, 9 or 25 heated rods; the reason for varying the number of heated rods was to test whether the group was adequate as a guard for the centre rod. The heated rods were constructed from thin walled brass tube, the thickness of which was adjusted to give a uniform heat release per unit length when heated by passing an electric current in the axial direction. A set of heated uniform diameter tubes corresponding to the undeformed fuel pins was also tested.

The assembly was placed in a 310mm square cross section wind tunnel, with provision for measuring the total flow through the assembly. A special fuel rod (of similar shape to the deformed heated rods) was used to measure the pressure distribution along the axis of the blocked region; the pressure distribution along the tunnel wall was also measured. Rod surface temperatures were measured by thermocouples soldered in small holes in the brass tubes, the leads being extracted through the tubes themselves. A thermocouple was also traversed axially along one of the central passages in the blocked region; whilst this would not register the bulk mean air temperature, it gave an indication of the axial gradient of air temperature. All temperatures were measured with respect to the air inlet temperature as datum. The probable errors in heat transfer coefficient resulting from various errors of measurement were: temperature, 3%; flow 2%; heat input 2%.

## 4. EXPERIMENTAL RESULTS AND THEIR INTERPRETATION

Space does not permit presentation of more than a fraction of the measurements that have been made. Data have therefore been selected to illustrate the following points

a) the determination of the pressure drop-flow characteristics of the passages in the partially blocked region of the assembly;

b) the pressure distribution in the partially blocked region;

c) surface temperature and air temperature measurements with deformed rods.

### 4.1 Friction factor for channels in the blocked region

A separate experiment using a flow of water in a model of a single passage gave the results shown in Fig 2. It will be seen that the friction factor is well described by the expression

$$f_r = 0.028\, Re_r^{-0.2} \qquad (1)$$

(Note: the Reynolds Number $Re_r$ is based on the usual "effective diameter" - i.e. 4 x cross sectional area/total perimeter. Since the passage shape is far from circular, it is not surprising that the coefficient 0.028 differs considerably from the usual value, 0.046).

### 4.2 Pressure distribution along axis of blockage

This is illustrated in Fig 3, together with the pressure distribution along the tunnel wall. The aim is to deduce from these results a method of predicting the pressure gradient through this type of blockage. Comparison of the two distributions in Fig 3 suggests that the partial stagnation of the flow upstream of the blockage more than compensates for the pressure drop required to accelerate the flow into the restricted passage; and this in spite of the fact that the velocity in the restricted passages is greater than that in the undistorted channels. The reason is probably the reduced frictional pressure gradient in the low velocity stagnation region. Following this interpretation, the pressure difference $(\Delta p)_1$, between the pressure at inlet to the blocked region (extrapolated from the linear gradient through that region) and the corresponding pressure at the tunnel wall, may be described as follows:

$$(\Delta p)_1 = 1.34\, \frac{\rho u_o^2}{2} - \frac{\rho u_r^2}{2} \qquad (2)$$

The pressure drop in the undistorted region around the blockage, $(\Delta p)_o$, can be calculated from standard data for pressure drops in turbulent flow (the effective diameter concept works quite well for longitudinal flow through clusters in which the pitch/diameter ratio is similar to that in a reactor core). Thus

$$(\Delta p)_o = \frac{\rho u_o^2}{2} \cdot \frac{4 f_o \ell}{d_o} \qquad (3)$$

$$\text{where } f_o = 0.046\, Re^{-0.2} \qquad (3a)$$

(Note: the pressure drop in the experiment is greater than would be calculated by this method because the blockage occupies a significant fraction of the complete cluster and thus increases the velocity in the outer unblocked region. When comparing the model with experiment this is accommodated by using an experimentally determined friction factor instead of $f_o$).

At the downstream end of the blockage there is a pressure recovery $(\Delta p)_2$. In most cases of interest, the area increase from blocked to unblocked passages will be large, so that the pressure recovery may be expressed in terms of the velocity head in the blocked passages. One would expect it to depend also on the precise shape of the downstream end of the blockage; however, one could certainly not depend upon the recovery being greater than determined in this experiment. From the results we deduce

$$(\Delta p)_2 = 0.37\, \frac{\rho u_r^2}{2} \qquad (4)$$

Finally, we compare the two expressions for the pressure drop over the blocked region, ie.

$$(\Delta p)_r = \frac{\rho u_r^2}{2} \cdot \frac{4 f_r \ell}{d_r} \qquad (5)$$

$$(\Delta p)_r = (\Delta p)_1 + (\Delta p)_o + (\Delta p)_2 \qquad (6)$$

(The friction factor $f_r$ for the shape tested here is described by equation (1); the selection of values for other shapes is discussed below in section 4.)

From equations (2,3,4,5, 6) we may obtain the following expression for the velocity $u_r$ in the blocked region in terms of that in the unblocked region. Thus

$$\frac{u_r}{u_o} = \left[\frac{1.34 + 4\,f_o \ell/d_o}{0.63 + 4\,f_r \ell/d_r}\right]^{\frac{1}{2}} \qquad (7)$$

A knowledge of the upstream flow conditions and the geometry of the blocked core now enables the flow and heat transfer conditions within the bundle to be calculated, as follows:

Given values of $u_o$, and the geometrical parameters $\ell$, $d_o$ and $d_r$ we may then use equation 7 in conjunction with equation 1 and 3a to determine $u_r$, the Reynolds Number in the blocked region, and $f_r$. As shown below, this enables heat transfer to be predicted.

### 4.3 Temperature distributions

Temperature measurements have been made for a range of heat inputs and air flow rates. Over the range of Reynolds Number tested (approx $10^4$ to $4.10^4$) there was no significant effect of either parameter on the dimensionless temperature distribution defined below (see Fig 4)

$$\theta = \frac{T - T_a}{T_b - T_a} \qquad (8)$$

Fig 4 illustrates how the dimensionless surface temperature $\theta_s$ and coolant temperature $\theta_c$ in the blocked region are determined from the following three deductions:

i) from Reynolds analogy it may be shown that

$$\frac{\Delta T}{T_2 - T_1} = \frac{\rho u_r^2}{p_1 - p_2} \cdot Pr^{0.6} = \frac{d_r \; Pr^{0.6}}{2\,f_r\,\ell} \qquad (9)$$

ii) by a heat balance, noting that the heat input per unit length of rod, whether distorted or not, is constant

$$\frac{T_2 - T_1}{T_b - T_a} = \frac{\ell}{L} \cdot \frac{u_o A_o}{u_r A_r} \qquad (10)$$

iii) since the temperature at inlet to the blocked region will be approximately the same as that at the same axial position in the unblocked region,

$$\frac{T_1 - T_a}{T_b - T_a} = \frac{\ell_1}{L} \qquad (11)$$

Using these relationships we find

$$\theta_c = \frac{\ell_1}{L} + \frac{x}{L} \cdot \frac{u_o A_o}{u_r A_r} \qquad (12)$$

$$\theta_s = \frac{\ell_1}{L} + \frac{x}{L} \cdot \frac{u_o A_o}{u_r A_r} + \frac{\ell}{L} \cdot \frac{u_o A_o}{u_r A_r} \cdot \frac{Pr^{0.6}\, d_r}{2 f_r \ell} \qquad (13)$$

Experimental values of $\theta_c$ and $\theta_s$ are compared with equations (12) and (13) in Fig 4.

## 5. APPLICATION OF THE RESULTS

The above work is limited to one particular passage shape in the blocked region - that which is obtained when the cladding deforms so that neighbouring fuel elements just touch each other. Further deformation will, of course, reduce the effective diameter, $d_r$, and there will be considerable fractions of the cladding perimeter that are inoperative for heat transfer.

As an approximation one might assume that the shape remains the same, so that equation (1) applies, and that the effective diameter varies as the square root of the cross sectional area for flow.

The above technique has been used to investigate the consequencies of this type of blockage in a hypothetical reactor fault. It is assumed that a group of neighbouring rods have ballooned over a length of 200mm. It appears from the tests using varying numbers of heated rods that a 5 x 5 group is sufficiently large for the model to apply, although further work is required to assess the relative importance of lateral heat transport from the overheated region. The other parameters governing the problem are given below:

| | | |
|---|---|---|
| Diameter of rods | = | 9 mm |
| Pitch of rods | = | 12 mm |
| Steam velocity in unblocked passages | = | 15 m/s |
| Steam pressure | = | 0.3 MN/$m^2$ |
| Heat generation in fuel rods | = | 1 kw/m |

These data may be used, together with the results given in section 3, to predict the temperature increase caused by the postulated blockage. The analysis is largely independent of the position of the blockage in the core, except insofar as the heat generation rate in the region of the blockage has been taken as 1 kW/m. In general, temperature increases will be greatest at the downstream end of the blockage. Both the coolant and the sheath temperatures will be raised above the values they would have in the unblocked core; in the following table these increases are designated as $\Delta T_c$ and $\Delta T_s$ respectively, and they are given for a range of blockages. (The degree of blockage is expressed either as the ratio of the flow passage area, $A_r$, to that without blockage, $A_o$, or by the % Blockage which is equal to $100(1-A_r/A_o)$).

| $A_r/A_o$ | 0.384 | 0.311 | 0.249 | 0.187 | 0.124 |
|---|---|---|---|---|---|
| % Blockage | 61.6 | 68.9 | 75.1 | 81.3 | 87.6 |
| $\Delta T_c$ (°C) | 64.4 | 96.9 | 140.8 | 219.4 | 392.1 |
| $\Delta T_s$ (°C) | 111.0 | 171.6 | 248.6 | 377.7 | 638.9 |

## 6. CONCLUSIONS

It must again be stressed that deformations of the ballooning type are necessary for the heat transfer deteriorations described here to occur; moreover the deformations must be sufficiently systematic for a group of adjacent rods to be affected. It is clear from the work that a group of 25 rods would provide complete thermal guarding for the centre rod in the group; however it is also true that a group of 9 rods would provide considerable guarding and might give temperatures at the centre rod which approached those for a larger group.

Many factors not considered in this paper must be accounted for in a full safety analysis: time dependence of temperatures; the possibility that cooling may be by a spray of water and steam; variations in the length and position of the blocked region. However, it appears from the results that temperatures sufficiently high to cause concern could result from this type of blockage; an analysis in which the deformation characteristics of the cladding are included would therefore appear to be necessary.

## 7. ACKNOWLEDGEMENTS

The authors wish to acknowledge the financial help they have received from the Inspectorate of Nuclear Installations in this work.

## 8. REFERENCES

1. Cermak, J.O., Leyse, R.H. and Dominicis, D.P. "Effect of flow blockage on bottom flooding heat transfer effectiveness (FLECHT)". Nuclear Technology, Vol.11, August 1971.

2. Davis, P.R. "Experimental studies of the effect of flow restrictions in a small rod bundle under emergency core coolant injection conditions". Nuclear Technology, Vol.11, August 1971.

3. Gambill, W.R. "Estimate of effect of localised flow blockages on PWR clad temperatures during reflood". Proc. A.N. Soc. Water Reactor Safety Meeting. Salt Lake City, March 1973.

4. Hall, P.C. and Duffey, R.B. "A method of calculating the effect of clad ballooning on Loss-of-Coolant Accident temperature transients". Nuclear Science and Engineering, August 1975.

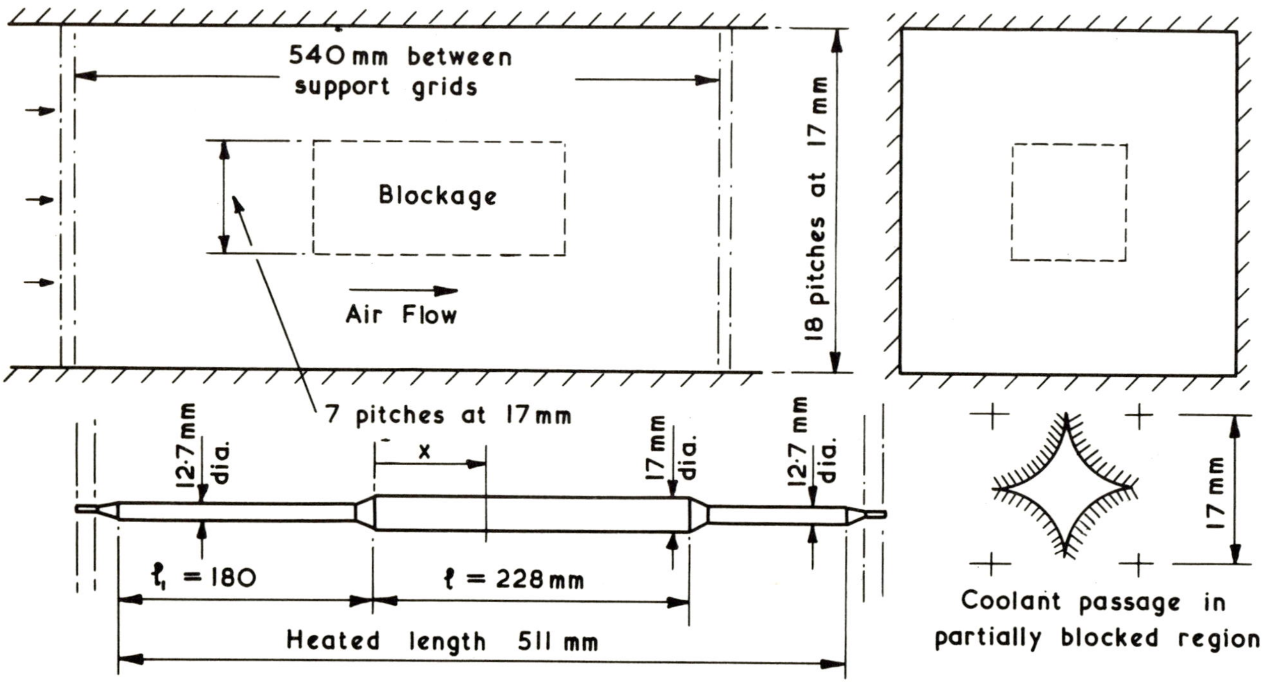


Fig. 1: Sketch of experimental rod cluster and deformed fuel rod

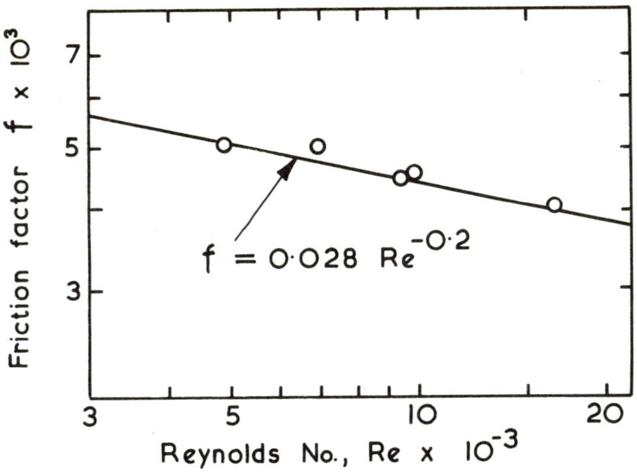


Fig. 2: Determination of friction factor for passage shape in partially blocked region

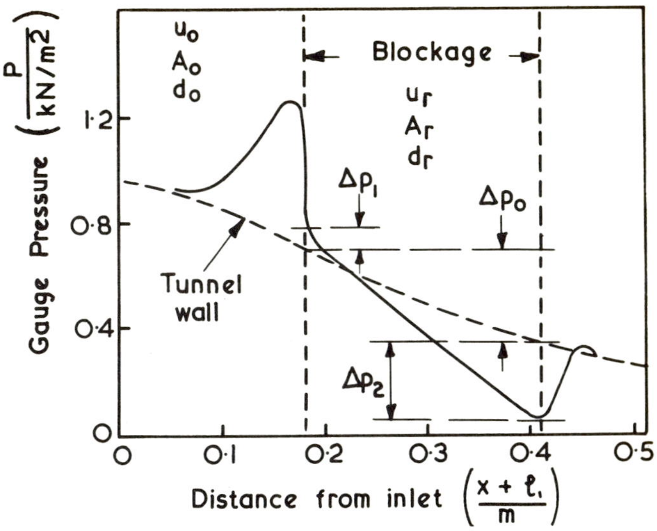


Fig. 3: Experimental pressure distribution along axis of cluster (through blockage) and along tunnel wall

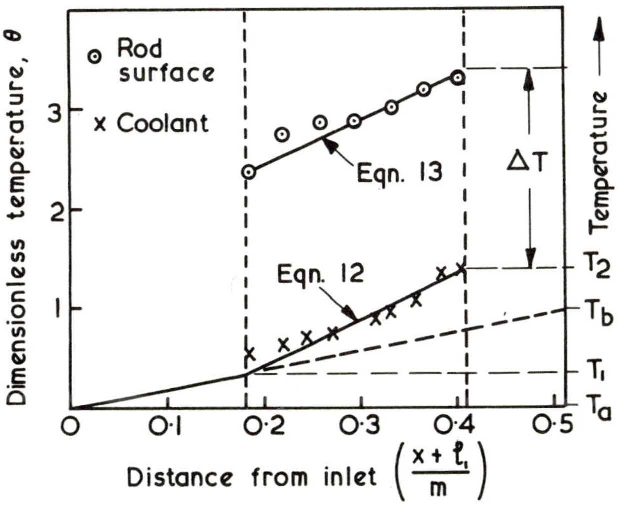


Fig. 4: Measured and calculated rod surface and coolant temperature distributions

C213/77

# REFLOOD EXPERIMENTS WITH BUNDLES OF BALLOONED RODS

P. IHLE, S. MALANG, K. RUST
Institut für Reaktorbauelemente, Kernforschungszentrum Karlsruhe, GFR

SYNOPSIS Experiments simulating the reflood phase of a loss-of-coolant accident in a pressurized water reactor have been performed. The particular aim of the project is to investigate the influence of size and shape of coolant channel blockages upon the cooling conditions. In a first step a row of five heater rods was used for base line tests without coolant channel blockages and with a perforated plate as blockage simulator. The effect of droplet dispersion in the two-phase flow caused by the blockage is described. The transients of heat transfer coefficient, heat flux, and stored heat are compared with the results of tests with unblocked coolant channels.

## INTRODUCTION

Most of the reflood experiments for pressurized water reactors are performed using an undisturbed bundle geometry. It is possible however, that the fuel rods balloon during a loss-of-coolant accident (LOCA) due to the internal pressure and the high sheath temperature.

Integral experiments with partly blocked coolant channels indicated that the heat transfer is improved in the region downstream of the blockage (ref. 1). Investigations of an unheated bundle with one-phase flow were showing the hydraulic principle (ref. 2). However, it is necessary to study in more detail the heat removal in the vicinity of blockages in heated bundles cooled by a two-phase flow. Therefore a separate effect experimental program is under way to investigate the influence of size and shape of coolant channel blockages on the cooling conditions during the reflood phase of a LOCA (ref. 3). The fuel rods are simulated by electrically heated rods with exchangeable blockages attached at the outer surface.

## EXPERIMENTAL EQUIPMENT

The test rig is designed for a separate effect program with constant flooding rate and constant back pressure to allow the investigation of the influence of coolant channel blockages independent of system effects (see Fig. 1a). A constant flooding rate is obtained with a pressure head of the pump much larger than the sum of pressure in the outlet plenum. There is a large buffer volume with a pressure regulating valve connected to the outlet plenum where a large massflow of saturated steam is coming in from a separate steam supply in order to keep changes in the back pressure, caused by variations in flow at the outlet of the test section, as small as possible. A steam-water separator is located at the outlet plenum to measure the water content of the two-phase flow. As fuel rod simulators electrically heated rods with full length are used. The axial cosine power profile is approximately realized by 7 steps of different specific rod power. The heated coil is embedded in MgO. In the sheath of 1 mm thickness up to 8 thermocouples of 0.5 mm diameter are embedded in grooves, milled into the outer surface. The grooves are closed by brazing over the total length to avoid any disturbance of the coolant flow. PWR-grid-spacers are located in 500 mm distance.

In a first step, grid plates as blockages are attached at the outer surface of the rods in the midplane of the heated region (see Fig. 1b). To ensure that the coolant passes through the holes of the grid plates situated in the center of each coolant channel the gaps between grid plate and rods as well as between grid plate and housing are sealed.

Three observation windows in different axial positions upstream and downstream of the blockage allow an optical observation of the two-phase flow. A special quartz glass with high infrared absorption is used to avoid a too big heat loss through the windows. Care has been taken to have the inner surface of the windows at the same plain as the test section wall in order to avoid any disturbance of the geometry.

Temperatures, pressures, pressure differences, flow rates, and rod power are recorded digitally with a scan frequency of 10 cycles/s. 14 channels are recorded in parallel on an anologue tape in order to get some informations about higher frequencies and in addition some redundancy.

In all tests the rods are heated up in stagnant steam to the desired initial clad temperature, using a small rod power. The test housing is heated up to approximately 500 deg. C by radiation of the rods. This leads to a heat content in the wall (6.5 mm thick) of approximately the same as in half a row of heater rods. To start the reflood phase, a bypass valve at the lower plenum is closed and the water level rises in this plenum. As soon as the water contacts the heated zone of the rods, the power is rised to the value given by the decay heat of a fuel rod.

## EXPERIMENTAL PROGRAM AND RESULTS

### 1. Parameters

The following values were kept constant for all the tests:

| | |
|---|---|
| - rod diamter | 10,75 mm |
| - rod pitch | 14,3 mm |
| - heated length | 290 cm |
| - axial power profile | cosine-profile approximated by 7 steps with a peak to average factor of 1,33 |
| - shape of the power transient | ANS-Standard (ref. 4) |

Varied from test to test are the following parameters:

| | |
|---|---|
| - flow rate (constant for each test), | 1,5 - 10 cm/s |
| - initial clad temperature, | 600 - 800 deg. C |
| - back pressure (constant for each test), | 1,0 - 4,5 bar |
| - level of decay heat, | 100 - 120 % |
| - flooding water temperature (constant for each test), | 40 - 80 deg. C |
| - type of the blockage, | grid plate |
| - blockage ratio, | 0, 62, 90 % |

The parameters of three typical tests with a row of 5 rods discussed in the following are listed in Tab. 1. This geometry has been chosen to allow a better optical observation. Later on tests with a 5 x 5 rod bundle will be performed.

Table 1

Varied parameters of the Tests shown in Fig. 2 and Fig. 3

| test number | | 30 | 77 | 70 |
|---|---|---|---|---|
| max. rod power at t = 0 s | (W/cm) | 27,5 | 27,5 | 27,5 |
| max. rod temperature at t = 0 s | (deg.C) | 700 | 750 | 750 |
| temperature of the flooding water | (deg.C) | 62 | 60 | 40 |
| cold flooding rate | (cm/s) | 6,4 | 6,6 | 2,1 |
| system pressure, abs. | (bar) | 3,5 | 4,6 | 3,5 |
| type of the blockage | - | no bl. | plate | plate |
| blockage factor | % | 0 | 62 | 62 |

The sheath temperatures of the rods are measured at the following levels:
-50 (50 mm upstream of the blockage), +35, +85, +135, +235, +335, +870 and +1120 (mm downstream of the blockage).

The fluid temperatures and pressures are measured in the lower plenum, in a cooling channel of the rod row and in the upper plenum. The steam temperature in the upper plenum is measured after the steam-water-separator.

### 2. Unblocked Geometry

Various tests in undisturbed geometry were made in the same test section to have a basis for the tests with blocked cooling channels. No qualitative differences were observed by the comparison with known experiments in larger bundle geometries. As example some measured values of test no. 30 are chosen to be compared later with values measured in blocked geometries (see Fig. 2a). In Fig. 2a the transients of four sheath temperatures near the midplane of the heated length are shown, all in the region of the maximum rod power. After beginning of the flooding at time = 0 (s) all temperatures are increasing during the first 10 - 15 seconds and are decreasing with nearly the same gradient. One temperature transient has a break down between 40 and 70 seconds after start, but it is rising up to the value measured at the same axial level of the adjacent rod. This effect is due to a local forced cooling which has not much influence on the average temperature in the cross section of this part of the rod. All measured levels of the rods are showing a quenching accordingly their axial position downstream.

### 3. Blocked Geometry

In these tests all coolant channels were blocked with the same blockage ratio. No coolant flow from unblocked channels or the upper plenum could reach the region downstream of the blockages as it would be the case in a reactor with partly blocked channels.

The ballooning of the rods was simulated by a stainless steel plate of 23 mm thickness. The gap between the housing and the plate was tightened by lip-seals and the gap between the plate and the rods of nominal 0.8 mm was sealed with graphite over a length of 10 mm. At the start of the reflood phase the average temperature of the grid plate was 100 to 150 deg C lower than the rod temperature at this level.

In all tests with this geometry, under variation of the parameters mentioned above, a second quenching front occured beginning just downstream of the blocked area. Its influence on the quenching of the downstream positions is reduced with growing distance from the blockage.

A comparison of the tests no. 30 and 77 shows that the temperatures measured at the same axial position have a quite different transient behaviour (see Fig. 2b). Only the temperature transient at the position -50 upstream of the blockage has the same tendency as the temperature transients in the unblocked geometry. All other temperatures downstream of the blockage are decreasing immediately.

At low flow rates, as it was the case in test no. 70 the cooling mechanism remains similar to the test no. 77 (see Fig. 2c). But in this case the rod temperatures downstream of the blockage increase in the first few seconds after start as for the unblocked geometry. However, they decrease earlier and do not reach the maximum value of a comparable test with unblocked geometry.

### 4. Optical Observation

At the beginning of the flooding high flow velocities occur. Bizarrely shaped droplets of a mean diameter of 1 mm are entrained with the steam, quenching immediately the spacers in the bundle. Inspite of a large buffer volume and a pressure regulating valve pressure pulsations are measured at the outlet plenum (see Fig. 3a, PUP). The same pulsations are measured at the inlet plenum and observed at the windows as mass flow pulsations. A comparison of the two pressure traces at the outlet and inlet plenum shows that phase and amplitude are approximately the same (see Fig. 3b, PUP, PLP). The pressure difference increases during the test from zero to the pressure corresponding to the height of the water column in the test section.

Two different hydraulic mechanisms were observed at the beginning of the reflood phase: During the first 10 - 20 seconds water droplets of much smaller size than in the unblocked channels are leaving the blocked area as jets which are dispersed 5 to 10 cm downstream. Later on, larger water droplets fall back in the region of low velocities outside the jets filling the space downstream of the blockage. This water 'layer' growing in thickness is forced to move by the steam production on the hot surfaces and the jets leaving the blockage. The jets disappear in the growing water "layer" before quenching is reached in this part of the rod row.

## ANALYSIS

The quantitative heat transfer analysis is made using a computer code developed on the basis of the HETRAP-code (ref. 5). Surface temperatures are calculated from the measured sheath temperatures. The stored energy is determined by temperatures, rod power and physical properties of the heater rods. The heat transfer is related to the temperature of the flooding water and not, as usual, to the saturation temperature. The reason is, that by cooling with subcooled water some parts of rod surfaces may have lower temperatures than the corresponding saturation temperature (see Fig. 2b), resulting in negative heat transfer coefficient. In Fig. 3 the heat transfer analysis of the two tests no. 30 and no. 77 is shown. In Figs. 3a und 3b fluid data are plotted, i.e. the pressure in the upper plenum (PUP), and in the lower plenum (PLP), the temperature in the upper plenum (TUP), and the fluid temperature in the midplane of the heater rod row (TF).
In Figs.3c and 3d the transients of the stored heat (SH), the heat flux (q"), and the heat transfer coefficient ($\alpha$) of the two tests can be compared.
The temperatures of the cladding (TC) at the level +35 mm in Figs.3a, 3b and 3c and at the level +870 mm in Fig. 3d are illustrating the reflood phase. In the case of unblocked channels the heat transfer increases slightly after the initial value of the reflood phase with a sudden peak in the moment of the quenching. The stored heat grows with increasing rod temperature. The heat flux keeps its mean value until quenching.

In blocked geometries the heat transfer downstream of the blockage increases to higher values as in unblocked geometries and then decreases, ending with a similar peak at the moment of the quenching of the rods.

## CONCLUSION

Reflood experiments under PWR conditions with a row of five electrically heated rods and uniform blocked cooling channels, show downstream of the blockage a good heat removal at the beginning of the reflood phase. This effect caused by droplet dispersion and water concentration downstream of the blocked area remains existant down to low flow rates. It remains to be seen if this effect is relevant for all types of blockages and if it is outweighed by the flow redistribution in a partly blocked bundle.

## REFERENCES


1. F.F. CADEK et al.
   PWR-FLECHT, Final Report
   WCAP-7665, April 1971

2. J.M. CREER, J.M. BATES
   Effects of Sleeve Blockages on Air Velocity Distributions in a Unheated 7 x 7 Rod Bundle
   BNWL-1975, Jan. 1976

3. G. HOFMANN et al.
   Projekt Nukleare Sicherheit
   Halbjahresbericht
   KFK-2050, Aug. 1974, S. 190 - 195
   KFK-2130, Mai 1975, S. 231 - 239
   KFK-2195, Sept.1975, S. 284 - 304

4. Proposed ANS Standard Energy Release Rates Following Shutdown of Uranium-Fueled Thermal Reactors
   ANS-5.1, October 1971

5. S. MALANG
   HETRAP: A Heat Transfer Analysis Program
   ORNL-TM-4555, Sept. 1974

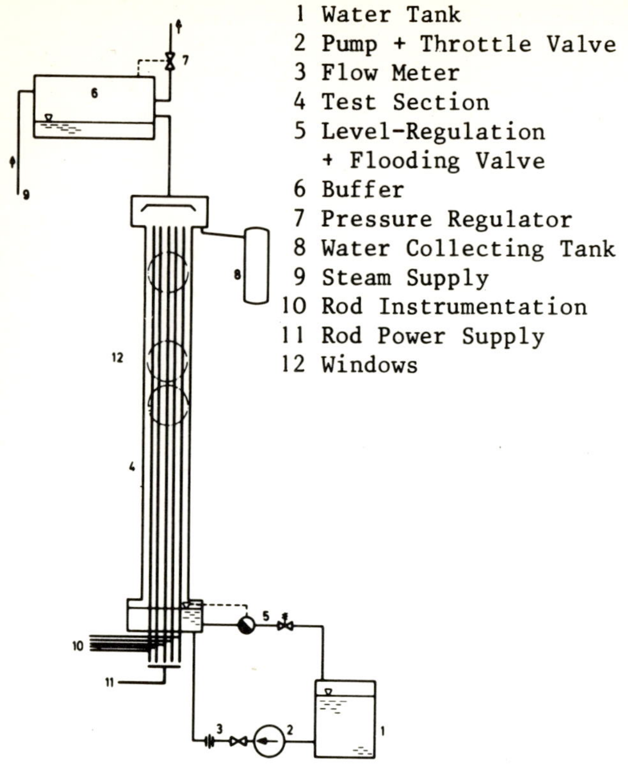


(a) Test rig (schematic)

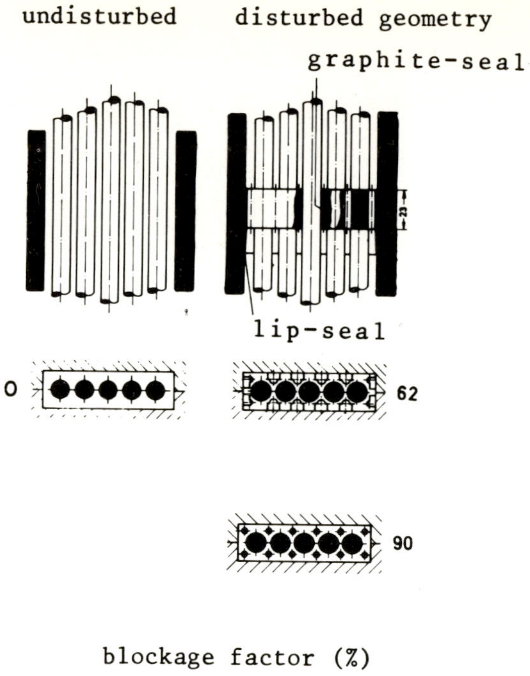


(b) Types of blockages

Fig. 1: Experimental equipment

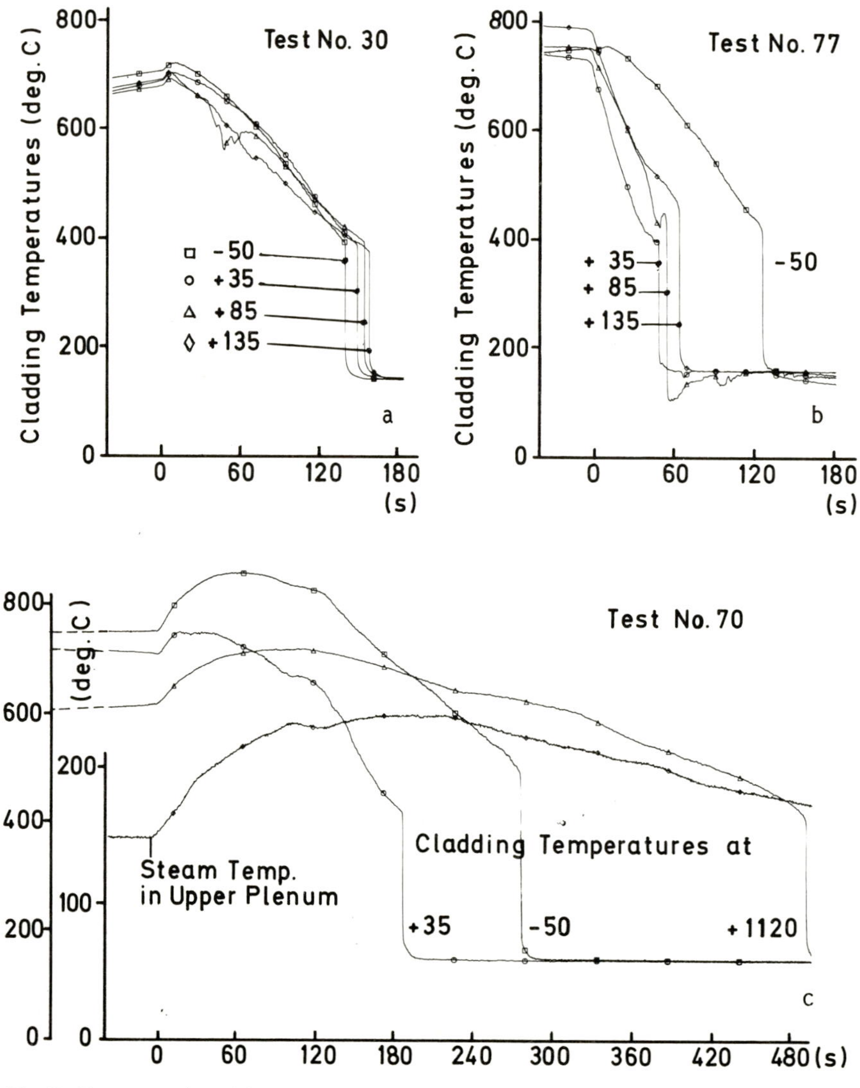


Fig. 2: Temperatures of 3 typical reflood tests

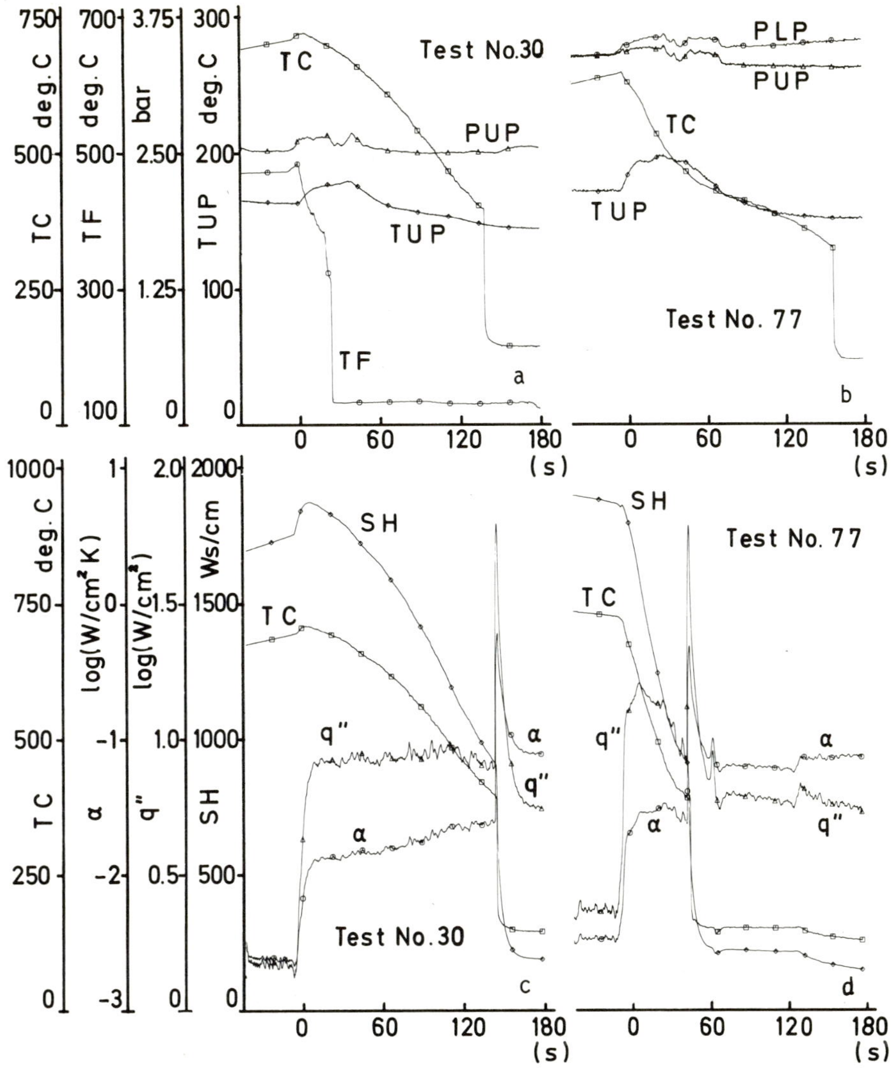


Fig. 3: Heat transfer analysis of reflood tests

# C214/77

# REFLOOD HEAT TRANSFER IN THE SEMISCALE MOD-1 CORE DURING FORCED FEED REFLOOD TESTS

A. C. PETERSON, PhD and G. G. LOOMIS, MS
EG&G Idaho, Inc., Idaho Falls, Idaho 83401, USA

ABSTRACT Experimental data and results drawn from experimental data obtained during forced feed reflood tests in the Semiscale Mod-1 system are analyzed. These tests provided information on the reflood portion of the loss-of-coolant accident. Heat transfer coefficients calculated from experimental data are compared with those calculated using selected heat transfer correlations from the literature to evaluate the capability of the correlations to describe the reflood heat transfer process. The test data and the core thermal response calculated with the REFLUX computer code are compared and the capability of the code to predict the thermal-hydraulic behavior occurring in the core is evaluated.

## INTRODUCTION

1. The reflooding phase of the loss-of-coolant accident (LOCA) starts when the coolant liquid level reaches the bottom of the rods. The heat transfer and two-phase flow phenomena that occur as the coolant penetrates the heated core during the reflood phase are complex processes. The hydraulic coupling between the vessel downcomer and the heated core in a pressurized water reactor (PWR) during the reflood process increases the complexity of the flow phenomena. To investigate the heat transfer characteristics without the complexity of the hydraulic coupling of the vessel downcomer and core, a study was performed in the Semiscale Mod-1 system in which the emergency core coolant (ECC) was injected into the core inlet at a constant rate (forced feed). The Semiscale Mod-1 facility is a small nonnuclear experimental facility modeled from a PWR; however, for these tests the vessel downcomer was isolated from the lower plenum and core inlet by a manifold. The objective of the tests was to provide experimental data that could be used to provide a basis for analytical model assessment.

2. Results calculated from experimental data are presented in the paper to illustrate the heat transfer processes that occurred in the Semiscale Mod-1 core during forced feed reflood. Several different heat transfer regimes were inferred from the calculated results. The heat transfer coefficients calculated from existing heat transfer correlations that are used to calculate reflood heat transfer for film and transition boiling are compared with those calculated from test results to evaluate the capability of the correlations to describe reflood heat transfer (Ref. 1, 2, and 3). The experimental core temperature data and the core temperature response calculated with an existing available computer code, REFLUX, are compared and discussed. REFLUX was developed to calculate the complex reflood process during forced feed reflood (Ref. 2). The purpose of presenting these results is to add to the reflood data and analysis obtained from other reflood heat transfer experiments conducted with longer (12 ft) rod lengths (Ref. 4). A second purpose is to evaluate the capability of an existing analytical model to predict the temperature response of the core during the reflood portion of the LOCA.

## SYSTEM DESCRIPTION AND TEST PROCEDURE

3. The Semiscale Mod-1 system configuration shown (Figure 1) was used to investigate reflood heat transfer with the ECC pumped into the core inlet. The system consisted of a pressure vessel with simulated reactor internals; an intact loop with phase separator, steam generator, and pump; a broken loop arranged in the cold leg break configuration but with the hot leg capped off at the vessel outlet and a rupture assembly with the rupture discs removed from the cold leg; a pressure suppression system with header, suppression tank and steam supply system to control the system pressure; and an ECC injection pump. A manifold was attached at the core inlet to direct ECC into the core.

4. The Mod-1 core simulator in the Semiscale system consists of 40 electrically heated rods of typical PWR fuel rod diameter (0.442 in.). The axial power distribution has a half cosine shape with an axial peaking factor of 1.58. The rod heating element is constructed of constantan wire (55% copper, 45% nickel) coiled with a varying pitch and sized to develop a specified power. Compacted boron nitride surrounds the element and insulates it from a composite sheath. The composite sheath was manufactured from 316L stainless steel. Four 0.025-in.-diameter Chromel-Alumel thermocouples swaged between inner and outer sheaths provide cladding temperature measurements for each rod at different axial locations along the 5.5-ft heated length of the heater rod. The overall length of the rod is about 17.2 ft and the bottom of the 5.5-ft heated section is about 11.9 ft below the cold leg center line. The rods extend from the bottom of the heated section to the upper plenum and pass out through the vessel upper head. The 40 rods are positioned and held in the core with ten grid spacers which maintain the heater rods on a typical PWR pitch (0.563 in.).

5. Measurements taken in the core included heater rod cladding temperatures and fluid temperatures. The core fluid temperatures were measured with thermocouples attached at the grid spacers located along the heated length of the rods. The core inlet flow rate and intact loop flow rate were obtained from a volumetric flow rate measurement obtained by a turbine

flowmeter. A shunt circuit technique provided core current, a digital voltmeter measured core voltage, and a Hall-effect multiplier device measured total core power. The core power was controlled to simulate the decay heating process during the reflood portion of a LOCA. The American Nuclear Society (ANS) +20% decay curve was used (Ref. 5).

6. The following procedure was used in establishing the initial conditions at the initiation of reflood: (a) the initial system pressure was established by controlling the pressure in the pressure suppression system, (b) the water level in the lower plenum was established up to the beginning of the heated length, (c) the core power was increased until the maximum specified power level was attained and (d) the ECC injection was initiated when the maximum specified rod cladding temperature was attained.

7. The test results discussed in this document include data at system pressures of 60 and 20 psia, core inlet flooding rates of 0.6 and 1.0 in./sec, and initial rod maximum cladding temperature of 1400°F, a maximum initial peak power of 0.7 kW/ft and an inlet subcooling of 140°F. Tests at core inlet flooding rates up to 2.5 in./sec, initial maximum cladding temperatures of 1100°F, a maximum peak power of 1.2 kW/ft, and a peak radial power profile were also conducted (Ref. 3). The results for these conditions are not included in this paper; however, the results that are presented are typical of the results for the entire range of test conditions.

## TEST RESULTS

8. On the basis of visual observations from the PWR-FLECHT tests, the heat transfer regimes that occur during the reflood process have been identified. At a typical axial location, the heat transfer regimes change from steam cooling, to dispersed flow film boiling, to film boiling, to transition boiling which ends with the rod quenching (Ref. 4). Since the Semiscale Mod-1 core was inside a core barrel and metal vessel, the heat transfer regimes could not be visually identified. However, the similarity of plots of the heat transfer coefficient versus the difference in the wall and saturation temperature ($\Delta T_{sat}$) shown (Figure 2) to classical pool boiling curves indicated that these type of plots could be used to identify the heat transfer regimes that occurred in the Semiscale Mod-1 core.

9. The heat transfer coefficients shown (Figure 2) were calculated using an inverse heat conduction technique from the measurement of a thermocouple and are based on the saturation temperature of the fluid. The inverse heat conduction technique solves for the surface heat flux that minimizes the sum of the squared differences between the calculated and measured temperature for the present time step and a few future time steps (Ref. 6). Through use of the calculated heat flux as a boundary condition, the radial temperature distribution, including the surface temperature, is calculated. The calculated surface temperature and heat flux are used in combination with the system saturation temperature to calculate the heat transfer coefficient. The heat transfer coefficient shown (Figure 2) was calculated from the thermocouple measurement at the core hot spot (29-in. elevation) for a test (Test S-03-A) conducted at a pressure of 20 psia and a flooding rate of 1.0 in./sec.

10. The heat transfer regimes that were inferred to occur in the Semiscale Mod-1 core are identified (Figure 2). Each symbol plotted (Figure 2) represents a 0.1-sec time step in the inverse conduction calculation so that the figure shows not only the magnitude of the heat transfer coefficient but the concentration of symbols indicates the duration of a heat transfer regime.

11. At the initiation of reflood the difference between the rod surface temperature and the saturation temperature was about 1150°F and the heat transfer coefficient was low. During the early portion of the reflood process the rod cladding temperature increased, which resulted in an increase in the temperature difference from about 1150°F to above 1300°F, and the heat transfer coefficient increased from about 1 to 10 Btu/hr-ft$^2$-°F. During this period, the heat transfer regime was initially forced convection to steam and changed to a dispersed-phase regime before the peak temperature occurred. After the peak temperature occurred, which corresponds to the maximum value of $\Delta T_{sat}$, the heat transfer coefficient continued to steadily increase and the temperature difference decreased to a value of about 500°F. The exact boundaries for the change in heat transfer regimes shown (Figure 2) were not directly observed during these tests, but were inferred from the general shape of the plot and the observations from other reflood heat transfer tests (Ref. 7). In addition an estimate of the injected mass that was entrained was calculated from a mass balance on the vessel based on measured inlet and exit mass flow rates and on the quench front velocity. This estimate indicated that a significant fraction of the injected mass was entrained above the quench front. The estimate of significant entrainment also implies the existence of a dispersed flow or film boiling regime during the period indicated (Figure 2). At temperature differences which were less than about 500°F, but greater than 150°F, transition boiling occurred and the heat transfer coefficient was typical of a transition boiling regime. The maximum heat transfer coefficient occurred at a temperature difference of about 150°F, and indicated the occurrence of critical heat flux (CHF). For temperature differences that were less than 150°F, the heat transfer regime changed from nucleate boiling to forced convection to liquid at the lower temperature differences. The spacing of the symbols plotted (Figure 2) indicates that for the bulk of the reflood time dispersed and film boiling regimes occurred at the 29-in. elevation whereas the transition boiling regime only ooccurred for about 3 seconds.

12. The heat transfer regimes inferred from the results presented (Figure 2) were used to determine the time span over which the various correlations could be reasonably compared with the test data. Since nonequilibrium thermal-hydraulic measurements could not be made in the Semiscale Mod-1 core and since the potential accuracy of the calculated heat transfer coefficient was uncertain following the quenching of the rod, comparisons of the test data with existing heat transfer coefficient correlations were limited to the film boiling, transition boiling, and the calculation of critical heat flux heat transfer regimes. The heat transfer coefficient correlations for film boiling and critical heat flux were developed from nonflow pool boiling conditions, whereas the Semiscale Mod-1 reflood tests were typically of convective flow conditions (Ref. 8, 9 and 10). However, since the reflood process in these tests and other tests involves relatively low liquid flow rates, a pool boiling description is considered valid (Ref. 7). A heat transfer coefficient correlation for the reflood process that was applicable for transition and film boiling was developed from PWR-FLECHT data by Hsu (Ref. 10). The combined heat transfer coefficient during reflood was calculated by adding a transition boiling heat transfer coefficient to a modified Bromley film boiling heat transfer coefficient.

13. Film and transition boiling correlations from the literature that have been used in computer codes to calculate reflood heat transfer were compared to results calculated from Semiscale Mod-1 reflood data. As indicated (Figure 3), at a system pressure of 20 psia the Berenson film boiling correlation appears to calculate the Semiscale Mod-1 film boiling regime more accurately than the modified Bromley film boiling correlation. The Hsu correlation which combined a transition boiling correlation developed from PWR-FLECHT data with a modified Bromley film boiling correlation, predicted the transition boiling heat transfer coefficients quite accurately for $\Delta T_{sat}$ values from about 450 to 150$^{o}$F, but not accurately at $\Delta T_{sat}$ values from about 450$^{o}$F to the minimum film boiling temperature at a $\Delta T_{sat}$ of about 550$^{o}$F. The higher values of $\Delta T_{sat}$ would correspond to lower values of local quality than the smaller values, which indicates that this correlation predicts the heat transfer coefficients more accurately at high quality than at low quality. However, since the majority (about two-thirds) of the time in transition boiling occurred at $\Delta T_{sat}$ values from about 550$^{o}$F to 450$^{o}$F, the Hsu correlation appears to generally calculate higher transition boiling heat transfer coefficients than were measured in the Semiscale Mod-1 core. The heat transfer coefficient in the film boiling regime was overcalculated by the Hsu correlation as indicated (Figure 3). These results for a system pressure of 20 psia were similar to the results obtained at a system pressure of 60 psia (Ref. 3).

14. The Semiscale Mod-1 data and results from the Zuber CHF correlation are compared (Figure 4) (Ref. 11). The CHF values calculated from the test results were significantly higher than the predictions of the Zuber correlation. The higher values of CHF calculated from the test data are partially attributed to the effect of axial conduction in the heater rod cladding on the calculation of heat flux using a one-dimensional inverse conduction solution. Axial conduction occurred because of the steep axial temperature profile near the quench front. The one-dimensional inverse conduction solution cannot include the effect of axial conduction and, therefore, includes the axial heat flux component as part of the calculated radial heat flux. The inclusion of the axial heat flux component in the calculated radial heat flux results in an overestimation of the heat flux during the latter part of transition boiling (near the CHF). Therefore, the difference between the CHF calculated from the Zuber correlation and the CHF calculated from the inverse conduction calculation from test data would be reduced if the axial component of the heat flux were not included in the inverse conduction calculation.

15. To evaluate the capability of a computer code which was developed primarily from PWR-FLECHT data to calculate the thermal-hydraulic response of the shorter Semiscale Mod-1 system, the experimental rod cladding temperatures and rod cladding temperatures calculated by the REFLUX computer code were compared (Ref. 2). The REFLUX computer code couples a core hydraulics model with a one-dimensional rod heat transfer model to calculate the temperature history of a rod undergoing the reflood process at a known positive inlet flow rate. On the basis of a physical description of the thermal-hydraulic interactions involved, criteria were developed to distinguish seven different heat transfer regimes in the REFLUX computer code. The heat transfer regimes identified in the REFLUX code were: forced convection to liquid, nucleate boiling, forced convection to annular flow, inverted annular film boiling, forced convection to vapor, and dispersed flow film boiling. The heat transfer coefficients for most regimes were calculated using standard heat transfer coefficients from the literature. The five parameters used to discriminate among the different heat transfer regimes were the critical vapor velocity, the calculated continuous liquid level, the system saturation temperature, the temperature at the critical heat flux, and the minimum temperature for film boiling. Axial heat conduction along the rod is not included in the rod model. Comparisons between the REFLUX calculation of rod cladding temperatures and PWR-FLECHT data indicated that the high flooding rate (2 to 6 in./sec) and high pressure (60 psia) tests were predicted well whereas the low pressure (20 psia) and low flooding rate (less than 1.5 in./sec) tests were not predicted well; in fact, considerable error was evident, particularly in the quench time (Ref. 2).

16. The rod cladding temperatures calculated by the REFLUX code and measured for a Semiscale Mod-1 test (Test S-03-A) at a system pressure of 20 psia and a flooding rate of 1.0 in./sec at three axial elevations (14, 29, and 39-in.) are shown (Figure 5). The predicted heat transfer regimes at the rod hot spot (29-in. elevation) are also shown on this figure. The calculated temperatures agree best with measured temperatures near the top of the core (39-in. elevation), with progressively worse agreement at the lower elevation.

17. The heat transfer regimes and heat transfer coefficients calculated by the REFLUX computer code were analyzed to investigate the reasons for the differences between the calculated and measured rod cladding temperatures. At the highest elevation investigated (39-in.), the predicted heat transfer regime oscillated between forced convection to vapor and dispersed flow film boiling several times during the first 50 seconds after reflood initiation. The switching between these heat transfer regimes was a result of changes in the vapor velocity which was near the threshold for liquid entrainment calculated from the Plummer criteria (Ref. 12). These oscillations of the heat transfer regime are not expected to have occurred during the test. During the periods when dispersed flow was the calculated heat transfer regime at this elevation, entrainment and carryover were calculated. During these periods the heat transfer coefficient was calculated by adding a wall to droplet heat transfer coefficient to the heat transfer coefficient calculated from the Dougall Roshenow correlation (Ref. 13). This calculation resulted in a significant increase in the heat transfer coefficient. The higher calculated heat transfer coefficients resulted in lower calculated rod cladding temperatures than were measured. From about 50 seconds after reflood initiation until the rods were calculated to begin to quench (at about 200 seconds), an inverted annular heat transfer regime was the dominant regime. The heat transfer coefficient for this regime which was calculated using a modified Bromley heat transfer coefficient and an experimental curve fit of PWR-FLECHT data, generally resulted in lower rod cladding temperatures than were measured.

18. At the rod hot spot (29-in. elevation), the initial difference between calculated and measured rod cladding temperatures occurred at about 12 seconds after reflood initiation when entrainment and carryover are calculated to occur as a result of exceeding the critical vapor velocity. Exceeding the critical vapor velocity results in a change in the calculated heat transfer regime from forced convection to vapor to dispersed flow, with a significant increase in the calculated heat transfer coefficient. From about 20 to 186 seconds after initiation of reflood, inverted annular film boiling was predicted at the 29-in. elevation. The calculated heat transfer coefficient during inverted annular film boiling was significantly higher than during

dispersed flow film boiling, which resulted in a further increase in the difference between the calculated and measured rod cladding temperatures. The calculation of an extended period of inverted annular film boiling and the associated high heat transfer coefficient during this period resulted in the calculation of a lower rod cladding temperature than was measured. The initiation of transition boiling was calculated at about 188 seconds after reflood initiation at the 29-in. elevation, which was about 8 seconds later than was calculated for the 39-in. elevation. The calculation of the initiation of transition boiling at a later time at the 29-in. elevation than at the 39-in. elevation resulted in a slightly later calculated rod quench at the 29-in. elevation than at the 39-in. elevation, whereas the measured rod quench time was about 50 seconds earlier at the 29-in. elevation.

19. Similarly to the 29-in. elevation, an inverted annular film boiling heat transfer regime was calculated at the 14-in. elevation for the majority (from 6 to 96 seconds after reflood initiation) of time until the rod quenched. The measured rod cladding temperatures were higher than the calculated temperatures for about the initial 50 seconds after the initiation of reflood. However at about 50 seconds after reflood initiation, lower rod cladding temperatures were measured than were calculated. Therefore, for the 14-in. elevation, the correlation for inverted annular film boiling calculated higher heat transfer coefficients than were measured early in the reflood process and lower heat transfer coefficients than were measured later in the reflood process, which resulted in a later calculated rod quench time than was measured. The calculation of the inverted annular film boiling heat transfer regime and the use of the applicable heat transfer correlations for this regime for an extended period of time at the 29-in. elevation and below resulted in the large difference between the measured and calculated rod cladding temperature.

20. In general, the REFLUX prediction of the Semiscale Mod-1 test at a pressure of 20 psia and a flooding rate of 1 in./sec did not exactly follow the trend of the data. The temperature rise and, therefore, the peak temperature were significantly underpredicted. The REFLUX code predicted later rod quench times and lower peak temperatures than were measured; for the PWR-FLECHT tests, the REFLUX code generally predicted higher peak temperatures and earlier rod quench times than were measured at 20 psia (Ref. 2). The calculations with REFLUX at 60 psia were somewhat better in predicting the bottom quenching behavior for the Semiscale Mod-1 core; however, the differences in the calculated and measured rod cladding temperatures were similar to the results shown (Figure 5) (Ref. 3).

## CONCLUSIONS

21. The thermal-hydraulic response of the Semiscale Mod-1 system during forced feed reflood tests was investigated to increase the understanding of the heat transfer phenomena during reflood heat transfer. The analysis of tests in the Semiscale Mod-1 system led to the following conclusions:

1. The models of forced feed reflood heat transfer that have the reflood heat transfer process proceed from forced convection to steam, to a dispersed flow heat transfer, to a film boiling regime, to transition boiling, and a final rod quench represent the phenomena occurring in the Semiscale Mod-1 system.

2. Dispersed flow and film boiling occurred for the bulk of the time at elevations near the center of the 5.5-ft Semiscale Mod-1 heated core. The transition boiling regime occurred for a total of about 3 seconds.

3. The Berenson film boiling correlation calculated the heat transfer coefficient in the Semiscale Mod-1 core quite accurately from the time the peak temperature was reached until the rods began to quench.

4. The Hsu transition boiling correlation predicted the heat transfer coefficient quite accurately at high quality but predicted higher values than were measured at lower qualities. When the combined Hsu correlation was used during film boiling, it predicted higher heat transfer coefficients than were measured.

5. The calculation of critical heat flux from the Zuber correlation was lower than the critical heat flux obtained from test data.

6. The REFLUX computer code did not calculate the peak rod cladding temperature or the trend of the progression of the quench front accurately. The differences between the test data and the calculated rod cladding temperatures generally resulted from an incorrect calculation of the time of entrainment. A more accurate representation of the heat transfer coefficient in the Semiscale Mod-1 core for inverted annular film boiling would also improve the agreement between REFLUX calculations and Semiscale Mod-1 data.

## REFERENCES


1. G. E. McCreery, "SUPERH – A Computer Model of Emergency Core Cooling Reflood Heat Transfer Assuming Constant Core Injection Rate," *Two-Phase Flow and Heat Transfer Symposium – Workshop, Proceedings of Condensed Paper, Univ. of Miami (October 1976) pp 109-111.*

2. W. L. Kirchner, "Reflood Heat Transfer in a Light Water Reactor," *PhD Thesis, MIT (January 1976).*

3. A. C. Peterson, G. G. Loomis, L. L. Chen, *Thermal and Hydraulic Response of the Semiscale Mod-1 Core During Forced Feed Reflood Tests,* TREE-NUREG-1001, (October 1976) pp 63-65.

4. G. Yadiganoglu et al, *Heat Transfer During the Reflood Phase of the LOCA-State of the Art,* EPRI 248-1 (September 1975).

5. American Nuclear Society, "Decay Energy Release Rates Following Shutdown of Uranium Fueled Thermal Reactors," *Proposed ANS Standard-5.1 (October 1971).*

6. J. V. Beck, "Nonlinear Estimation Applied to the Nonlinear Inverse Heat Conduction Problem," *International Journal of Heat and Mass Transfer, 13* (1970) pp 703-716.

7. F. F. Cadek et al, *PWR-FLECHT (Full Length Emergency Cooling Heat Transfer) Final Report,* WCAP-7665 (April 1971).

8. P. J. Berenson, "Film Boiling Heat Transfer from a Horizontal Surface," *Transactions of the American Society of Mechanical Engineers,* (August 1961) p 351.

9. L. A. Bromley, "Heat Transfer in Stable Film Boiling," *Chemical Engineering Progress, Vol. 46, No. 5* (1950) pp 221-226.

10. Y. Y. Hsu, "Tentative Correlations of Reflood Heat Transfer," *U. S. Nuclear Regulatory Commission Memorandum* (June 1975).

11. N. Zuber et al, "The Hydrodynamic Crisis in Pool Boiling of Saturated and Subcooled Liquids," *International Developments in Heat Transfer, Part II, No. 27 International Heat Transfer Conference, Boulder, Colorado (August 1961) pp 230-236.*

12. D. N. Plummer et al, *Post Critical Heat Transfer to Flowing Liquid in a Vertical Tube,* MIT Mechanical Engineering Report No. 72718-91 (September 1974).

13. R. S. Dougall and W. M. Roshenow, *Film Boiling on the Inside of Vertical Tubes with Upward Flow of the Fluid at Low Qualities,* MIT No. 9079-26 (1963).

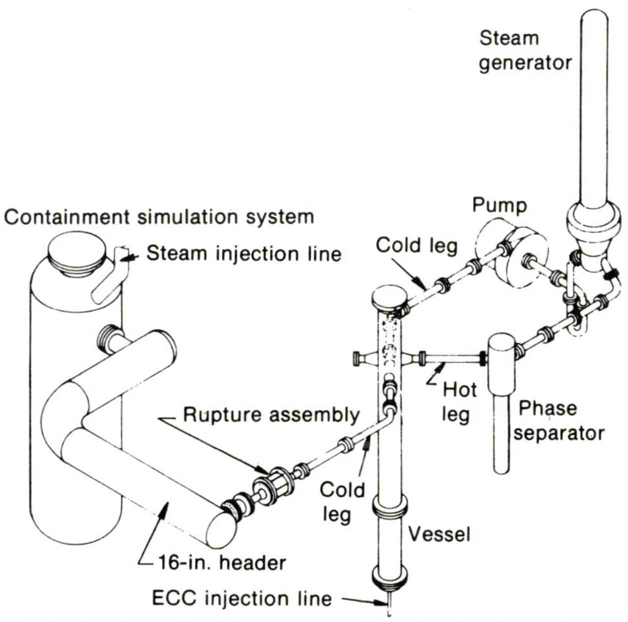


**Fig. 1: Semiscale Mod-1 system configuration for forced feed reflood heat transfer tests**

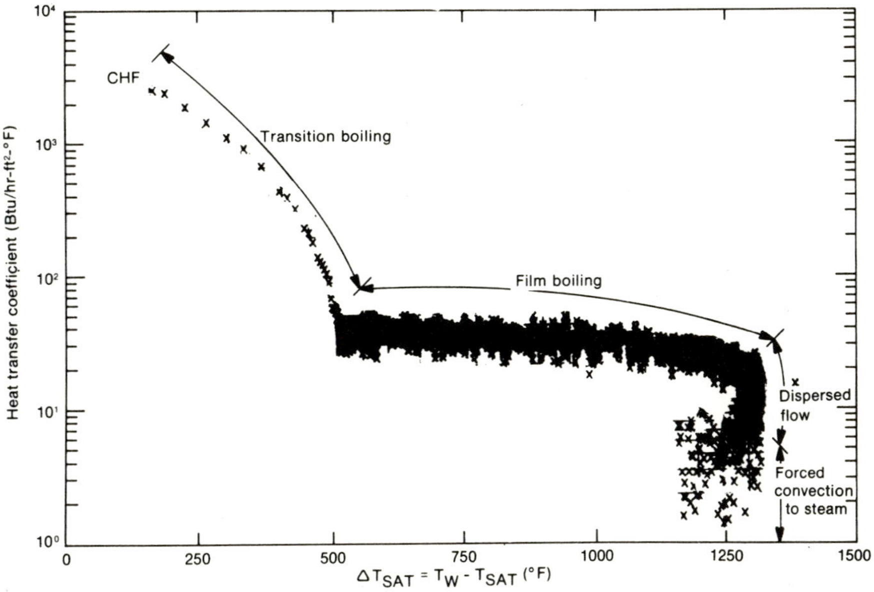


**Fig. 2: Typical plot of heat transfer coefficient versus the difference in the wall and saturation temperatures**

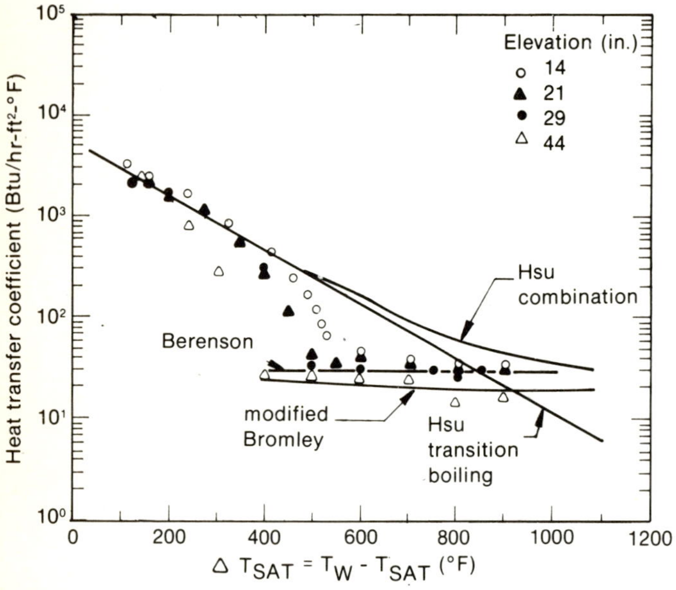


**Fig. 3: Comparison of film and transition boiling heat transfer coefficients at 20 psia — Hsu, modified Bromley, and Berenson correlations**

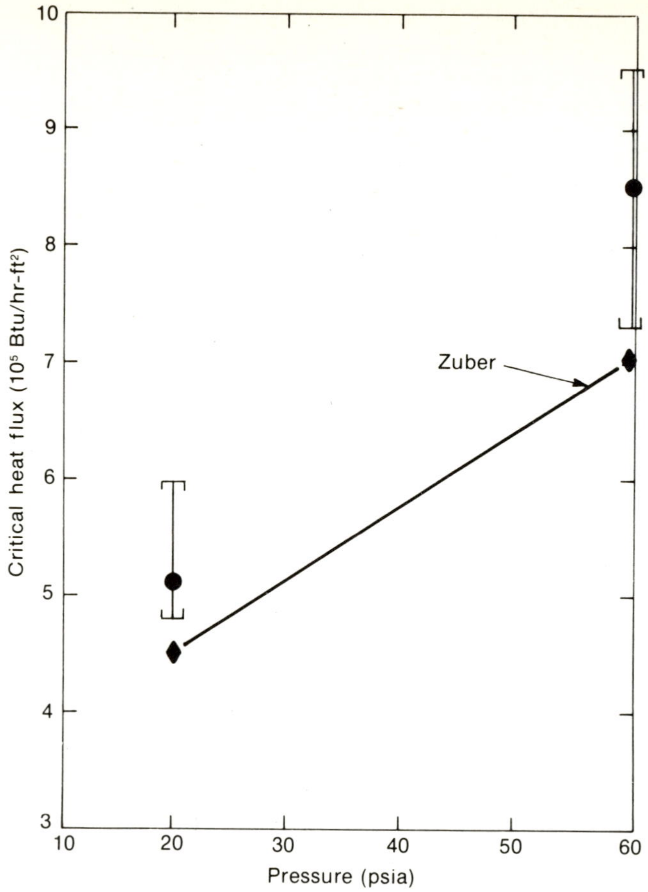


**Fig. 4: Comparisons of calculated and experimental critical heat fluxes**

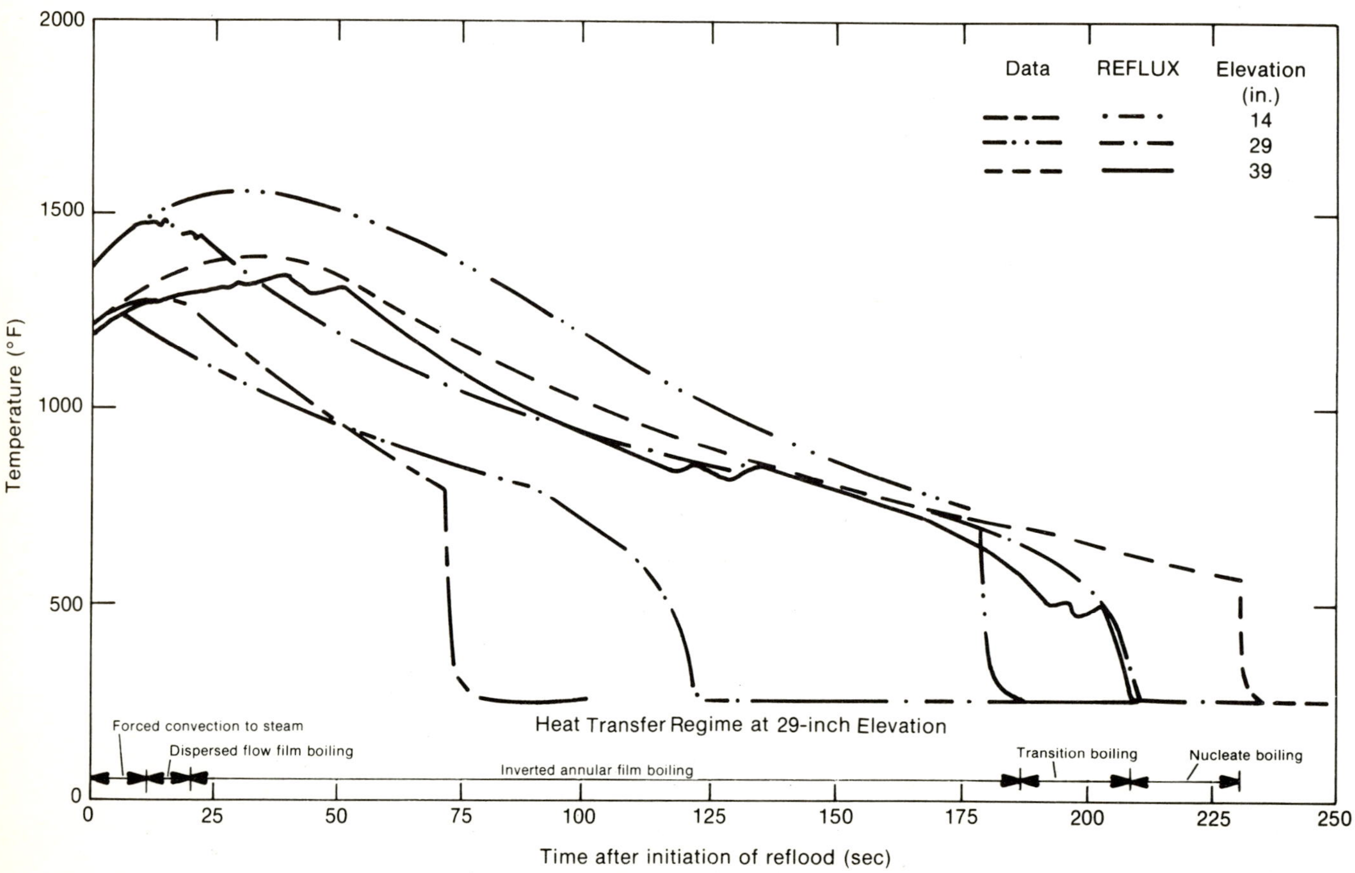


**Fig. 5: Comparison of measured and calculated rod cladding temperature response**

C215/77

# FLECHT-SET SYSTEMS: SYSTEMS EFFECTS ON REFLOODING BEHAVIOR

J. P. WARING and L. E. HOCHREITER
Westinghouse Electric Corporation, Pittsburg, Pa 15230, USA

SYNOPSIS The reflooding transient is currently the most limiting phase of a hypothetical loss of coolant calculation for a PWR. The FLECT-SET (Full Length Emergency Core Heat Transfer-Systems Effects Tests) experiments were performed to examine the influence of the system behavior on the reflooding heat transfer and flooding rate performance during a postulated loss of coolant accident for a PWR.

This paper describes the FLECHT-SET facility, the resulting data, and the data analysis which is discussed in detail particularly contrasting what was observed to occur in the tests against how the reflooding portion of the transient is modelled with present day safety analysis methods.

## INTRODUCTION

1. In present safety analyses, the reflood calculation is the most limiting factor for a PWR Loss Of Coolant Accident (LOCA). During reflood the potential for cooling the core is limited by the amount of downcomer head available to force the emergency core cooling flow throughout the primary system. The models and assumptions used to calculate how the downcomer driving head is utilized, and the resulting system feedback, significantly influence the core reflooding rate and peak clad temperature. The Full Length Emergency Cooling Heat Transfer Systems Effects Test (FLECHT-SET) are designed to provide experimental information on how the downcomer, core, and primary system interact during the reflood phase of a postulated LOCA. The data from these experiments are useful for assessing the system behavior during reflood, the amount of cooling received by the core, and the interaction among the system components which determines the core flooding rate. These data will be used to verify and improve PWR reflood calculational techniques.

## TEST DESCRIPTION

1. The FLECHT-SET program used two test loop facilities, a Phase A facility which was a one-loop simulation of a four-loop PWR, and the FLECHT-SET Phase B facility which was a two-loop simulation of a four loop PWR.

The FLECHT-SET Phase B test facility was designed such that its response simulated the reflood phase of a PWR during a hypothetical LOCA. Therefore, the rod bundle simulated the average fuel assembly in the core such that the steam generation rate and liquid entrainment would be representative of the average core behavior. Flow blockage was not simulated in these experiments. Simulated intact and broken loops with active steam generators were added to the existing FLECHT-SET Phase A(1) downcomer, test section and single loop design. The facility design, is described in detail in the FLECHT-SET Phase B System Design Description(2). The physical arrangement of facility components and instrumentation is shown in Figure 1.

One of the prime requirements for the facility design was to maintain typical PWR flow resistances, elevations, and lengths and to scale the flow areas by the ratio of a typical PWR core flow area to the flow area of the FLECHT 10 x 10 rod bundle. This scaling preserved the local steam velocity in the bundle and loops. While the hydraulic resistance of each component in the FLECHT-SET Phase B loops may differ from those of the PWR component, the overall resistance is adjusted to equal the total PWR resistance by a suitable choice of orifice plate characteristics in the simulated pump suction piping. In general, the scaling logic used would preserve the downcomer fill-up rate, exhaust velocities, transport times, and delays caused by mass storage and venting flow resistance such that they would be similar to those expected in a four-loop PWR during the reflood phase of a hypothetical LOCA.

As with any experiment, some scaling compromises did occur. The upper plenum in the FLECHT-SET tests was the original upper plenum used in the original FLECHT program and it did not preserve the proper flow area or height. Therefore, an extension pipe was added to obtain the proper elevation and the larger plenum volume was reduced by filling it with metal blocks. The flow area of the plenum was also treated as a test parameter and was varied by filling the plenum with additional metal blocks. The PWR lower plenum was simulated using the bundle lower plenum and a circular crossover pipe which provided the additional scale volume needed. The lower plenum turning loss and other resistances were simulated with an orifice plate.

Instrumentation of the 10 x 10 heater rod bundle is similar to that of previous FLECHT(3) bundles and contained 138 heater rod thermocouples. Important differences are the wall thermocouples on two of the simulated control rod thimbles, the eight thermocouple heater rods providing better definition of quench front elevation vs time. The rod bundle consisted of 91, 1.072 cm diameter stainless steel clad, boron nitride insulated, indirectly heated, heater rods and nine simulated control rod thimbles. The rods and thimbles were arranged in a square pitch of 1.430 cm and the rods had a 1.66 axial cosine power shape and were arranged in three radial power zones with 0.95, 1.0 and 1.11 relative powers. In addition to the heater rod instrumentation there was a substantial amount of loop instrumentation such that mass storage effects and loop flows rates could be measured as well as the steam generator performance. The void fraction distribution in the bundle was measured using differential pressure cells which were attached every 0.61 meters over the bundle length. The flooding rate into the bundle (inlet flow rate) was calculated from a mass balance on the downcomer.

The system was heated by steam and strip heaters on the piping to preset initial temperatures to simulate the conditions of a PWR at the beginning of reflood. Since the piping surface area to volume was not scaled, the pipe wall temperatures were adjusted such that they would be passive, i.e., neither a heat source or sink to the coolant. This is a good approximation since the PWR metal heat release downstream of the core is negligible compared to core and steam generator heat release. The steam generators, secondary side was filled and heated uniformly until they reached 268°C. When initial conditions of the facility met the specifications within the allowable tolerances power was applied to the heater rods. When the specified initial temperature was reached, injection of coolant was begun. Decay of the power input to the rods, began when the water reached the bottom of the heated length. The specified power decay was the ANS + 20% standard beginning at 30 seconds after the postulated break had occurred. As the injected coolant reached the heated portion of the rods, steam was generated by the energy released from the rods. The steam and entrained water exited the test section through the two loops. Each loop had an active steam generator where the bundle effluent was generally superheated, and an orifice plate in the simulated pump suction piping to provide a measurement of mass flow rate and to provide the correct flow resistance. The steam then flowed directly to the containment tank (for the broken loop) and via the top of the downcomer and overflow tanks for the simulated unbroken loop. The containment tank pressure was maintained constant by a pressure control valve located in the exit line.

## TEST RESULTS AND ANALYSIS

Several tests were conducted in the FLECHT-SET phase B facility to investigate the effects of different test bundle and loop parameters on the reflooding performance of the facility. Two tests were analyzed in detail while the other tests were analyzed in a parametric fashion to examine loop or bundle effects relative to the two reference tests.

Test 3105B was a reference test and was conducted with the following initial conditions:

- Containment Pressure = 0.407 MPa
- Initial Clad Temperature = 593.3°C
- Peak Power = 2.756 kW/m
- Coolant Temperature = 66.7°C
- Average Housing Temperature = 152.2°C
- Injection Flow Rate into the downcomer = 5.58 kg/sec for 14 sec., variable thereafter
- Steam Generator's Secondary Level (Cold) = 7.31 m (Hot $\cong$ 10.67 m)
- Steam Generator's Secondary Temperature = 255.7°C

In this test, the injection flow rate into the downcomer was varied by hand to prevent spillage of the downcomer fluid. Earlier tests indicated that downcomer spillage into the downcomer collection tanks and mixing with the intact loop steam flow would result in a non-typical condensation transient. If the nominal injection rate had been used, the downcomer would have overflowed for most of the test giving a maximum driving head of 4.88 m. Since the test had a driving head of about 0.76 m less, the resulting flooding rate into the bundle is lower than if the full downcomer head had been available.

As the injected water builds up a static head in the downcomer, the coolant flows into the heated bundle. The water penetrates approximately 38 cm into the bundle before steam generation begins. As the steam is formed, the ECC water continues to penetrate into the bundle at a decreasing rate and the pressure in the test vessel increases due to the steam generation. Since the steam flow, and entrainment is restricted from flowing to the containment tank by the loop pressure drop, the inlet flow into the bundle is stalled and reverses as the steam flow vents out through the system resistance. This initial injection water surge followed by the steam generation initiates the three-second manometer tube flow oscillation obtained in both FLECHT-SET Phase A and FLECHT-SET Phase B. The rod bundle was observed to oscillate in a one-dimensional fashion with little radial diffusion. Manometer tube flow oscillations have also been observed in semi-scale(4) gravity reflood tests and Kraftwork Union's PKL loop gravity reflood tests(5). As the water oscillates between the downcomer and the core, significant amounts of liquid are entrained and carried the full length of the rod bundle. The entrained water and pulsating steam flow increase the local rod heat transfer on every positive oscillation. These three-second flow oscillations have been modeled by White and Duffy(6) and independently by Yeh(7) to find out what mechanism was primarily responsible for the initiation and duration of the oscillations. It was found by Yeh that the oscillations could only occur if some disturbance was applied to the system. Examination of the FLECHT-SET data indicated that the most likely disturbance was the rapid increase in the steam generation rate and upper plenum pressure. The resulting pressure increase in the test section retards the inlet flow rate. The result is usually a large initial flooding rate which decreases as both the steam generation in the bundle increases and the driving head decreases. Both the upper plenum pressure build-up and the large increase in the steam generation rate was most likely due to the quenching of the lower portions of the heater rods which released large amounts of initial stored energy. These three-second oscillations eventually died out as the bundle filled and the steam flow out of the loops became were constant. The three- second oscillations died out after approximately 30 seconds in test 3105B. The effect of the oscillations on the heat transfer and bundle void fraction was pronounced as shown in Figure 2 and indicates that significant entrainment is being carried up the bundle. It was also found that lower pressure, lower subcooling, increased loop resistance and increased rod power would increase the magnitude and duration of the three second manometer type oscillations.

As the flooding process continues and the three-second manometer oscillations die out, longer period flooding rate variations occur which cause the flooding rate into the bundle to change with time. The effect of these longer term system variations can be seen in Figure 3 from 70 to 100 seconds which shows the bundle flooding rate history and the measured void fraction distribution.

The two-phase flow which leaves the bundle exit, separates to a large degree in the test section upper plenum, upper plenum extention, and the steam generator lower plenum. This phase separation decreases the mass flow through the simulated loops but increases the static pressure head downstream of the bundle quench front. The increase of static pressure drop above the quench front offsets partically the effective downcomer driving head (downcomer head - core elevation head) which forces the water into the bundle. Therefore, the bundle flooding rate decreases. As the flooding rate decreases, the saturation line ($\alpha = 0$) moves down the bundle toward the entrance and more steam is generated in the bundle between the saturation line and the quench front. This steam generation further retards the bundle flooding rate front and the quench movement. The increased steam flow out of the bundle entrains some of the liquid stored in the bundle, upper plenum, and steam generator lower plenums resulting in an increased loop pressure drop, but a reduced elevation head downstream of the quench front. With the increased loop pressure drop and the stalled flooding rate, the injected ECC water is stored in the downcomer, or if the downcomer is full it spills to the overflow tanks, until it has a sufficiently large head that it can force water into the bundle again. As the water, which has accumulated in the downcomer is forced into the bundle, it collapses the voids below the quenched front and the saturation line moves further up into the bundle increasing the bundle mass inventory. Collapsing the void fraction below the quench front reduces the steam generation and loop pressure drop, both of which increase the flooding rate into the bundle. This behavior is shown in Figure 3 from 70 to 125 seconds. In time, the downcomer driving head will diminish and the flooding rate will decrease and the cycle will begin again. This type of oscillation was observed only once in test 3105B; but lower system pressure, lower injection water, subcooling, and increased loop resistance tests indicated that this long term oscillation occurred several times before bundle quench and caused long term variations in the inlet flooding rate.

When comparing the measured heat transfer in the FLECHT-SET tests with the predicted heat transfer using the FLECT correlation(1,8), the measured heat transfer was always greater even when the variable flooding rate in the FLECHT-SET was accounted for. This indicates that the bundle heat transfer is quite sensitive to not only the instantaneous flooding rate but the flooding rate history of the bundle. In most cases the steam flow out of the bundle would be approximately constant while the flooding rate would vary indicating that the mass within the bundle was alternately being accumulated and boiled-off. The boiling- off period provides cooling of the rods by steam and entrained liquid despite the low or even negative flooding rate at these times.

This effect is not modeled accurately when, as is currently done, the rod heat transfer is considered to be directly related to inlet flooding rate. Proper two-phase modeling of the core which would calculate the local heat transfer as a function of local flow and quality, would account for this effect.

As mentioned earlier, several system parameters and bundle parameters were varied to see the effect on the bundle flooding rate, reflood heat transfer, and resulting temperature rise. The majority of the main parameters and their effect on the mid plane maximum temperature, turnaround time and quench time are summarized in Table 1. System pressure and subcooling were the two parameters which gave the largest change in the bundle and system thermal hydraulic performance.

When comparing the results of the FLECHT-SET tests and the current PWR safety reflood evaluation model predictions, several differences are noticed. The PWR evaluation models(9,10,11) do not predict either the three-second manometer tube oscillations or the longer term flooding rate variations described above. Normally these codes only predict a continually decreasing flooding-rate as the core is reflooded. The flow oscillations and the longer term flooding rate variations improve the heat transfer when compared to the FLECHT correlation which is used in the safety evaluation models. While it is not clear that a real PWR would oscillate in the same manner as FLECHT-SET, some type of flow oscillation would be expected. This data indicates that the proper modeling of this type of flow would yield higher calculated heat transfer coefficients than are currently used for safety analyses.

Significant two phase separation was found to occur in the FLECHT-SET test facility which reduced the total mass flow through the simulated pump, which is the largest restriction in the system, and through the heated steam generators, and hence decreased the steam binding effects. Current reflooding evaluation models do not account for phase separation and assume that all mass flow leaving the quench front is superheated in the steam generators and passes through the pump. These assumptions maximize the steam binding effects and are not consistent with the FLECHT-SET data and require further examination. It is felt that proper modeling of the de-entrainment, phase separation, and possible liquid fallback in the core will significantly improve the current PWR flooding rates and the calculated heat transfer.

It was also observed in the FLECHT-SET tests that the entrained liquid is not totally vaporized in the FLECHT-SET steam generators. Calculations using different drop sizes, and accounting for the drop motion within the steam generators also showed that only the smallest (0.254mm) drops would have significant evaporation as they passed through the steam generators(8). These calculation and observations are similar to those observed by Lee and Ryley(12), who studied single drop evaporation rates. Again the current reflood evaluation models assume total evaporation in the steam generators thereby maximizing the steam binding effect. It is felt that modeling of the non-equilibrium steam-droplet heat transfer will decrease the steam binding effects and thereby improve the calculated PWR core reflooding rates.

## CONCLUSION

The FLECHT-SET Phase B data indicate several thermal hydraulic phenomenan: variable flooding rate, phase separation, flow oscillations, and incomplete superheating, which are not presently accounted for in the current reflood evaluation models which act to minimize core heat transfer. Inclusion of these effects, through proper two phase modeling, should result in a significant decrease in the PWR steam binding effects and thereby improve greatly the calculated core reflooding rate and calculated heat transfer. The FLECHT-SET data should prove useful to these analyst's who wish to develop these improved reflood models as well as to those who wish to verify their present reflood model.

## REFERENCES


1. Blaisdell, J. A., Hochreiter, L. E., and Waring, J. P., "PWR FLECHT-SET Phase A Report", WCAP-8238, December 1973.

2. Cleary, W. F., et.al., "FLECHT-SET Phase B System Design Description", WCAP-8410, October 1974.

3. Cadek, F. F., Dominicis, D. P., Yeh, H. C., and Leyse, R. H., "PWR FLECHT Final Report Supplement", October 1972.

4. Crapo, H. S., Jensen, M. F., and Sockett, K. E., "Experimental Data Report for Semi-Scale Mod 1 Tests, S-03-5, S-03-6, S-03-7, and S-03-8 (Reflood Heat Transfer Tests)", ANCR-NUREG-1308, June 1976.

5. Riedle, K., Gaul, H. P., Ruthrof, K., and and Sarkar, J., "Reflood and Spray Cooling Heat Transfer in PWR and BWR Bundles", Presented at 1976 AIChE-ASME National Heat Transfer Conference, St. Louis, Mo.

6. White, E. P., and Duffy, R. B., "A Study of the Unsteady Flow and Heat Transfer in the Reflooding of Water Reactor Cores", Annals of Nuclear Energy, Vol. 3, pp. 197-210, (1976).

7. Yeh, H. C. and Hochreiter, L. E., "An Analysis of Oscillations in Simulated Reflood Experiments", ANS Transactions, Vol 22, 490, (1975)

8. Waring, J. P. and Hochreiter, L. E., "PWR FLECHT-SET Phase B1 Evaluation Report", WCAP-8533, August 1975.

9. Collier, G., Kelly, R. D., Spencer, A. C., and Waring, J. P., "Calculational Model for Core Reflooding After a Loss of Coolant Accident", WCAP-8171, June 1974.

10. Combustion Engineering, "COMPERC-II, A Program for Emergency Refill - Reflood of the Core", CENPD-134, Rev. 01, August 1974.

11. Bingham, B. E. and Shieh, K. C., "REFLOOD Description of Model for Multinode Core Reflood", BAW-10093, July 1974.

12. Lee, K., and Ryley, D. J., "The Evaporation of Water Droplets in Superheated Steam", Journal of Heat Transfer, Vol 90, pg. 445 (1968).

TABLE 1

FLECHT-SET PHASE B1 PARAMETER EFFECTS SUMMARY

| Parameter | Run No. | Variation | $T_{max}$ midplane | $\Delta T_{rise}$ | $t_{turn}$ midplane | RESULTS $t_{quench}$ midplane | Comments |
|---|---|---|---|---|---|---|---|
| Pressure | 2714B | 0.138 MPa | 776.1°C | 215.6°C | 121 sec | 324 sec | Effects are same as in Phase A. Higher pressure results in faster reflood due to higher steam density and larger coolant subcooling. Lower steam velocities at higher pressure result in greater separatic of entrained liquid above bundle. |
| | 3105B | 0.414 MPa | 593.9°C | 41.7°C | 27 sec | 101 sec | |
| Upper Plenum Flow Area | 2714B | 21:1 | 776.1°C | 215.6°C | 121 sec | 324 sec | Upper plenum volume and flow area reduced by using aluminum filler plates. Change has little effect on overall results, except at 0.414 MPa the entrained liquid reaches the hot leg and steam generator plenum faster. |
| | 2818B | 2.5:1 | 772.8°C | 204.4°C | 118 sec | 331 sec | |
| | 3105B | 21:1 | 593.9°C | 41.7°C | 27 sec | 101 sec | |
| | 3019B | 2.5:1 | 593.9°C | 37.8°C | 12 sec | 100 sec | |
| Steam Generator Secondary Water Level | 3105B | 9.45m | 540.6°C | 36.1°C | 27 sec | 76 sec | Rod power is slightly higher in 3105B causing larger temperature rise and turn-around time. Loop flow rates are 10% greater in 2011B due to lower steam temperatures. Large change in water level does not cause large changes of overall response. This is consistent with observation that only bottom 1.83 m of steam generator cools down. |
| | 2011B | 2.44 m | 583.9°C | 20.6°C | 11 sec | 82 sec | |
| Loop Resistance | 2714B | $K_{loop}$=31.40 | 776.1°C | 215.6°C | 121 sec | 324 sec | Loop resistances increased by 35%. As expected, flooding rate and loop flow rates decreased by ∿15%. Temperature rise, turnaround and quench time increase slightly. |
| | 3215B | $K_{loop}$=42.54 | 792.2°C | 227.2°C | 133 sec | 329 sec | |
| Test Section Inlet Resistance | 3105B | $K_{inlet}$=11 | 540.6°C | 36.1°C | 27 sec | 76 sec | Rod power higher in 3105B causing larger temperature rise and turnaround time. Average flooding rates are approximately equal but initial oscillations are decreased greatly. (Maximum velocities are reduced by factor of 3). |
| | 1607B | $K_{inlet}$=91 | 602.2°C | 31.7°C | 13 sec | 82 sec | |
| Initial Housing | 2714B | $T_{avg}$ 120.6°C | 776.1°C | 215.6°C | 121 sec | 324 sec | As in Phase A, housing heat releases occurs too fast due to low temperature of housing walls. Additional energy input results in slower reflood. Oscillations are larger and last longer if housing is heated above $T_{sat}$ (50 vs. 320 seconds at 0.138 MPa 20 vs. 40 seconds at 0.414 MPa) |
| | 2617B | $T_{avg}$ 233.9°C | 912.8°C | 321.1°C | 169 sec | 481 sec | |
| | 3105B | $T_{avg}$ 152.2°C | 593.9°C | 41.7°C | 27 sec | 101 sec | |
| | 1001B | $T_{avg}$ 205.6°C | 637.2°C | 65.0°C | 44 sec | 117 sec | |
| Coolant Subcooling Effects | 2714B | $T_{coolant}$ 67.2°C | 776.1°C | 215.6°C | 121 sec | 324 sec | Decreasing subcooling caused water to boil sooner and lower in the bundle resulting in more steam generation. This results in lower flooding rates, less mass stored in the bundle, and higher loop flow rates. |
| | 3316B | $T_{coolant}$ 110.0°C | 790.6°C | 229.4°C | 115 sec | 387 sec | |

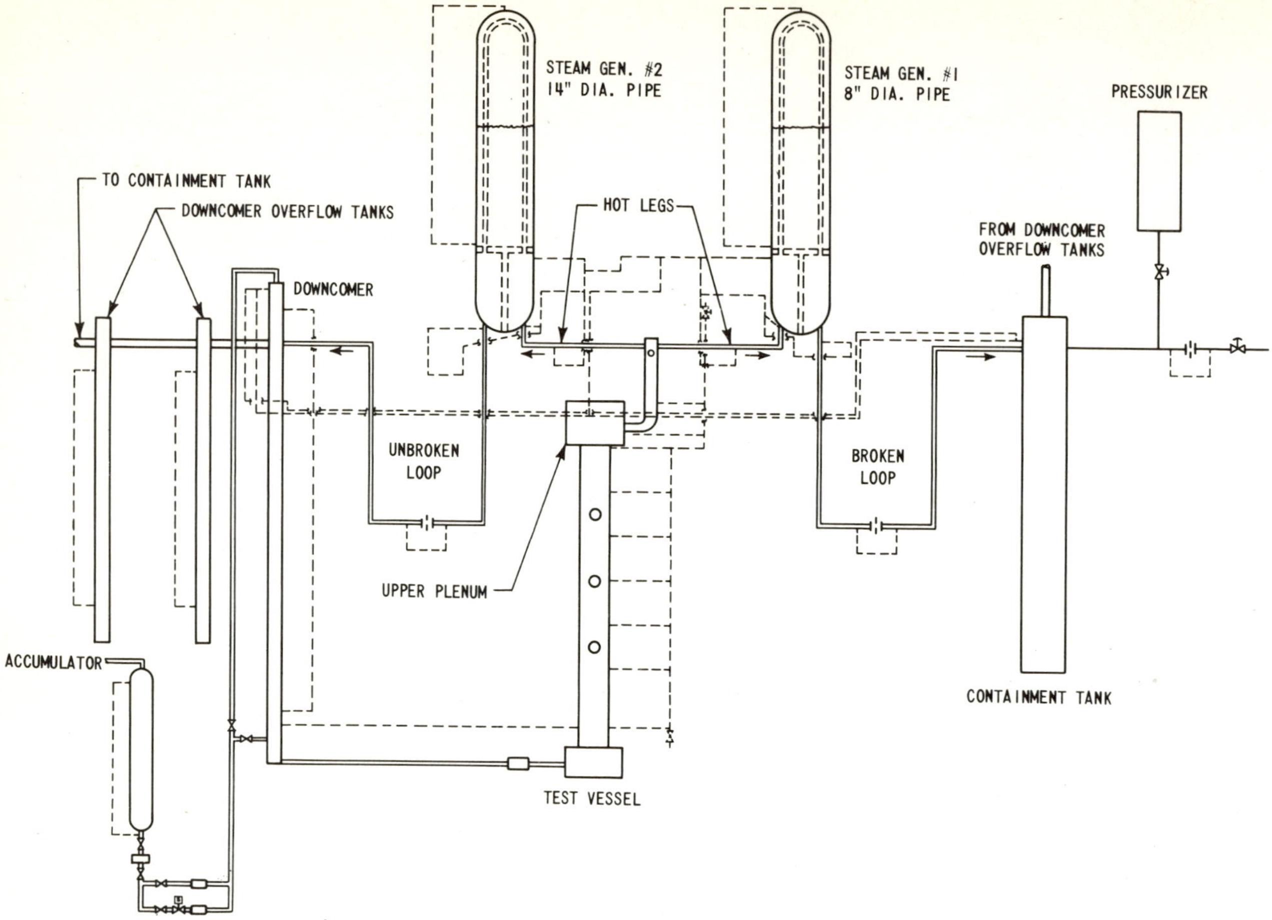


Fig. 1: FLECH–SET phase B loop schematic

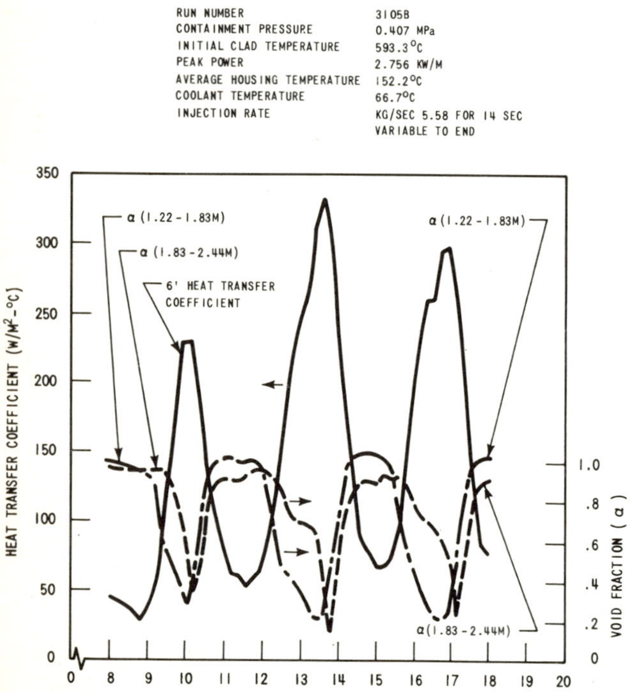


Fig. 2: Comparison of calculated void fractions and heat transfer coefficient for run 3105B

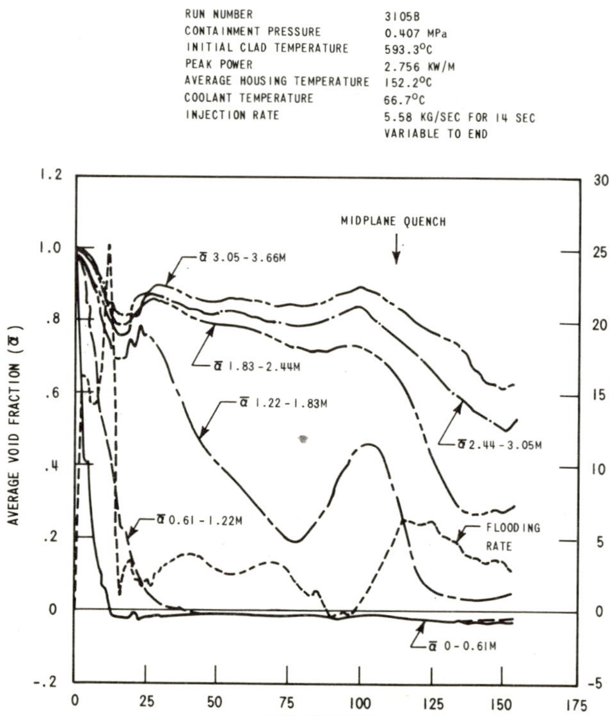


Fig. 3: Test section average void fractions and flooding rate for run 3105B

# C216/77

# PWR REFILL STUDIES

D. H. ROONEY, BSc, PhD, ARCST, CEng, MIMechE, MIED
H. C. SIMPSON, BSc, SM, ScD, FIMechE, FInstP, FRSE and
T. CAMPBELL, BSc
University of Strathclyde, Glasgow G1 1XJ, UK

SYNOPSIS An apparatus has been constructed, and tests carried out, to study the performance of the emergency core cooling system (ECCS) of a pressurised water reactor (PWR) during the refill stage of a loss of coolant accident (LOCA). In particular, the occurrences in the PWR downcomer and the effectiveness with which the coolant reached the lower plenum were studied.

The working fluids were steam and water, at pressures around atmospheric, with a range of subcooled liquid conditions at the coolant entry. The downcomer test sections were transparent, 1/10 full scale and the tests reported here were carried out in two phases. Phase 1 tests involved a developed downcomer annulus with tangential coolant entry whilst Phase 2 used a developed downcomer annulus but with normal coolant entry.

The test results indicate that liquid bridging (i.e. by-passing of the coolant from the inlet to the outlet pipe without reaching the lower plenum) occurred under certain critical conditions, the transition being well defined and easily measured. A number of factors affecting the critical conditions are discussed. The important effect of coolant entry conditions on the events leading up to, and evaluation of, the critical conditions for liquid bridging is also examined.

## INTRODUCTION

1. This work relates to the safety aspects of pressurised water reactors (PWR) in the event of a loss of coolant accident (LOCA) caused by a fractured cold leg entry pipe.

2. When such a LOCA occurs, the subsequent behaviour of the reactor system can be classified generally into three stages (i) the blowdown or depressurisation stage (ii) the refill stage, during which the emergency core coolant system (ECCS) comes into operation with a view to flooding the lower plenum (iii) the reflood stage, during which the emergency coolant level rises through the reactor core and quenches it. In this paper, the interest is confined to the refill stage and, in particular, to the effectiveness with which the emergency coolant (water) traverses the PWR downcomer annulus and reaches the lower plenum.

3. Previous work carried out in the USA (ref 1)* has indicated that under certain conditions of water downflow and steam (or gas) upflow, in situations similar to those of PWR downcomers, a "liquid bridging" or "by pass" effect occurs whereby the water (emergency coolant) entering the top of the downcomer, via an entry pipe, is bridged or by-passed to the fractured coolant pipe and does not reach the lower plenum. Defining the conditions under which the liquid bridging takes place is clearly desirable.

4. At the initiation of the project, data on liquid bridging had been presented by Wallis and co-workers (ref 1), the data being taken mainly from a 1/30, but also from a 1/15, scale model of a PWR downcomer annulus using air-water and steam-water mixtures. A correlation for predicting the critical conditions for liquid bridging was also presented by these authors and involved a relationship between a dimensionless parameter for each phase $j_s^*$ and $j_w^*$ (these effectively representing a balance between the inertial and hydrostatic forces) and a thermodynamic balance between the component phases. The characteristic defining the critical conditions for liquid bridging over a range of flowrates was divided into 3 regions. Region 1 was defined by a flooding type relationship, previously deduced for non-evaporating or non-condensing flows, using the annulus gap size as the characteristic dimension. Region 2 was defined by a line representing thermodynamic equilibrium between the component phases i.e. the flowrate of water with a particular subcooling which was just sufficient to condense all of the steam flowing. Region 3 was defined by a horizontal line, on a plot of $j_g^*$ versus $j_w^*$, emanating from the point of intersection of Region 2 and a line defined by the same relationship as for Region 1 except that the characteristic dimension was taken as the annulus width.

*This reference is quoted as being representative of a number of progress reports and items of literature emanating from the same sources, namely Dartmouth and Creare.

5. A satisfactory physical appreciation of the correlation was not easy, hence the present work was started in order to obtain a better understanding of the liquid bridging phenomenon and to substantiate the Wallis findings on a larger scale test section. A 1/10 scale simulation of a PWR downcomer was designed, for operation with steam-water mixtures, having transparent test sections to allow for the study of events leading to liquid bridging and with facilities for the further study of (i) the direction of the inlet coolant injection (ii) the effect of inlet subcooling of the coolant relative to the steam (iii) the effect of heated downcomer walls.

## DESCRIPTION OF TEST RIG

6. The test rig was designed to simulate the conditions in a PWR downcomer during the refill stage of a LOCA and used 1/10 scale transparent test sections manufactured from polycarbonate. The working fluids were steam and water and the use of polycarbonate necessitated operation at pressures close to atmospheric.

7. A line diagram of the test rig and instrumentation is shown in Fig.1. Steam was introduced into a large tank, representing the lower plenum of a PWR, the exit path of the steam being via the test section (representing the PWR downcomer annulus) and the outlet pipe (representing the fractured cold leg). Concurrent with the steam entry, water was supplied at twin entry points on either side of the test section (representing cold leg entries adjacent to the fractured cold leg) and either passed to the lower plenum or was forced through the outlet pipe according to the flow conditions pertaining. Any liquid reaching the lower plenum tank was measured in terms of an increase in water level. The steam, or steam/water, leaving the test section through the outlet pipe passed to a separator tank and condenser.

8. Measurements of pressure, temperature and flowrate were taken at the points in the flow path indicated in Fig. 1, with additional pressures and temperatures measured in the test section.

9. The steam flowrate entering the lower plenum was measured using an orifice plate in conjunction with a differential pressure transducer and the (matched) water flowrates entering the test section by turbine flowmeters. The water level variation in the lower plenum was measured by a float gauge in conjunction with a displacement transducer. Pressure measurements were made with transducers and temperature measurements with small silicon diodes. All of the measurements were transduced to give electrical outputs which were fed to a data acquisition system, thus providing for either automatic or manual collection of data. Cine films of the events occurring in the test section were also taken.

### Test Sections

10. The test sections used in the work presented here are also shown on Fig. 1 and had rectangular cross sections, representing a developed or unwrapped PWR annulus. In each case, the gap size was 25.4 mm (1") and the width 690 mm (27"), these representing a 1/10 scale development of a PWR with the width corresponding to twice the circumferential pitch of the cold leg entry pipes.

11. One test section had side, or tangential, water entry ducts of rectangular cross section and was used in Phase 1 tests, the other had normal water entry through half diameter pipes and was used in Phase 2 tests. Heaters were fitted to the steel back plate of each test section to simulate the heating effect of the reactor core.

### Test Section Dimensions

12. Gap size 25.4 mm; test section width 690 mm; outlet pipe diameter 76.2 mm.
Phase 1 tests: Inlet pipes rectangular 25.4 mm broad x 73 mm high.
Phase 2 tests: Inlet pipes half circular, radius 38.1 mm
Total height of test section 848 mm
Height from bottom of test section to outlet pipe ℄ 650 mm

### Range of Conditions

13. Water (coolant) flow rates: up to 0.0135 $m^3/s$ (180 gal/m)
(on the basis of Froude number similarity, this is equivalent to 4.27 $m^3/s$ or 57,000 gal/min under reactor conditions)
Steam flow rates: up to 0.63 kg/s (5000 lb/h) $\approx$1 $m^3/s$ (35 $ft^3/s$)
Steam supply conditions: around atmospheric, normally 1.1 bar, 110°C (16 psi a, 230°F)
Inlet subcooling (i.e. steam temperature-water temperature): 30 K (54F°) to 90 K (162F°)
Heated downcomer wall temperatures: up to 150°C (302°F)

### Accuracy of Measurements

14. Allowing for all known factors, the estimated accuracies of the various measurements were:

Steam supply flow rates - to nearest $3 \times 10^{-3}$ $m^3/s$
Water supply flow rates - to nearest $1.7 \times 10^{-5}$ $m^3/s$
Pressure difference (lower plenum-outlet pipe) - to nearest 0.2 $kN/m^2$
Temperatures - to nearest °C.

### Test Procedure

15. For each particular setting of inlet subcooling and heated downcomer wall temperature, the form of tests involved -

(i) "Water first" tests - i.e. a steady water flow was introduced into the test section, evenly distributed between the two entry pipes, and the steam flow rate increased, in steps, until liquid bridging occurred. The procedure was repeated for several different water flow rates.

(ii) "Steam first" tests - i.e. a steady steam flow was introduced into the test section, via the lower plenum, and the water flow rate increased, in steps, until liquid bridging ceased.

## EXPERIMENTAL RESULTS

16. The events leading up to the critical conditions for liquid bridging can be described briefly as follows:

(i) In Phase 1 with water first tests, the water initially flowed down the test section like two waterfalls, filling the gap but rarely filling the test section width. At low steam flowrates, the steam travelled up the centre of the test section in a pulsating manner, causing the waterfalls to flap sideways in doing so. Increasing the steam flowrate increased the pulsating effects until a rate was reached where water hold-up occurred i.e. the water was carried out through the outlet pipe with the upflowing steam and only an insignificant amount reached the lower plenum. This was accompanied by severe swirling in the test section around and above the outlet pipe although the liquid bridge (or jet) could be seen to extend part way across the test section from inlet to outlet pipe.

(ii) In Phase 1 with steam first tests, the established initial steam flowrate caused the water at low flowrates to be passed through the outlet pipe without reaching the lower plenum, the conditions in the test section being similar to those described above. Increasing the water flowrate eventually produced a sharp transition whence all of the inlet water did reach the lower plenum.

(iii) The main difference during the Phase 2 tests, with the water entering the test section at right angles to the plane of the section, was that the water impinged on the back plate of the test section and flowed down into the lower plenum as a liquid film on the back plate, the film seldom filling the gap. Particularly at high water flowrates, the liquid "jet" produced a wide film of liquid across the test section.

17. Typical characteristics obtained from "water first" and "steam first" tests are shown in Fig. 2. These indicate that the critical conditions for liquid bridging are sharp and clearly defined. It was also noticed that when a particular water flowrate was established, the steam flowrate necessary to produce liquid bridging was greater than that required when the steam flowrate was established first. Hence, for the remainder of this paper, the emphasis will be placed on the steam first test results since these (i) are more representative of the reactor situation during a LOCA (ii) provide the more conservative conditions to be catered for in the safety case.

18. Data plots of the loci of the critical conditions for bridging are shown in Fig. 3, these referring to Phase 1 and Phase 2 tests and typifying the effects of inlet subcooling and heated downcomer wall temperature. The area defined as "safe region" coincides with conditions where the inlet coolant reaches the lower plenum.

19. Several effects are apparent from Fig. 3, not the least significant being the effect of coolant (water) entry conditions. A comparison between the Phase 1 and Phase 2 $Q_s$ versus $Q_w$ characteristics, for any particular inlet subcooling and downcomer wall temperature, reveals (i) the Phase 1 characteristics quickly become linear whereas the Phase 2 are decidely curved (ii) the Phase 1 characteristics are higher than the Phase 2. The different shape of characteristic is important since it appears to indicate different mechanisms for effecting liquid bridging with the different inlet conditions. This was borne out, to some extent, in the experimental observations described previously e.g. in Phase 1 tests the gap was filled whereas in Phase 2 tests a film flow down the back plate was more common. The different magnitudes of the characteristics suggests that tangential coolant supply would be more effective than the usual normal type entry.

20. Fig. 3 also indicates that a lower water supply flowrate is necessary to penetrate to the lower plenum at high inlet subcoolings and with low downcomer wall temperatures. This is conducive to condensation playing some part in the mechanism leading to liquid bridging since high inlet subcooling and low wall temperature represent a greater capacity for condensing the inlet steam. During these tests, the effect per degree change in temperature indicated that inlet subcooling was the major effect but this will depend to some extent on the thermal capacity of the downcomer wall and the energy source supplying it.

21. A comparison with the data of Wallis and co-workers (ref. 1) is shown in Fig. 4 using the Wallis dimensionless parameters $j_s^*$ and $j_w^*$ defined by

$$j_s = j_s \rho_s^{\frac{1}{2}} \left[ g D_H (\rho_w - \rho_s) \right]^{-\frac{1}{2}}$$

and $$j_w = j_w \rho_w^{\frac{1}{2}} \left[ g D_H (\rho_w - \rho_s) \right]^{-\frac{1}{2}}$$

where $j_s = Q_s/A$ and $j_w = Q_w/A$ and the characteristic dimension is taken as the hydraulic diameter $D_H = 4A/P$.

22. The comparison relates to conditions of $\Delta T$ subcooling of 90K and wall temperature of 20°C, this being the closest set of matching conditions between the two experiments. The "Wallis" data were taken from Fig. 18 in reference 1. As can be seen, the data from the Phase 2 tests agree remarkably well with the Wallis data considering the respective scales of the apparatus used were 1/10 and 1/30. The type of coolant entry condition was virtually

matched also. The Phase 1 test data do not agree either qualitatively or quantitatively with the Wallis data, except at low water flowrates, and this difference emphasises the effect of liquid entry conditions on bridging.

23. Fig. 5 shows a typical pressure difference characteristic spanning the critical conditions for bridging, the pressure difference being measured between the lower plenum and the outlet pipe. As can be seen, there is a considerable increase in the pressure difference required to maintain bridging relative to the condition where downcomer penetration occurs. A significant proportion of this increase in pressure difference will be due to the momentum flux, or kinetic head, of the two phase mixture in the outlet pipe after bridging has been established. Other contributions will come from friction and other losses in the test section, entry losses to the test section and outlet pipe, and gravity effects due to the "liquid bridge" formation.

## DISCUSSION

24. Whilst this paper is confined to a presentation and discussion of experimental results, it is the intention that the information given here will form the basis of a theoretical model, defining the bridging process, which will be presented at a later date.

25. Returning to the comparison of the Phase 1 and Phase 2 data, typified in Fig. 3, it has already been stated that the most striking difference is the shape of the $Q_S$ versus $Q_W$ characteristics, the Phase 2 being curved whereas the Phase 1 appear linear down to low liquid flowrates. This raises the question as to whether or not the characteristics should pass through the origin since both sets of data indicate the possibility of an intercept on the $Q_S$ axis. Such an intercept could be attributed to (i) condensation in the lower plenum (ii) steam leakage or (iii) a change in events in the test section at low liquid flowrates. Heat loss calculations could not support the (sometimes) high condensation rates suggested by the intercepts, no steam leakage was apparent during the tests, and hence item (iii) would seem most likely. At this stage of experimental data presentation, this point was considered not too important, since the region of main interest lies in the higher flowrates where these effects would be small. Hence the experimental curves were not extrapolated beyond the lowest flowrate data point. However, in subsequent theoretical modelling, this point would require to be resolved.

26. It also appears that different models will be required for the different liquid entry conditions. The Phase 1 model must accommodate the raising of an initial waterfall type liquid flow, with any consequential acceleration or choking effects (but where the water fills the annulus gap), whereas the Phase 2 will describe the steam flow over a liquid film, perhaps similar to a flooding type effect.

27. As previously stated, the agreement between the Strathclyde Phase 2 data and the data from ref. 1, shown in Fig. 4, is very good considering the Strathclyde rig was 3 times larger than the other. The agreement suggests that the co-ordinates used in Fig. 4 are reasonable for presenting data on critical conditions and that, perhaps, families of such curves could be used to define the corresponding reactor conditions. However, in the absence of a satisfactory theoretical model to permit true scaling to be made, it must be pointed out that there is a big step between the scale factor of 3, shown in Fig. 4, and the further factor of 10 necessary to extrapolate to reactor conditions. It should also be stated that the agreement shown in Fig. 4 does not necessarily subscribe to agreement with the essentially empirical correlation defined in the introduction.

## CONCLUSIONS

28. On the basis of the experimental data obtained from Phase 1 and Phase 2 tests, the following conclusions can be drawn.

(i) Under particular conditions of steam and water flowrates, the water injected into the downcomer did not reach the lower plenum. Instead, a liquid bridging, or bypass, action occurred between the liquid entry point and the outlet.

(ii) The critical conditions at which bridging took place were sharp and well defined and indicated either total or virtually no penetration of the inlet water to the lower plenum.

(iii) Condensation of the upflowing steam occurred and should be included in any meaningful theoretical model.

(iv) For any particular test condition, the characteristics describing the critical conditions for steam first and water first tests were not coincident, with steam first tests providing the more conservative conditions to be catered for.

(v) Decreasing the inlet subcooling or increasing the downcomer wall temperature required a higher water supply flowrate to ensure penetration to the lower plenum.

(vi) The different liquid entry conditions represented by the Phase 1 and Phase 2 tests produced different characteristics for the critical conditions and suggested different bridging mechanisms for the two test series. This was confirmed to some extent by visual and photographic measurements.

(vii) Very good qualitative and quantitative agreement between the Phase 2 data and the results of Wallis et al, for a similar set of conditions, was obtained, despite the scale factor of 3 between the experiments.

(viii) The production of a theoretical model (or models) to describe the bridging mechanisms is desirable to allow confident extra polation of the results to the reactor situation.

ACKNOWLEDGEMENTS

This work was carried out for the Nuclear Installations Inspectorate of the UK Government and the authors are grateful for their financial and other support and for permission to publish these data. Support is continuing for a Phase 3 test series involving a 1/10 scale model of a complete PWR downcomer and theoretical studies.

REFERENCES

1. CROWLEY, C J, WALLIS, G B, LUDWIG, D L, Steam/Water Interaction in a Scaled Pressurised Water Reactor Downcomer Annulus. Report No. COO-2294-4. Dartmouth College, Thayer School of Engineering, New Hampshire, USA, September 1974.

NOMENCLATURE

| | |
|---|---|
| A | Cross sectional area of downcomer test section |
| $D_H$ | Hydraulic diameter or characteristic dimension |
| g | Gravitational acceleration |
| $j_s^*$ | Dimensionless steam flux defined in Section 3 |
| $j_w^*$ | Dimensionless water flux defined in Section 3 |
| $M_s$ | Mass flowrate of steam entering test section |
| $M_w$ | Mass flowrate of water entering test section |
| p | Pressure |
| $\Delta p$ | Pressure difference between lower plenum and outlet pipe |
| P | Perimeter of test section |
| $Q_{LP}$ | Volume flowrate of water reaching lower plenum |
| $Q_s$ | Volume flowrate of steam entering test section |
| $Q_w$ | Volume flowrate of water entering test section |
| $T_s$ | Steam temperature |
| $T_w$ | Water temperature |
| $T_{wall}$ | Downcomer wall temperature |
| $\Delta T_{SUB}$ | Inlet subcooling $= T_s - T_w$ |
| $\rho_s$ | Steam density |
| $\rho_w$ | Water density |

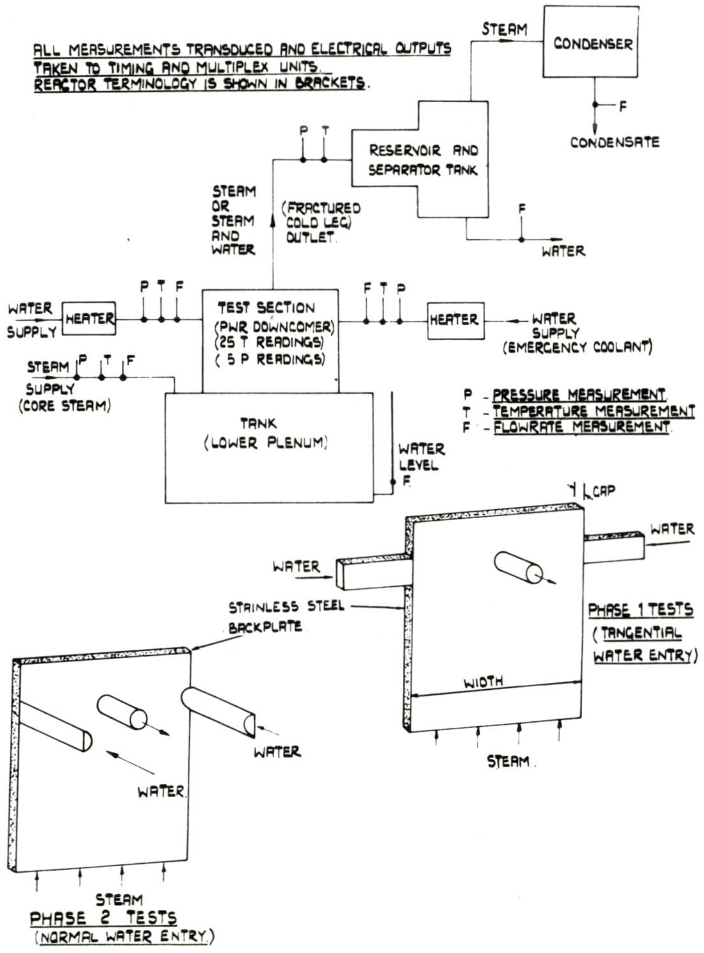


**Fig. 1: Line diagram of refill test rig showing instrumentation and test sections**

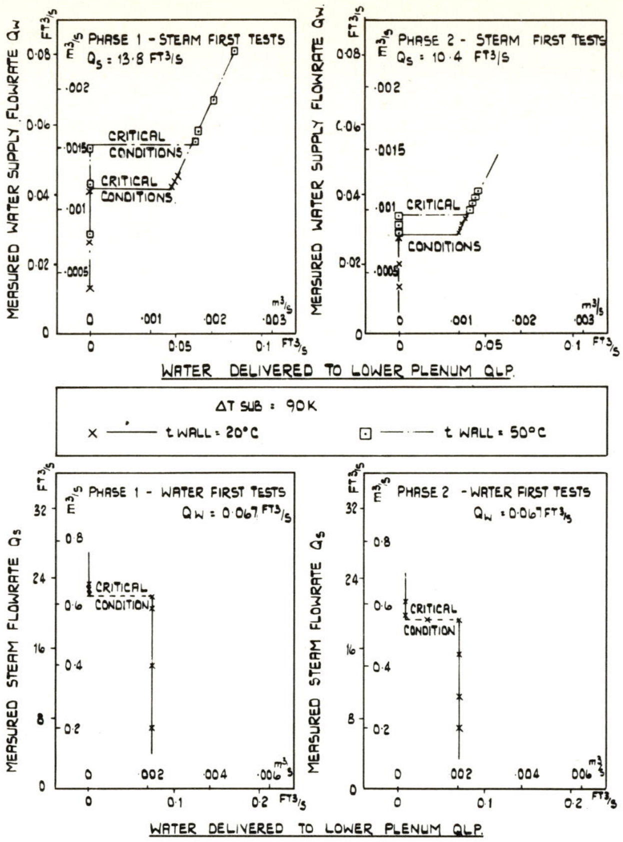


**Fig. 2: Typical water penetration characteristics**

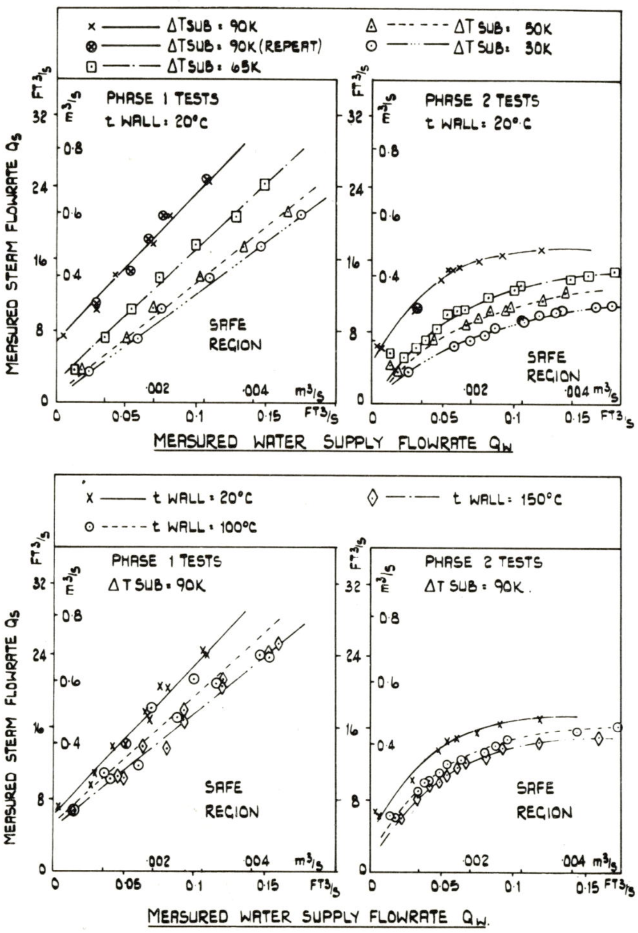


**Fig. 3: Critical conditions for liquid bridging (steam first)**

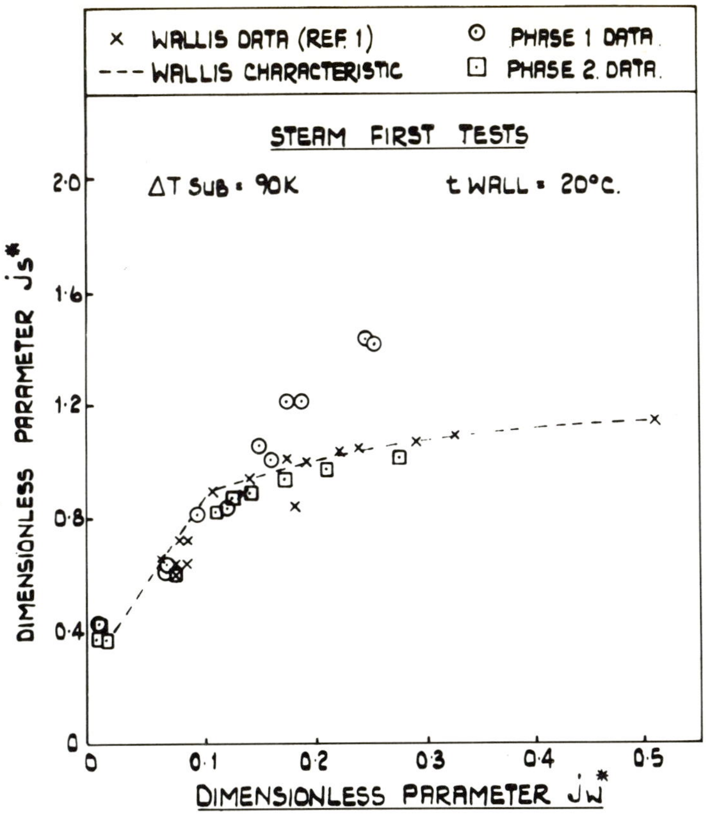


Fig. 4: Comparison with data of Wallis et al

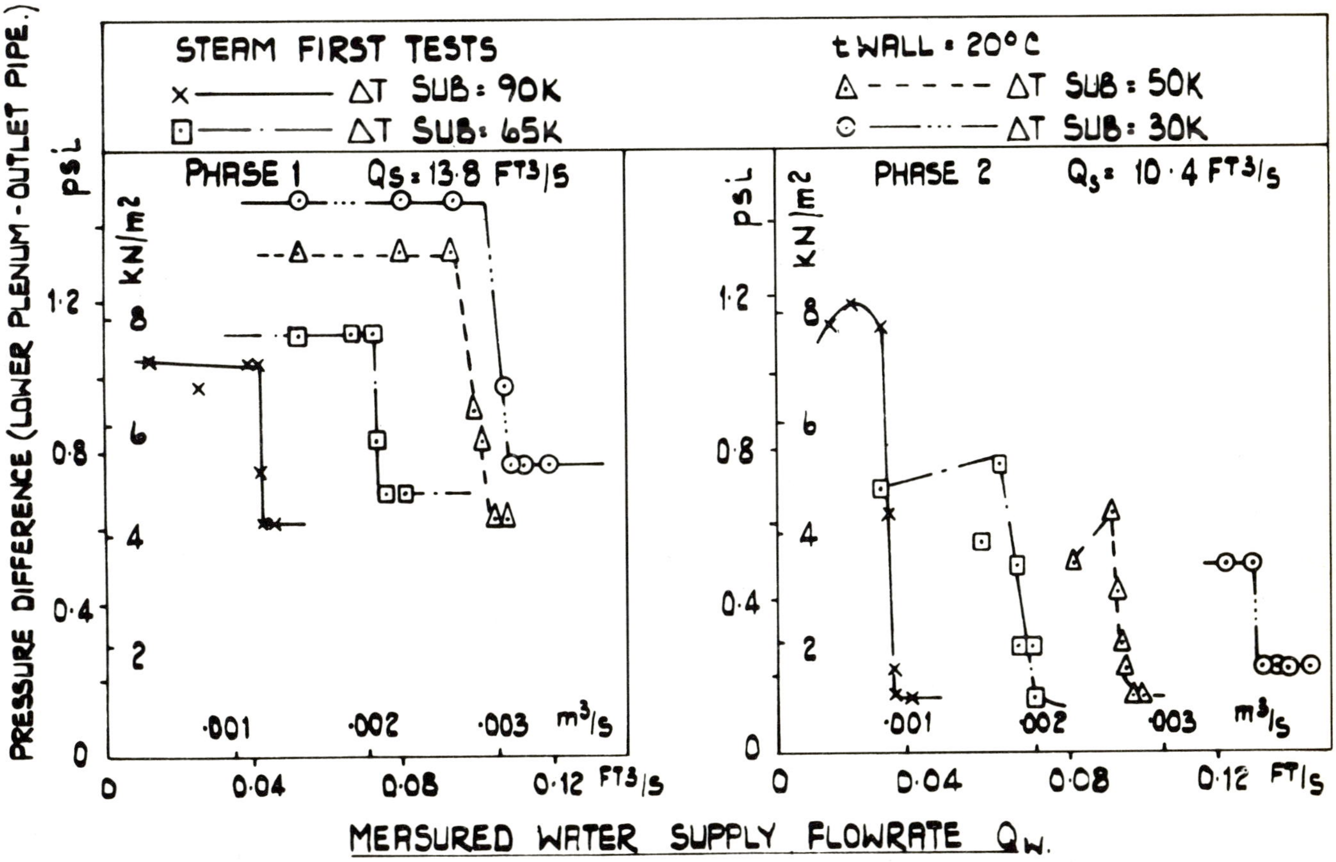


Fig. 5: Typical pressure difference characteristics during liquid bridging

C217/77

# A NEW MIXED FLOW CLUSTER DRYOUT CORRELATION FOR PRESSURES IN THE RANGE 0.6–15.5 MN/m² (90–2250 psia) – FOR USE IN A TRANSIENT BLOWDOWN CODE

R. W. BOWRING, MA
UKAEA, AEE Winfrith, Dorset

SYNOPSIS A mixed-flow dryout correlation has been developed from steady-state experimental cluster data covering the following ranges: geometry - representative of Pressure Tube Reactors, BWRs and PWRs; pressure 0.6-15.5 $MN/m^2$; mass velocity 50-400 $kg/m^2s$; the low flow and low pressure conditions are particularly relevant in Blowdown applications. The correlation contains terms to account for axial and radial non-uniform heat flux and a notional correction for steam at channel inlet. The RMS error of the new correlation is 12% over the whole pressure range compared to 33% for the correlation originally programmed in RELAP-3; at 7 $MN/m^2$, the RMS error is 11% compared to 18% and 21% for the Macbeth and Barnett correlations The correlation has been adopted in RELAP-UK.

NOTATION

A, B, C Terms in correlation

$D_h$, $D_w$ Heated and wetted equivalent diameter of channel

$f_g$ Effective fraction of supplemental flow affecting dryout

$F_p$ Radial heat flux peaking factor = $\phi_p/\phi$

$F_1$, $F_2$ Pressure terms in correlation

G Mass velocity

$G_s$ Contribution to mass velocity due to spray supplement entering between inlet and point considered

G* Used instead of G in correlation when there is supplemental flow; based on mass velocity at channel inlet ($G_c$) and $G_s$ - see para 13.

$\Delta H_i$ Enthalpy subcooling of water at channel inlet

$\Delta H_i$* Used instead of $\Delta H_i$ in correlation when there is supplemental flow; based on subcooling of flow at channel inlet ($\Delta H_{i,c}$) and subcooling of spray supplement ($\Delta H_{i,s}$) - see para 13.

p System pressure

X Local steam quality ($0 \leq X \leq 1.0$)

$X_i$ Steam quality at channel inlet = $-\Delta H_i/\lambda$

Y Axial heat flux factor - see para 6.

Z Distance of dryout position from channel inlet

$\phi$ Radial - average of heat flux on rods at given axial position, Z

$\phi_{do}$ Radial - average dryout heat flux at axial position, Z

$\phi_p$ Maximum heat flux on rods at given axial position, Z

$\lambda$ Latent heat of evaporation

INTRODUCTION

1. A new dryout correlation was required for the transient blowdown code RELAP-UK (Ref 1) since experience had shown the correlation programmed in RELAP-3 (Ref 2) to be inaccurate when applied to Steam Generating Heavy Water Reactor (SGHWR) and other Pressure Tube Reactor (PTR) geometries.

2. SGHWR emergency cooling during blow-down is effected by spraying water into the fuel cluster from small holes at discrete intervals up a central sparge pipe; this supplemental flow could be on continuously during normal reactor operation. Experimental dryout results had recently been obtained in an out-of-pile rig for conditions of continuous supplemental flow, from 0 - 100% of the total flow in a cluster simulating a SGHWR fuel element (Ref 3). The parameter ranges in these experiments had been extended down to low pressure (0.6 $MN/m^2$) and low mass velocity (50 $Kg/m^2s$). It was required to be able to represent these results in RELAP-UK calculations.

3. Furthermore, analysis of possible Loss-of-Coolant Accidents (LOCA) for various reactors had indicated that flashing might occur during rapid depressurisation, giving steam at channel inlet. It was known from earlier work (Refs 4, 5) that when there was quality at inlet, the dryout heat flux could not be extrapolated from, and thus predicted by, correlations developed from data obtained with subcooled water at inlet. It was required to represent any possible deleterious effect on dryout, even if only in a notional way, prior to a more rigorous experimental investigation of the problem.

4. There was thus a need for a new dryout correlation for RELAP-UK which could be applied accurately to SGHWR but which would also be a general purpose correlation for use with Boiling Water Reactor (BWR) and Pressurised Water Reactor (PWR) geometries. It would be complementary to the simpler correlation developed (Ref 6) for the SGHWR Prototype Reference Design.

5. In this paper, the new correlation is given first followed by brief descriptions of its development, its accuracy and finally the method of using it in RELAP-UK.

## CORRELATION EQUATIONS

### The heat balance axial heat flux parameter Y

6. The parameter Y has not been used in previous published correlations hence its function is described in some detail. It is a measure of the non-uniformity of the axial heat flux profile and also provides a means of converting from the inlet subcooling ($\Delta H_i$) to the local quality (X) form of the correlation via the heat balance equation.

7. It is defined as:

$$Y = \frac{\text{average flux from entry to Z}}{\text{local radially averaged flux at Z}}$$

$$= \left[\int_o^Z \phi dZ\right] \Big/ (\phi Z) \qquad (1)$$

8. The heat balance equation may thus be written as

$$\Delta H_i + \lambda X = 4\phi YZ/(GD_h) \qquad (2)$$

For a cluster with uniform axial heat flux, Y = 1 at all Z. For non-uniform heat flux, Y varies along the length; for example, for a chopped-cosine profile, Y < 1 over the first part of the channel, = 1 at about 2/3 of the length and > 1 near the exit.

### Units

9. For convenience of programming in RELAP-UK, the correlation was developed with the parameters specified in British Units. Hence when using the correlation, the units should be as follows:

| | | | |
|---|---|---|---|
| $D_h$, $D_w$, Z | : inches | (1 m | = 1/0.0254 in) |
| G | : $Mlb/ft^2hr$ | (1 $kg/m^2s$ | = 0.0007373 $Mlb/ft^2hr$) |
| $\Delta H_i$, $\lambda$ | : Btu/lb | (1 kj/kg | = 0 4299 Btu/lb) |
| p | : psia | (1 $MN/m^2$ | = 145.07 psia) |
| $\phi$ | : $MBtu/ft^2hr$ | (1 $kW/m^2$ | = 0.0317 $MBtu/ft^2hr$) |

### Correlation for Pressure Tube Cluster

10. Pressure = 90 - 1250* psia (0.6 - 8.6 $MN/m^2$)
No Spray Cooling Supplement
Subcooled at Inlet

$$\phi_{do} = \frac{A + B\Delta H_i}{C + ZY} \qquad (3a)$$

$$= \frac{A - B\lambda X}{C + ZY[1 - 4B/(GD_h)]} \qquad (3b)$$

where

$$p_T = 0.001p \qquad (4)$$

$$F_1 = \left[1.0 - 0.04p_T\,(1 + 0.47p_T^2)^{\frac{1}{2}}\right]^2 \qquad (5)$$

$$F_2 = \begin{cases} 0.45 + 1.25p_T & : p \leq 415 \\ 0.424 + 1.959p_T - 1.556p_T^2 & : 415 < p \leq 650 \\ (3.2 - p_T)(0.32 + 0.135p_T) & : p > 650 \end{cases} \qquad (6)$$

$$A = \frac{242.4\,F_1GD_h}{1 + \dfrac{1.52\,(F_pD_h)^2G}{F_2D_w^{1.3}\left\{1 + G\,(0.8F_pD_h/D_w - 1)\right\}}} \qquad (7)$$

$$B = 0.25GD_h\,e^{-0.2G} \qquad (8)$$

$$C = 60D_w^{0.57}\,G^{0.27}\left\{1 + \frac{Y - 1}{G + 1}\right\} \qquad (9)$$

### Correlation for Square Lattice Clusters (eg PWR, BWR)

11. Pressure = 500 - 2250 psia (3.4 - 15.5 $MN/m^2$)
No Spray cooling supplement
Subcooled at Inlet

p ≤ 1250 : Use PTR correlation (para 10, equations 3-9)

p > 1250 : Use PTR correlation with new A, namely instead of equation 7,

$$A = A_2 + (2.250 - 0.001p).(A_1 - A_2) \qquad (10)$$

where

$A_1$ = PTR value of A at p = 1250, ie with $F_1$ = 0.8726, $F_2$ = 0.953 in equation 7,

$$A_2 = 18.0G + \frac{9.5GD_h}{0.1 + G} \qquad (11)$$

* The PTR correlation may well be valid at higher pressure since the pressure functions $F_1$ and $F_2$ extrapolate to the critical pressure without obvious breakdown. The limit of 1250 psia given represents the highest pressure tested.

Correlation for Quality at Inlet

12. Pressure Tube and Square Lattice Clusters (as paras 10, 11)
No Spray Cooling Supplement
Steam Quality at Inlet

$$\phi_{do} = \frac{A}{C + ZY}\left[\frac{1 - X_i}{1 + bX_i}\right] \qquad (12)$$

where

$$b = (2.5B\lambda/A) - 1$$

A, B, C as defined in paras 10, 11

Correlation for 36-rod PTR Cluster with Supplemental Spray Cooling

13. Equations 3-9 in para 10 are used with $\Delta H_i^*$, $G^*$ instead of $\Delta H_i$ and G, where (see also Notation)

$$\Delta H_i^* = \Delta H_{i,c}\, G_c + \Delta H_{i,s}\, G_s \qquad (13)$$

$$G^* = G_c + f_g\, G_s \qquad (14)$$

$f_g$ = Fraction depending on geometry, nominal value = 0.6

14. It should be noted that when equation 3b is used, X = true local radial-average quality, taking into account the spray enthalpy. In the denominator, $G = G_c + G_s$, not $G^*$, since it is derived from equation 3a, via the heat balance.

Correlation Range

15. The correlation has been tested over the following parameter ranges, although the ranges were not investigated simultaneously:

| | |
|---|---|
| Geometry | : PTR, BWR, PWR |
| Pressure | : 0.6 - 15.5 $MN/m^2$ (90 - 2250 psia) |
| Mass Velocity | : 50 - 4000 $kg/m^2s$ (0.04 - 3.0 $Mlb/ft^2hr$) |
| Heated Diameter | : 7.6 - 36.6 mm (0.3 - 1.4 in) |
| Channel Length | : 1.5 - 4.6 m (60 - 180 in) |
| Radial Peaking Factor | : 1.0 - 1.32 |
| Axial Form Factor | : 1.0 - 1.38 |

CORRELATION DEVELOPMENT

16. Experimental steady-state dryout data from a number of rod-clusters, cooled by water in vertical up-flow had already been analysed in the development of the subchannel dryout correlation WSC-2 (Ref 7) and a representative sample were available on punched cards. A subset was chosen as being amenable to correlation on a mixed flow basis, ie clusters in which the radial enthalpy gradients were not excessive. Data from the 34 clusters used to develop the new correlation are described in Table 1, divided into 3 groups, namely by PTR, BWR and PWR geometries. The new Winfrith data pertaining to low flow and to the lower pressure range are contained in the block "New". The various types of cluster cross-sections represented are shown in Figure 1.

Optimisation of Subcooled - Inlet Correlation

17. The basic form of the correlation for uniform axial heat flux (Y = 1) in terms of the parameters A, B and C was that pioneered by Macbeth (Ref 8) and Barnett (Ref 9). The use of Y with C and Z in the denominator had been identified in the development of the subchannel correlation (Ref 7). The values of B and C (at Y = 1) were taken to be those found by Macbeth (Ref 8). Values of A could then be computed for each data point from the experimental power. The parameter A was correlated assuming a functional shape similar to that already found for subchannels (Ref 7) and round tubes (Ref 10).

Optimisation of Inlet Quality Correction

18. Experimental dryout data obtained by CISE for round tubes, annuli and small clusters (Ref 11) were examined to identify the effect of quality at inlet. There were two main effects. First, hydrodynamic instability could occur under certain conditions but, in principle, this could be represented in a transient code and the effect on dryout allowed for. A second effect was to modify the dryout characteristics under stable conditions and the trend of these data was represented by the equation in para 12.

19. The equation gives a dryout heat flux with continuity at $\Delta H_i = X_i = 0$ and decreasing to zero at $X_i = 1.0$, as found experimentally. The term in $bX_i$ represents the shape of the curve at intermediate values of inlet quality. The constant 2.5 represents the change in slope of $d\phi/dX_i$ at $X_i = 0$ and was a mean value for the experimental data. It should be emphasised that the correction for inlet quality in general reduces the dryout heat flux and is included in the interests of greater conservatism in reactor assessment. It should be regarded as tentative until its reactor application has been substantiated by experimental measurement.

Optimisation of Supplemental Spray Correction

20. The correlation was found to overpredict the dryout heat flux if the local value of mass velocity was used. It was as if the effect of the extra flow, entering at the centre, was not immediately felt in the dryout region on the outside of the cluster. This was expressed by defining an effective mass velocity G* for use in the correlation. This device was found to be satisfactory over the entire range of spray - fraction of the total from 0 to 100%.

CORRELATION ACCURACY

21. The prediction error (E) for an individual data point is defined as $E = (Q_p - Q_e)/Q_e$ where $Q_p$, $Q_e$ = predicted and experimental channel powers. The RMS error for a number (N) of points is defined as $(\Sigma E^2/N)^{\frac{1}{2}}$; this is a measure of the overall spread of the correlation and thus is the main index of its performance.

22. Table 2 shows the errors of the correlation when applied to all data. For the 970 data points, the RMS error was 11.9% and average error 4.4%. This is very much better than the

behaviour of the composite correlation in RELAP-3 (Ref 2) for which the RMS error was 33.1% and average error 16.9% when applied to the same data.

23. The Macbeth (Ref 8) and Barnett (Ref 9) mixed flow cluster correlations are applicable to data at 1000 psia (6.9 $MN/m^2$). For the 417 points at this pressure in the data set, the performance of the new correlation was superior; the errors were: New RMS = 10.8%, Average = 4.8%; Macbeth RMS = 17.7%, Average 6.5%; Barnett RMS = 20.7%, Average = 2.6%.

24. Table 3 shows the performance of the correlation when applied to the SGHWR-type clusters. The overall RMS error is low, ie 5%, and the data with supplemental flow are as well correlated as those without.

USE IN RELAP-UK

25. A number of experiments reported in the literature indicate that the performance of a steady-state dryout correlation tends to be marginally conservative when used under reactor-type transient conditions (eg Refs 12, 13). However, this requires more detailed investigation and to further this, the correlation has been programmed in RELAP-UK. For subcooled inlet, it is applied on a local conditions basis, ie using the $\emptyset$ vs. X form of the equation in para 10. However for quality at inlet, the inlet conditions seemed to be more important hence the equation of para 12 is coded in the $\emptyset$ vs. $X_i$ form.

26. In RELAP-UK, the channel length is divided into a number of nodes with the heat flux assumed constant radially and axially over the nodal volume. The calculation is "driven" by the driver-pins for which the heat transfer mode changes when the node, "on average", undergoes dryout. This is interpreted as applying to the average rod half-way along the node, ie $F_p$ is set to 1.0 and the conditions at the mid-point of the node are used in the correlation. A "slave-pin" model is also incorporated in RELAP-UK; it does not affect the calculation but allows the first onset of dryout to be determined. For this pin, the cluster value of $F_p$ is used with the length and driver-coolant conditions at the beginning or end of the node as required. Cumulative effects eg G*, Y and Z are integrated from channel (not node) inlet which is taken to be the bottom of the cluster for up-flow but the top for down-flow.

SUMMARY AND CONCLUSIONS

27. A mixed-flow dryout correlation for water in vertical up-flow has been developed from steady-state experimental data from 34 sets of rod cluster data. The geometries were representative of PTR, BWR and PWR clusters, the pressure range was 0.6 - 15.5 $MN/m^2$ and the mass velocity range 50 - 4000 $kg/m^2s$. The correlation contains terms to account for axial and radial non-uniform heat flux, a notional correction for quality at inlet and can be used for conditions of supplemental flow from a central sparge pipe.

28. The RMS error of the correlation was 12% for the whole pressure range (12% for PTR, 9% for BWR and 14% for PWR geometries) compared to 33% for the correlation originally programmed in RELAP-3. It was nearly half that of the Macbeth and Barnett correlations for relevant data at 7 $MN/m^2$.

29. The correlation has been adopted in the RELAP-UK code for use under transient conditions and its method of application is described. Low flow rates and low pressures are likely to be reached in a reactor blowdown, hence the matching of the correlation to the new Winfrith experimental data at these conditions increases the confidence with which the correlation may be applied to Blowdown calculations.

REFERENCES


1. BRITAIN I., BRYCE W.M., GREEN C. and HOLMES J.A. "The Status of RELAP-UK Mk III at July 1976 - A program for transient thermal hydraulic analysis" UKAEA Report AEEW - R 1083 (1976).

2. RETTIG W.H., JAYNE G.A., MOORE K.W., SLATER C.E. and UPTMOR M.L. "RELAP3 - A computer program for reactor blowdown analysis" Idaho Nuclear Corp. IN-1321 (1970).

3. BOWDITCH F.D., HOWE W.D. and LEE D.H. "Experimental measurements of dryout power and pressure drop in an SGHWR 36-pin cluster with differential enrichment, non-uniform axial heat flux, with and without continuous spray flow" UKAEA Report AEEW - R 1085, 1086 - to be published.

4. MACBETH R.V. Advances in Chemical Engineering 1968, Volume 7, pages 208-293.

5. COLLIER J.G. Convective boiling and condensation 1972, McGraw-Hill, New York 1972, pages 280-284.

6. BAILEY N.A. and LEE D.H. "A review of dryout data in SGHWR fuel bundle simulations" Paper 29, I Mech E TDFM Conf on "Heat and Fluid Flow in Water Reactor Safety", Manchester, September 1977.

7. BOWRING R.W. Unpublished work.

8. MACBETH R.V. "Burnout Analysis: Part 5, Examination of Published World Data for Rod Bundles". UKAEA Report, AEEW - R 358 (1964).

9. BARNETT P.G. "A correlation of Burnout Data for Uniformly Heated Annuli and its Use for Predicting Burnout in Uniformly Heated Rod Bundles". UKAEA Report AEEW - R 463 (1968).

10. BOWRING R.W. "A Simple but Accurate Round Tube, Uniform Heat Flux, Dryout Correlation over the Pressure Range 0.7 - 17 $MN/m^2$ (100-2500 psia)" UKAEA Report, AEEW - R 789 (1972).

11. ADORNI N., GASPARI G.P., GERMANI G., PETERLONGO G., RAVETTA R. and ZAVATTARELLI R. "Heat transfer crisis with steam-water mixtures in complex geometries: experimental data in annuli and clusters" Centro Informazioni Studi Esperienze (CISE) Report R - 123 (1964) This report refers to other relevant CISE Reports, namely R-63, 69, 74, 78, 86, 90 and 99.

12. CERMAK J.O., FARMAN R.F., TONG L.S., CASTERLINE J.E., KOKOLIS S. and MATZNER B. "The departure from nucleate boiling in rod bundles during pressure blowdown". Paper 70-HT-12 ASME Heat Transfer Conf. Detroit, May 1970.

13. MOXON D. and EDWARDS P.A. "Dryout during flow and power transients". UKAEA Report, AEEW - R 553 (1967).

14. Unpublished UKAEA Data. Experimental Data given on Pages A27 - 8 of AEEW - R 358 (Ref 8).

15. HESSON G.M., FITZSIMMONS D.E. and BATCH J.M. "Experimental Boiling Burnout Heat Fluxes with an Electrically Heated 19-rod Bundle Test Section". Battelle Report, BNWL - 206 (1965).

16. MATZNER B. and CASTERLINE J. E. "The Effect of Length and Pressure on the Critical Heat Flux for a closely-spaced 19-rod Bundle in Forced Convection Boiling". TID 22539 (1965).

17. MATZNER B. "Critical Heat Flux in a 9' long, closely spaced, 19-rod Bundle with an Unheated Central Rod". TID 24956 (1966).

18. NYLUND O. et al. "Hydrodynamic and Heat Transfer Measurements on a Full-Scale Simulated 36-rod Marviken Fuel Element with Non-Uniform Radial Flux Distribution". ASEA-ATOM, Sweden FRIGG-3 (1969).

19. JANSSEN E. "Nine-rod Critical Heat Flux Investigation, Steam-Water at 600 - 1400 psia; Final Summary Report". GEAP 5616 (1968).

20. Unpublished Data.

21. JANSSEN E., SCHRAUB R.A., NIXON R.B., MATZNER B. and CASTERLINE J.E. "Sixteen-rod Heat Flux Investigation, Steam-Water at 600-1250 psia". Symposium on two-phase flow and heat transfer in rod bundles, ASME Winter Meeting, Los Angeles, November 1969.

22. JANSSEN E. "Nine-rod Critical Heat Flux Investigation, Steam-Water at 600-1400 psia". Final Summary Report GEAP-5616 (1968).

23. ISRAEL S., CASTERLINE J. and MATZNER B. "Critical Heat Flux Measurements in a 16-rod Simulation of a BWR Fuel Assembly". Trans. ASME, Journal of Heat Transfer, 355 (August 1969).

24. GELLERSTEDT J.S., LEE R.A., OBERJOHN W.J., WILSON R.H. and STANEK L.J. "Correlation of Critical Heat Flux in a Bundle Cooled by Pressurised Water". Symposium on Two-Phase Flow and Heat Transfer in Rod Bundles. ASME Winter Meeting, Los Angeles, November 1969 - Page 63.

25. WILSON R.H., STANEK L.J., GELLERSTEDT J.S. and LEE R.A. "Critical Heat Flux in a Non-Uniformly Heated Rod Bundle". Symposium on Two-Phase Flow and Heat Transfer in Rod Bundles. ASME Winter Meeting, Los Angeles, November 1969 - Page 56.

Table 1

Cluster Parameters

| Block | Ref. | No. Rods | Geom. | Length m | Rod dia. mm | Spacer | Axial Flux | $F_p$ | $D_h$ mm | Pressure $MN/m^2$ | No. Pts |
|---|---|---|---|---|---|---|---|---|---|---|---|
| 3 | 14 | 7/6 | A | 1.83 | 12.7 | Grid | U | 1.00 | 15.7 | 6.9 | 9 |
| 4 | 15 | 19/19 | B | 1.93 | 15.5 | Warts | " | 1.08 | 7.5 | 8.3, 6.9 | 16 |
| 7 | 16 | " | " | 1.83 | 19.8 | " | " | 1.12 | 8.6 | 3.4 - 6.9 | 25 |
| 8 | " | " | " | " | " | " | " | " | " | " | 15 |
| 9 | " | " | " | 2.74 | " | " | " | 1.14 | " | " | 22 |
| 10 | 17 | 19/18 | " | " | " | " | " | 1.13 | 9.1 | " | 24 |
| 13 | 6 | 37/36 | C | 3.66 | 15.9 | Grid | " | 1.21 | 13.5 | 4.5 - 6.8 | 34 |
| 14 | " | " | " | " | " | " | NU | 1.22 | " | " | 29 |
| 15 | " | " | " | " | " | " | U | 1.03 | " | " | 22 |
| New | 3 | " | " | " | " | " | NU | 1.06 | " | 0.6 - 6.8 | 178 |
| 16 | 3 | " | " | 4.57 | 18.0 | " | U | 1.23 | 7.8 | 3.4 - 6.8 | 60 |
| 27 | 18 | " | " | 4.37 | 13.8 | " | " | 1.00 | 36.6 | 5.0 | 9 |
| 28 | " | " | " | " | " | " | " | 1.18 | " | 3.0 - 8.7 | 22 |
| 29 | " | " | " | " | " | " | NU | 1.10 | " | 3.0 - 6.9 | 30 |
| | | | | | | | | | | | |
| 26 | 19 | 9/9 | G | 1.52 | 11.1 | Pins | U | 1.07 | 19.5 | 4.1 - 9.7 | 26 |
| 35 | 20 | " | " | 1.50 | 12.2 | Coil | " | 1.02 | 18.8 | 3.3 - 7.0 | 9 |
| 40 | 21 | 16/16 | I | 1.83 | 14.3 | Pins | " | 1.00 | 18.5 | 4.1 - 8.6 | 26 |
| 41 | " | " | " | " | " | BWR | " | " | " | " | 27 |
| 42 | " | " | " | " | " | " | " | " | " | 6.9 | 12 |
| 43 | " | " | " | " | " | Pins | " | 1.25 | " | " | 12 |
| 44 | " | " | " | " | " | BWR | " | " | " | " | 4 |
| 45 | 22 | 9/9 | G | " | 14.5 | Pins | " | 1.00 | 19.4 | " | 11 |
| 46 | " | " | " | " | " | " | " | " | 15.2 | " | 6 |
| 54 | 23 | 16/16 | I | " | 14.3 | Warts | " | 1.32 | 15.8 | " | 14 |
| 55 | " | " | " | " | " | " | " | " | " | 5.2, 6.9 | 26 |
| 56 | " | " | " | " | " | Egg Cr. | " | " | " | 6.9 | 21 |
| 57 | " | " | " | " | " | " | " | " | " | " | 18 |
| | | | | | | | | | | | |
| 47 | 24 | 9/9 | F | 1.83 | 10.7 | Ferrule | U | 1.00 | 13.5 | 6.9, 15.2 | 47 |
| 48 | 25 | " | " | " | " | " | NU | " | " | 13.8 - 16.6 | 42 |
| 52 | 12 | 21/21 | L | 1.52 | " | Simple | U | 1.01 | 18.8 | 4.6 - 10.3 | 17 |
| 58 | 20 | " | M | 1.83 | " | Wart | " | 1.09 | 16.0 | 6.9 - 15.5 | 27 |
| 59 | " | 21/20 | " | " | " | " | " | 1.10 | 16.4 | " | 28 |
| 63 | " | 21/21 | " | " | " | " | NU | 1.09 | 16.0 | " | 27 |
| 64 | " | 25/25 | K | 3.66 | " | Grid | U | 1.24 | 16.1 | 4.9 - 15.5 | 75 |

**Notes**

No. Rods = Number of rods in cluster; Total/Heated
Geometry : Letters refer to shape in Figure 1
Axial Flux : U = Uniform, NU = Non-Uniform
Pressure : $P_1$, $P_2$ indicates two pressures tested, $P_1 - P_2$ indicates range covering more than two pressures
No. Pts : Number of points used in correlation development

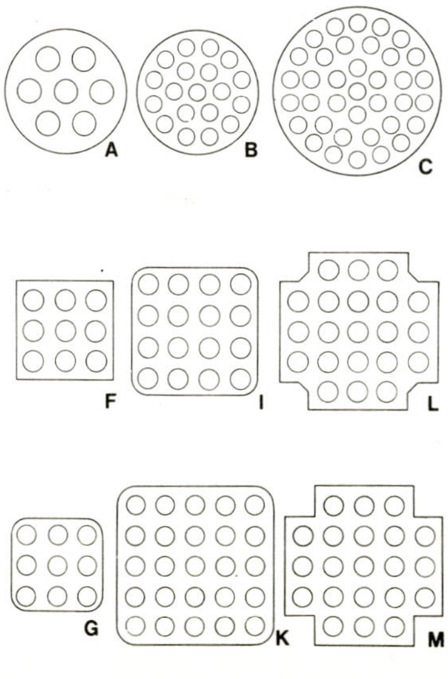


Fig. 1: Cluster cross-sections for data blocks in Table 1

Table 2

Correlation Comparison for all Data
(No Spray Supplement)

| Blk | No. Pts | Errors (%) | | | |
|---|---|---|---|---|---|
| | | New | | RELAP(Orig) | |
| | | RMS | Av | RMS | Av |
| 3 | 9 | 8.5 | 8.2 | 5.6 | -5.2 |
| 4 | 16 | 36.0 | 34.7 | 18.3 | 13.1 |
| 7 | 25 | 27.4 | 25.1 | 56.1 | 39.4 |
| 8 | 15 | 24.6 | 22.7 | 57.9 | 37.1 |
| 9 | 22 | 17.0 | 15.4 | 52.2 | 28.6 |
| 10 | 24 | 9.9 | 9.1 | 48.0 | 23.4 |
| 13 | 34 | 3.5 | -0.6 | 32.2 | 15.4 |
| 14 | 29 | 4.4 | 1.8 | 24.7 | 1.2 |
| 15 | 22 | 2.0 | -1.4 | 26.0 | 13.3 |
| NEW | 178 | 5.3 | -2.3 | 31.1 | 24.9 |
| 16 | 60 | 5.2 | 1.2 | 74.3 | 59.7 |
| 27 | 9 | 11.0 | 10.6 | 5.2 | -0.8 |
| 28 | 22 | 4.2 | -2.7 | 10.2 | 5.7 |
| 29 | 30 | 4.3 | -3.9 | 15.6 | -12.8 |
| ALL PTR | 495 | 11.7 | 3.5 | 41.0 | 23.2 |
| 26 | 26 | 6.0 | 0.7 | 9.8 | 7.0 |
| 35 | 9 | 8.9 | -3.7 | 12.5 | 7.2 |
| 40 | 26 | 7.2 | 0.5 | 8.5 | 1.1 |
| 41 | 27 | 8.7 | -6.4 | 9.5 | -5.6 |
| 42 | 12 | 4.1 | -0.2 | 6.7 | -5.5 |
| 43 | 12 | 9.3 | -9.0 | 5.8 | -2.8 |
| 44 | 4 | 2.4 | 2.3 | 9.9 | 9.5 |
| 45 | 11 | 2.9 | -2.0 | 9.5 | -9.0 |
| 46 | 6 | 5.8 | 5.1 | 7.9 | -4.9 |
| 54 | 14 | 15.0 | 13.7 | 20.4 | 19.1 |
| 55 | 26 | 11.9 | 11.2 | 24.8 | 21.2 |
| 56 | 21 | 6.6 | 4.8 | 13.1 | 10.5 |
| 57 | 18 | 8.6 | 7.2 | 14.1 | 11.7 |
| ALL BWR | 212 | 8.6 | 2.1 | 13.7 | 5.6 |
| 47 | 47 | 8.8 | -2.5 | 9.3 | -3.8 |
| 48 | 42 | 9.9 | 2.2 | 5.5 | -4.9 |
| 52 | 17 | 18.9 | 15.1 | 38.9 | 34.9 |
| 58 | 27 | 11.3 | 7.6 | 23.0 | 19.5 |
| 59 | 28 | 12.0 | 6.8 | 23.0 | 20.3 |
| 63 | 27 | 12.2 | 6.9 | 10.8 | -3.2 |
| 64 | 75 | 19.8 | 16.8 | 41.7 | 33.3 |
| ALL PWR | 263 | 14.4 | 7.9 | 27.1 | 14.1 |
| TOTAL | 970 | 11.9 | 4.4 | 33.1 | 16.9 |

Table 3

Accuracy of New Correlation for SGHWR Clusters

| Block | Spray? | Pressure $MN/m^2$ | No. Pts. | Errors (%) | | | |
|---|---|---|---|---|---|---|---|
| | | | | RMS | Av | Extremes | |
| 13 | No | 6.8 | 18 | 3.9 | 0.3 | 4.6 | -7.8 |
| | | 5.9 | 8 | 3.5 | -1.3 | 2.8 | -6.2 |
| | | 4.5 | 8 | 2.6 | -1.8 | 0.3 | -6.0 |
| | | ALL | 34 | 3.5 | -0.6 | 4.6 | -7.8 |
| 14 | No | 6.8 | 15 | 5.1 | 2.5 | 7.3 | -5.8 |
| | | 5.9 | 8 | 4.1 | 1.0 | 5.4 | -6.3 |
| | | 4.5 | 6 | 1.9 | 1.2 | 4.0 | -0.7 |
| | | ALL | 29 | 4.4 | 1.8 | 7.3 | -6.3 |
| 15 | No | 6.8 | 8 | 0.5 | 0 | 1.1 | -0.5 |
| | | 5.9 | 8 | 2.7 | -2.4 | 0.1 | -4.4 |
| | | 4.5 | 6 | 2.1 | -1.9 | -0.8 | -3.4 |
| | | ALL | 22 | 2.0 | -1.4 | 1.1 | -4.4 |
| New | No | 6.8 | 68 | 5.1 | -1.1 | 7.8 | -15.3 |
| | | 5.9 | 47 | 6.6 | -3.9 | 4.1 | -12.7 |
| | | 2.9 | 24 | 4.7 | -2.9 | 2.5 | -7.6 |
| | | 1.5 | 29 | 4.0 | -3.1 | 1.6 | -8.7 |
| | | 0.6 | 10 | 3.9 | 0.9 | 6.2 | -4.2 |
| | | ALL | 178 | 5.3 | -2.3 | 7.8 | -15.3 |
| New | Yes | 6.8 | 87 | 5.2 | 0.9 | 11.0 | -14.8 |
| | | 5.9 | 46 | 7.1 | 2.9 | 10.6 | -21.3 |
| | | 2.9 | 27 | 4.1 | 1.1 | 6.7 | -11.9 |
| | | 1.5 | 42 | 3.9 | -1.1 | 8.0 | -7.4 |
| | | 0.6 | 14 | 2.8 | 1.0 | 6.7 | -4.8 |
| | | ALL | 216 | 5.2 | 1.0 | 11.0 | -21.3 |
| ALL | - | TOTAL | 479 | 5.0 | -0.4 | 11.0 | -21.3 |

# C218/77

# UP-FLOW AND DOWN-FLOW BURNOUT

M. CUMO, R. BERTONI
University of Rome
R. CIPRIANI, G. PALAZZI
CNEN – CSN Casaccia

SYNOPSIS The effects of buoyancy force on the burn-out phenomenon have been studied with Freon 12 at low mass flow-rates. The critical heat flow is higher in up-flow than in down-flow (up to 35%) and this difference depends on the flow patterns occurring along the tube. The main influence of the buoyancy force seems to be localized in subcooled and bubble flow regions. A correlation between crisis qualities in down-flow and in up-flow has been obtained for these parameter ranges: - specific mass flow-rate G, 100 ÷ 1000 kg/m$^2$s; pressure $0.25 < p/p_{cr} < 0.60$; inlet quality $-0.40 < X_{in} < +0.20$

## 1. PREVIOUS LITERATURE ON THE SUBJECT

More and more consideration has been devoted in recent years to the difference between upflow and downflow burn-out heat fluxes;* mainly because of its importance for safety measures in nuclear plants. Papell et al.(1) investigated the buoyancy effect on the critical heat flux in forced convection vertical flow with nitrogen as fluid test. The main conclusions were that under certain conditions - low flows and high exit qualities - the critical heat flux for downflow is significantly lower than for upflow. Kirby et al.(2) experimented situations closer to those found in power plants. They noticed that by using water (25 p.s.i.) and changing the flow direction from upflow to downflow, the burn-out heat flux was reduced anywhere from 10 to 30%, the greatest reduction being at the lowest flowrate. In another experiment with a duct uniformly heated (20 mm I.D., 2.5 m long), with water boiling at 50, 70 and 90 bar and exit flow in subcooled or saturated conditions, Skaug (3) found slightly lower burn-out qualities in downflow as compared to upflow data (from 2% to 10% in the critical heat flux). The upflow critical heat fluxes were not experimented but calculated with the Becker correlation. Thorsen et al. (4), experimenting with both single-phase and boiling Freon 113 in upflow and downflow and visualizing the test section, revealed significantly different flow patterns for upflow and downflow during boiling. This difference was made evident by disparities in the pressure drop and heat transfer behaviours. There is a significant difference between upflow and downflow subcooled boiling as well as a lesser difference for saturated boiling. For a given wall superheat, larger heat fluxes are obtained for upflow than for downflow for both subcooled and saturated boiling. This behaviour, which is postulated as being due to the difference in the drag force on bubbles at the heating surface relative to the buoyancy force for upflow an downflow, is more pronounced in the subcooled regime.
Thorsen's paper strengthens the feeling that main differences between upflow and downflow behaviours lie in subcooled boiling and bubbly flow regimes.

2. Further experiments are reported in refs. (5) to (7). In the light of these conclusions and with the mind mainly to nuclear situations with safety problems (for instance the inversion of flow through PWR core during the loss of coolant accident), the authors decided to investigate systematically the differences between upflow and downflow burn-out, particularly at low flowrates. Freon 12 was chosen as fluid because of its similarity to water. Water tests in full PWR scale are also scheduled.

## 3. EXPERIMENTAL APPARATUS

The freon loop employed, sketched in fig. 1a, is described at length in ref (8). In order to have identical thermohydraulic conditions in both upflow and downflow, the test section is in the form of a U

* The phenomenon of thermal crisis is known under different terms. In this paper, 'burn-out', 'DNB' (Departure from Nucleate Boiling), 'dryout', and CHF (Critical Heat Flux) are used as synonyms, even if the more correct terms for the authors seems to be CHF.

shaped tube with the two heated lengths close together (fig. 1b). The parameters of the heated tubes are: 7.8 mm inner diameter and 2.00 m heated length; the arrangements of wall thermocouples are also clearly shown. Tests with Freon 12 were run both for subcooled and saturated inlet conditions, first heating the upflow side of the test section and then the downflow side, keeping the test section in vertical position. The burn-out condition is detected by a sudden rise in temperature of the wall thermocouple closest to the outlet.
The uncentainty of ± 7% in burnout measure is estimated to account for instrumentation errors and the different inlet geometry in the two test sections.

## 4. DISCUSSION OF RESULTS

Runs have been performed (see Table I) at different specific mass flowrates ($130 < G < 1000$ kg/m$^2$s), different pressures (p=10.5; 14.5; 17.5; 24 bar) and different inlet qualities ($-0.45 < X_{in} < +0.30$). In fig.2 the main results of the critical heat fluxes versus the inlet qualities are shown: upflow critical heat flux $q''_{CHF}$ is always higher than downflow one (clearly up to $X_{in} \simeq 0.15$ at least). In the cases where the inlet condition is a liquid vapour mixture, the trend is, sometimes, not completely linear as versus the negative quality. Since the steam freon mixture entry to the test section is generated in a low heat flux preheater, the calminglength ($L/D \simeq 25$) from the bend would be not sufficient to have a perfectly stable flow. In any case this fact does not seem to invalidate the comparison between the upflow and downflow CHF in the positive quality region too. The influence of mass flowrate, for some typical cases, both on $q''_{CHF}$ and $X_{B.O.}$, is made more evident by the diagrams of fig. 3. Again there is a sharp distribution between upflow (upper line) and downflow (lower line). Intuition would suggest that at even higher flowrates the $q''_{CHF}$ or $X_{B.O.}$ curves should merge as gravity becomes less a differentiating factor. However the graphs could seem to contradict such a conclusion, at least in the range of flowrates investigated. For both $q''_{CHF}$ and $X_{B.O.}$ curves, there are trends of much the same type as shown in fig.4 both for upflow and downflow: again the respective curves are clearly distinct and separated. An overall representation of the experimental data is shown in fig.5. The two burn-out qualities are directly compared and the statement that $X_{B.O.up}$ is greater than $X_{B.O.down}$ is immediately evident. Nevertheless there are a few cases at highly positive inlet qualities where an inversion of trend can be seen.

5. The influence of the inlet quality deserves particular consideration. The diagrams of fig. 6 consistently confirm that the difference between the two burn-out qualities decreases as the inlet quality increases, till positive inlet qualities of 15% about are reached and this difference disappears. This observation leads to the conclusion that the difference in up and down two-phase diabatic flows lies mainly in the quality region from $\simeq -5\%$ to $\simeq +15\%$ (subcooled boiling region and bubbly flow region).
The two-phase flow regimes at higher qualities and up to the burn-out quality seem to be quite independent of the direction of the gravity vector as far as the up and down burn-out quality values are concerned (at least at the flow-rates and pressures indicated).
Therefore the influence of the buoyancy force on the burn-out heat flux seems to be localized in subcooled and bubble regions. In the annular flow region the voidage is higher with downflow than with upflow, while the difference in the forces, due to the flow acceleration and to gravity, decreases the shear stress at the vapour-liquid interface in downflow, offering a less favourable condition for the onset of dry-out. Because of these two contrasting situations, we can conclude that the main influence on the critical heat flux difference is in the low quality region.

## 6. THEORETICAL REMARKS

Therefore, limiting ourselves to the low quality region (bubbly flow), let's consider the pressure forces acting on a bubble (9):

- buoyancy forces : $\frac{2}{3} g\, d\, (\rho_l - \rho_g)$
- inertia forces : $\frac{1}{2} \rho_l u^2 C_d$
- viscosity forces : $u_l \frac{u}{d}$
- surface forces : $\frac{\sigma}{\pi d}$

Neglecting the numerical constants and the coefficients, the respective ratios to the buoyancy force are the following:

$$\frac{\rho_l u^2}{g\, d\, (\rho_l - \rho_g)} \quad ; \quad \frac{u_l u}{g\, d^2 (\rho_l - \rho_g)} \quad ;$$

$$\frac{\sigma}{g\, d^2\, (\rho_l - \rho_g)}$$

In our experimental cases, viscous forces and surface tension forces are negligible with respect to the buoyancy forces. *

7. The main parameter seems to be the first ratio (inertial forces to buoyancy), i.e.:

$$\frac{\rho_1 \; u^2}{g \; d \; (\rho_1 - \rho_g)} \qquad (1)$$

Introducing the "surface" velocities of the two phases:

$$J_1 = \frac{Q_1}{A}, \quad J_g = \frac{Q_g}{A}$$

The fluid velocity u is : $u = \dfrac{Q_1 + Q_g}{A}$

and, substituting, the ratio (1) becomes:

$$\left( \frac{J_g}{\sqrt{g D \frac{\rho_1 - \rho_g}{\rho_1}}} + \frac{J_1}{\sqrt{g D \frac{\rho_1 - \rho_g}{\rho_1}}} \right)^2 = \left| J_g^* + J_1^* \right| = J^{*2}$$

Now $J^*$, the sum of the adimensionalized surface velocities, must be in someway related to the difference between $X_{B.O.up}$ and $X_{B.O.down}$, $\Delta X_{B.O.} = X_{B.O.up} - X_{B.O.down}$. Let's say, in a rather general way:

$$\frac{\Delta X_{B.O.}}{X_{B.O.up}} = \frac{f(X_{in}, G)}{J^*}$$

indicating that at increasing $J^*$ the $\Delta X_{B.O.}$ difference disappears and being $f(X_{in}, G)$ a function of inlet quality and mass flowrate.

In this way $X_{B.O.down}$ is correlated with $X_{B.O.up}$ which can be obtained with a burn-out correlation. In the case of Freon 12, the following (10) correlation may be adopted for $X_{B.O.up}$ (fig. 7d):

$$X_{B.O.up} = \frac{a}{1 + \frac{b}{L_b}} \cdot F \qquad (2)$$

where:

$$a = (1 - \pi) \left( \frac{G}{100} \right)^{-1/3} \quad \text{(C.G.S. system)}$$

$$b = 0.6 \left( \frac{1}{\pi} - 1 \right)^{0.4} D^{0.25} \; G$$

$$F = 1 - \left( \frac{1}{4} - \pi \right) \left( \frac{9}{4} - \pi \right)^{-1}$$

$$\pi = P/Pcr$$

From the experimental results we get:

$$f(X_{in} : G) = \left( \frac{G}{100} \right)^{3/2} (0.6 - X_{in}) \qquad \begin{cases} 100 < G < 1000 \text{ kg/m}^2\text{s} \\ -0.40 < X_{in} < +0.20 \\ 0.25 < \frac{P}{P_{cr}} < 0.60 \end{cases}$$

and fig. 7b provides an overall comparison between the experimental and predicted values of $X_{B.O.down}$. Most of dots are gathered in a ± 15 % band.

Similar calculations may also be made by employing the Worley and Griffith (11) formula:

$$\frac{X_{B.O.down}}{X_{B.O.up}} = \frac{-\frac{1}{F_2} + \left[ \frac{\rho_1 F_1}{\rho_g F_2^2} - \frac{100 g D (\rho_1 \; \rho_g) \rho_1^2 \; F_1 \; F_2}{G^2 \rho_g} + \frac{100 g D (\rho_1 \; \rho_g) \rho_1}{G^2 F_2} \right]^{1/2}}{-\frac{1}{F_2} + \left[ \frac{\rho_1 F_1}{\rho_g F_2^2} - \frac{100 g D (\rho_1 - \rho_g) \rho_1^2 \; F_1 \; F_2}{G^2 \rho_g} + \frac{100 g D (\rho_1 - \rho_g) \rho_1}{G^2 F_2} \right]^{1/2}}$$

where:

$$F_1 = \frac{1 + 75 \; (1 - \alpha)}{\alpha^{5/2}}$$

$$F_2 = (1 - \alpha)$$

The results are graphed in fig. 8, showing that the correlation (3) rather overpredicts the experimental values.

As said above, upflow burn-out happens at higher qualities, and therefore at higher vapour bulk velocity, than in downflow. The rise in upflow wall temperature is therefore lower than downflow because of different convection heat transfer in post-dryout region. This influence is shown in fig. 9.

## 8. CONCLUSIONS

Extensive tests with Freon 12 at reduced pressures of interest in nuclear reactor safety evaluations (after L.O.C.A.) have led to these main conclusions:

- burn-out heat fluxes (and burn-out qualities ) are greater (10 to 30%) in upflow than in downflow (fig. 5);
- this difference seems mainly related to buoyancy effects in the two-phase flow region at low qualities (-5% to 15%, fig. 6);
- downflow burn-out quality is related to upflow burn-out quality by a correlation of the type:

$$X_{B.O.down} = X_{B.O.up} \left[ 1 - \frac{\left( \frac{G}{1000} \right)^{3/2} (0.6 - X_{inlet})}{J^*} \right]$$

where $X_{B.O.up}$ may be predicted by the correlation (2).

* For the value of bubble diameter d it is possible to take the duct diameter, referring to the onset of slug flow.

9. ACKNOWLEDGEMENTS

The authors wish to thank for his advice Ing. E. Farello, leader of Heat Transfer Laboratory, and Mr. F. Fabrizi, undergraduating in Nuclear Engineering for his interest and helpful advice given for this work.

10. NOMENCLATURE

| | |
|---|---|
| A | duct cross section area |
| Cd | bubbles shape factor |
| D | duct inner diameter |
| d | bubbles diameter |
| G | specific mass flowrate |
| g | gravity acceleration |
| J | "surface" velocities |
| $J^*$ | adimensional "surface" velocities |
| $L_b$ | boiling length |
| p | pressure |
| Q | volumetric flowrate |
| q" | heat flux |
| T | temperature |
| u | fluid velocities |
| X | quality |

GREEK SYMBOLS

| | |
|---|---|
| $\alpha$ | void fraction |
| $\mu$ | viscosity |
| $\pi$ | reduced pressure |
| $\rho$ | density |
| $\sigma$ | surface tension |

SUBSCRIPT

| | |
|---|---|
| B.O. | refers to dry-out (burn-out) condition |
| CHF | critical heat flux |
| down | refers to downflow condition |
| g | refers to vapour |
| in | refers to inlet condition |
| l | refers to liquid |
| up | refers to upflow condition |
| sat | refers to saturation conditions |
| w | refers to duct wall |

REFERENCES

(1) S.S. Papell, R.J. Simoneau, D.D. Brown "Buoyancy effects on critical heat flux of forced convective boiling in vertical flow", NASA - THD - 3672 (1966)

(2) G.J. Kirby, R. Stainforth, J.H.Kinneir "A visual study of forced convection boiling. Flow patterns and burn-out for a round test section", AEEW Report 506, Winfrith, Dorchester, Dorset (1967)

(3) J.A. Skaug :"Downflow burn-out of a uniformly heated 20 mm round duct", European Two-Phase Flow Group Meeting, Karlsruhe, June 2-5, 1969

(4) R.S. Thorsen, F. Dobran, J.A. Alcorta: "A comparative study of vertical upflow and downflow in a uniformly heated boiling fluid", 5th International Heat Transfer Conf., Tokyo, Sept. 3-7, 1974

(5) R.M. Wright :"Downflow forced convection boiling of water in uniformly heated tubes", University of California, Lawrence Radiation Laboratory, UCRL 9744, TID 4500, 1961

(6) R.A. Smith :"CHF in flow reversal transients", MIT, P.H.D. Thesis in M. E., May 1975

(7) P. Griffith :"Transient CHF presentation", 3rd Water Reactor Safety Research Information Meeting, Washington, Sept. 29 - Oct. 2, 1975

(8) G.E. Farello :"Rapporto di sicurezza CF/1" (in italian), CNEN Report, STAV 11, 1968

(9) G.B. Wallis :"One dimensional two-phase flow", Mc Graw Hill, 1969

(10) M. Cumo, G. Ferrari, G. Urbani : "Prediction of burn-out power with Freon up to the critical pressure", 27th Meeting ATI, Napoli 27-29 September 1972

(11) L.C. Worley, P. Griffith :"Downflow void fraction, pressure drop and critical heat fluw", Department of Mechanical Engineering, MIT, Report No. 80620-83, 1973

TABLE I

| UPFLOW (inlet diameter = 0780 cm) | | | | | | | DOWNFLOW (inlet diameter = 0780 cm) | | | | | | |
|---|---|---|---|---|---|---|---|---|---|---|---|---|---|
| RUN No | Heated length [m] | Inlet pressure [bar] | Specific mass flow-rate [kg/m² sec] | Inlet quality | Critical heat-flux [W/m²]×10⁴ | Dry-out quality | RUN No | Heated length [m] | Inlet pressure [bar] | Specific mass flow-rate [Kg/m² sec] | Inlet quality | Critical heat-flux [W/m²]×10⁴ | Dry-out quality |
| 1 | 2.00 | 12.7 | 688 | .176 | 4.46 | .710 | 1 | 2.00 | 11.2 | 695 | .191 | 4.44 | .700 |
| 2 | 2.00 | 11.9 | 695 | .076 | 5.34 | .696 | 2 | 2.00 | 11.1 | 695 | .085 | 5.02 | .639 |
| 3 | 2.00 | 11.9 | 695 | -.073 | 6.60 | .693 | 3 | 2.00 | 11.1 | 695 | -.046 | 6.16 | .639 |
| 4 | 2.00 | 11.9 | 695 | -.211 | 7.80 | .694 | 4 | 2.00 | 10.7 | 691 | -.196 | 7.18 | .621 |
| 5 | 2.00 | 11.2 | 510 | .208 | 4.15 | .854 | 5 | 2.00 | 11.1 | 506 | .198 | 4.00 | .824 |
| | | | | | | | 6 | 2.00 | 10.7 | 502 | .225 | 3.79 | .814 |
| 7 | 2.00 | 11.2 | 506 | .023 | 5.42 | .853 | 7 | 2.00 | 10.5 | 510 | .620 | 3.61 | .617 |
| | | | | | | | 8 | 2.00 | 10.2 | 502 | .076 | 3.93 | .667 |
| 9 | 2.00 | 11.2 | 510 | .025 | 5.75 | .870 | 9 | 2.00 | 10.9 | 514 | -.041 | 5.51 | .805 |
| 10 | 2.00 | 11.1 | 510 | -.191 | 6.69 | .849 | 10 | 2.00 | 10.8 | 514 | -.201 | 6.41 | .782 |
| 11 | 2.00 | 10.7 | 310 | .279 | 2.10 | .799 | 11 | 2.00 | 10.6 | 302 | .258 | 2.46 | .892 |
| | | | | | | | 12 | 2.00 | 10.9 | 317 | .205 | 2.68 | .868 |
| | | | | | | | 13 | 2.00 | 11.0 | 317 | .194 | 2.69 | .864 |
| 14 | 2.00 | 11.0 | 340 | .130 | 3.53 | .876 | 14 | 2.00 | 10.9 | 302 | .082 | 2.22 | .662 |
| | | | | | | | 15 | 2.00 | 11.0 | 325 | .065 | 2.47 | .666 |
| | | | | | | | 16 | 2.00 | 11.2 | 314 | .114 | 2.44 | .721 |
| 17 | 2.00 | 10.7 | 302 | .029 | 3.94 | .996 | 17 | 2.00 | 11.0 | 325 | -.044 | 3.53 | .792 |
| 18 | 2.00 | 10.6 | 302 | -.035 | 3.93 | .987 | | | | | | | |
| 19 | 2.00 | 10.7 | 302 | -.189 | 4.51 | .983 | 19 | 2.00 | 10.8 | 325 | -.211 | 4.17 | .806 |
| 20 | 2.00 | 10.9 | 302 | -.208 | 4.49 | .967 | | | | | | | |
| 21 | 2.00 | 11.3 | 200 | .243 | 1.91 | .981 | 21 | 2.00 | 11.3 | 208 | .177 | 1.72 | .814 |
| 22 | 2.00 | 11.3 | 208 | .079 | 2.10 | .842 | 22 | 2.00 | 10.9 | 208 | .071 | 1.85 | .740 |
| 23 | 2.00 | 11.3 | 208 | -.056 | 2.63 | .952 | 23 | 2.00 | 11.0 | 208 | -.046 | 2.35 | .851 |
| 24 | 2.00 | 11.2 | 208 | -.191 | 2.98 | .947 | 24 | 2.00 | 11.0 | 208 | -.205 | 2.68 | .825 |
| 25 | 2.00 | 11.2 | 136 | .149 | 1.50 | .979 | 25 | 2.00 | 11.1 | 132 | .214 | 1.22 | .877 |
| 26 | 2.00 | 11.3 | 136 | .065 | 1.67 | 1.020 | 26 | 2.00 | 10.4 | 136 | .077 | 1.03 | .597 |
| 27 | 2.00 | 11.3 | 136 | -.050 | 1.76 | .956 | 27 | 2.00 | 10.9 | 136 | -.062 | 1.61 | .873 |
| 28 | 2.00 | 11.3 | 136 | -.205 | 2.00 | .967 | 28 | 2.00 | 11.0 | 132 | -.200 | 1.78 | .848 |
| | | | | | | | 29 | 2.00 | 14.7 | 1035 | .195 | 2.91 | .437 |
| | | | | | | | 30 | 2.00 | 14.7 | 1035 | .100 | 3.59 | .398 |
| | | | | | | | 31 | 2.00 | 14.7 | 1043 | -.071 | 5.25 | .366 |
| | | | | | | | 32 | 2.00 | 14.7 | 1043 | -.182 | 6.24 | .333 |
| 33 | 2.00 | 14.6 | 673 | .242 | 2.68 | .584 | 33 | 2.00 | 14.6 | 684 | .233 | 3.03 | .613 |
| 34 | 2.00 | 14.6 | 673 | .138 | 3.53 | .588 | 34 | 2.00 | 14.6 | 676 | .119 | 3.53 | .566 |
| 35 | 2.00 | 14.6 | 692 | -.040 | 4.79 | .555 | 35 | 2.00 | 14.6 | 691 | -.042 | 4.42 | .506 |
| 36 | 2.00 | 14.6 | 699 | -.174 | 5.90 | .550 | 36 | 2.00 | 14.6 | 691 | -.179 | 5.33 | .483 |
| 37 | 2.00 | 14.9 | 525 | .200 | 2.91 | .679 | 37 | 2.00 | 15.0 | 510 | .332 | 2.92 | .778 |
| 38 | 2.00 | 14.8 | 525 | .068 | 3.66 | .669 | 38 | 2.00 | 15.0 | 510 | .256 | 3.10 | .730 |
| 39 | 2.00 | 14.9 | 525 | -.056 | 4.42 | .672 | 39 | 2.00 | 15.2 | 510 | -.076 | 3.86 | .589 |
| 40 | 2.00 | 14.9 | 520 | -.189 | 5.02 | .645 | 40 | 2.00 | 15.0 | 510 | -.195 | 4.57 | .581 |
| 41 | 2.00 | 14.3 | 306 | .172 | 2.59 | .859 | 41 | 2.00 | 14.7 | 306 | .221 | 2.22 | .828 |
| 42 | 2.00 | 14.3 | 306 | .035 | 2.91 | .845 | 42 | 2.00 | 14.7 | 306 | .088 | 2.46 | .780 |
| 43 | 2.00 | 14.6 | 312 | -.078 | 3.66 | .928 | 43 | 2.00 | 14.9 | 306 | -.074 | 2.91 | .747 |
| 44 | 2.00 | 14.6 | 312 | -.174 | 4.00 | .924 | 44 | 2.00 | 14.9 | 306 | -.191 | 3.21 | .716 |
| 45 | 2.00 | 15.1 | 237 | .165 | 2.04 | .915 | 45 | 2.00 | 14.6 | 219 | .326 | 1.81 | .974 |
| 46 | 2.00 | 15.2 | 219 | .069 | 2.09 | .900 | 46 | 2.00 | 14.8 | 221 | .210 | 2.00 | .918 |
| 47 | 2.00 | 15.2 | 219 | -.070 | 2.51 | .927 | 47 | 2.00 | 14.6 | 221 | -.058 | 2.40 | .873 |
| 48 | 2.00 | 15.0 | 219 | -.197 | 2.85 | .931 | 48 | 2.00 | 14.8 | 221 | -.191 | 2.62 | .829 |
| 49 | 2.00 | 14.6 | 123 | .330 | 1.06 | no crisis | 49 | 2.00 | 15.0 | 128 | .400 | 1.04 | 1.030 |
| 50 | 2.00 | 15.0 | 123 | .165 | 1.28 | 1.060 | 50 | 2.00 | 15.0 | 123 | .270 | 1.27 | 1.070 |
| 51 | 2.00 | 15.0 | 123 | -.075 | 1.49 | .973 | 51 | 2.00 | 15.0 | 123 | -.075 | 1.42 | .928 |
| 52 | 2.00 | 15.3 | 117 | -.199 | 1.62 | 1.000 | 52 | 2.00 | 15.0 | 118 | -.155 | 1.55 | .947 |
| 53 | 2.00 | 18.2 | 1012 | .135 | 2.55 | .361 | 53 | 2.00 | 16.8 | 1005 | .128 | 3.05 | .401 |
| 54 | 2.00 | 17.7 | 997 | .040 | 4.09 | .407 | 54 | 2.00 | 16.6 | 997 | .047 | 3.61 | .370 |
| 55 | 2.00 | 18.4 | 1005 | -.077 | 5.25 | .413 | 55 | 2.00 | 17.4 | 1005 | -.034 | 4.02 | .352 |
| 56 | 2.00 | 18.5 | 997 | -.226 | 6.82 | .416 | 56 | 2.00 | 17.1 | 1005 | -.178 | 5.62 | .328 |
| 57 | 2.00 | 18.1 | 1016 | -.461 | 9.27 | .386 | 57 | 2.00 | 16.5 | 997 | -.409 | 7.61 | .277 |

TABLE I (continued)

| UPFLOW (inlet diameter = 0.780 cm) | | | | | | | DOWNFLOW (inlet diameter = 0.780 cm) | | | | | | |
|---|---|---|---|---|---|---|---|---|---|---|---|---|---|
| RUN No | Heated length [m] | Inlet pressure [bar] | Specific mass flow-rate [Kg/m² sec] | Inlet quality | Critical heat-flux [W/m²]×10⁴ | Dry-out quality | RUN No | Heated length [m] | Inlet pressure [bar] | Specific mass flow-rate [Kg/m² sec] | Inlet quality | Critical heat-flux [W/m²]×10⁴ | Dry-out quality |
| 58 | 2.00 | 17.2 | 695 | .198 | 2.56 | .532 | 58 | 2.00 | 17.8 | 684 | .186 | 2.62 | .518 |
| 59 | 2.00 | 18.1 | 691 | .051 | 3.23 | .483 | 59 | 2.00 | 17.5 | 684 | .042 | 3.15 | .460 |
| 60 | 2.00 | 17.7 | 725 | -.068 | 4.56 | .509 | 60 | 2.00 | 17.5 | 680 | -.057 | 3.71 | .430 |
| 61 | 2.00 | 17.3 | 695 | -.191 | 5.23 | .495 | 61 | 2.00 | 17.3 | 680 | -.185 | 4.64 | .435 |
| 62 | 2.00 | 16.9 | 673 | -.418 | 6.84 | .503 | 62 | 2.00 | 17.3 | 680 | -.424 | 6.09 | .398 |
| 63 | 2.00 | 17.8 | 510 | .224 | 2.41 | .652 | 63 | 2.00 | 17.5 | 518 | .153 | 2.62 | .615 |
| 64 | 2.00 | 17.7 | 510 | .093 | 2.88 | .611 | 64 | 2.00 | 17.4 | 510 | .060 | 2.90 | .580 |
| 65 | 2.00 | 17.5 | 510 | -.060 | 3.65 | .600 | 65 | 2.00 | 17.5 | 506 | -.049 | 3.43 | .569 |
| 66 | 2.00 | 17.8 | 510 | -.200 | 4.28 | .586 | 66 | 2.00 | 17.5 | 502 | -.201 | 3.95 | .516 |
| 67 | 2.00 | 17.6 | 529 | -.459 | 6.05 | .590 | 67 | 2.00 | 17.3 | 502 | -.432 | 5.04 | .486 |
| 68 | 2.00 | 17.2 | 501 | -.426 | 7.13 | .615 | 68 | 2.00 | 17.8 | 521 | -.499 | 5.73 | .422 |
| 69 | 1.85 | 17.7 | 501 | -.445 | 7.69 | .617 | 69 | 1.85 | 17.6 | 518 | -.482 | 6.59 | .459 |
| 70 | 1.70 | 17.5 | 505 | -.437 | 8.57 | .051 | 70 | 1.70 | 18.2 | 518 | -.500 | 7.03 | .453 |
| 71 | 1.50 | 18.6 | 496 | -.485 | 10.37 | .690 | 71 | 1.50 | 18.1 | 518 | -.494 | 7.69 | .473 |
| 72 | 2.00 | 17.8 | 496 | -.453 | 9.50 | .695 | | | | | | | |
| 73 | 2.00 | 16.7 | 501 | -.407 | 6.15 | .596 | | | | | | | |
| 74 | 2.00 | 17.1 | 302 | .287 | 1.95 | .837 | 74 | 2.00 | 17.4 | 302 | .201 | 1.81 | .738 |
| 75 | 2.00 | 17.9 | 302 | .067 | 2.62 | .846 | 75 | 2.00 | 17.3 | 302 | .027 | 1.98 | .607 |
| 76 | 2.00 | 18.1 | 302 | -.083 | 2.96 | .828 | 76 | 2.00 | 17.3 | 302 | -.046 | 2.39 | .657 |
| 77 | 2.00 | 18.4 | 302 | -.223 | 3.32 | .818 | 77 | 2.00 | 17.1 | 302 | -.176 | 2.84 | .676 |
| 78 | 2.00 | 18.5 | 302 | -.473 | 4.14 | .834 | 78 | 2.00 | 16.9 | 328 | -.439 | 4.10 | .668 |
| 79 | 2.00 | 17.7 | 315 | -.457 | 5.41 | .810 | 79 | 2.00 | 18.1 | 317 | .487 | 4.20 | .630 |
| 80 | 1.85 | 17.7 | 321 | -.453 | 5.90 | .817 | 80 | 1.85 | 18.1 | 317 | .487 | 4.78 | .641 |
| 81 | 1.70 | 18.1 | 321 | -.471 | 6.41 | .829 | 81 | 1.70 | 18.1 | 317 | .487 | 5.24 | .670 |
| 82 | 1.50 | 16.7 | 326 | -.413 | 8.08 | .917 | 82 | 1.50 | 17.7 | 317 | .467 | 5.57 | .663 |
| 83 | 2.00 | 17.2 | 326 | -.432 | 6.59 | .760 | | | | | | | |
| 84 | 2.00 | 17.1 | 219 | .394 | 1.48 | .986 | 84 | 2.00 | 17.7 | 219 | .332 | 1.40 | .919 |
| 85 | 2.00 | 17.4 | 219 | .085 | 2.14 | .950 | 85 | 2.00 | 17.7 | 219 | .126 | 1.67 | .820 |
| 86 | 2.00 | 17.6 | 219 | -.057 | 2.44 | .962 | 86 | 2.00 | 17.8 | 219 | -.054 | 1.99 | .786 |
| 87 | 2.00 | 17.9 | 219 | -.215 | 2.65 | .915 | 87 | 2.00 | 17.8 | 215 | -.198 | 2.32 | .794 |
| 88 | 2.00 | 17.7 | 223 | -.438 | 3.26 | .907 | 88 | 2.00 | 17.2 | 215 | -.449 | 3.14 | .760 |
| 89 | 2.00 | 18.1 | 219 | -.473 | 4.20 | .960 | 89 | 2.00 | 18.1 | 219 | -.483 | 3.14 | .728 |
| 90 | 1.85 | 17.7 | 219 | -.453 | 4.49 | .963 | 90 | 1.85 | 17.8 | 219 | -.471 | 3.65 | .767 |
| 91 | 1.70 | 17.4 | 219 | -.445 | 4.94 | .998 | 91 | 1.70 | 17.8 | 219 | -.471 | 3.92 | .772 |
| 92 | 1.50 | 17.7 | 219 | -.453 | 3.90 | 1.023 | 92 | 1.50 | 17.7 | 219 | -.463 | 4.34 | .813 |
| 93 | 2.00 | 17.5 | 136 | .196 | 1.28 | 1.050 | 93 | 2.00 | 17.2 | 132 | .210 | 1.01 | .870 |
| 94 | 2.00 | 17.7 | 125 | .155 | 1.20 | .998 | 94 | 2.00 | 17.3 | 129 | .131 | 1.15 | .950 |
| 95 | 2.00 | 18.0 | 132 | -.070 | 1.45 | .947 | 95 | 2.00 | 17.7 | 125 | -.048 | 1.24 | .842 |
| 96 | 2.00 | 18.4 | 132 | -.216 | 1.64 | .954 | 96 | 2.00 | 17.8 | 125 | -.190 | 1.44 | .860 |
| 97 | 2.00 | 18.2 | 132 | -.477 | 1.97 | .925 | 97 | 2.00 | 17.9 | 125 | -.477 | 1.72 | .755 |
| 98 | 2.00 | 17.7 | 132 | -.453 | 2.67 | 1.040 | 98 | 2.00 | 18.0 | 132 | -.477 | 2.13 | .882 |
| 99 | 1.85 | 17.7 | 132 | -.453 | 2.96 | no crisis | 99 | 1.85 | 17.7 | 132 | -.461 | 2.45 | .908 |
| 100 | 1.70 | 17.5 | 132 | -.445 | 3.26 | no crisis | 100 | 1.70 | 18.1 | 132 | -.481 | 2.67 | .932 |
| 101 | 1.50 | 17.8 | 139 | -.461 | 3.92 | no crisis | 101 | 1.50 | 18.1 | 132 | -.481 | 2.84 | .919 |
| 102 | 2.00 | 24.0 | 1340 | -.328 | 7.47 | .276 | 102 | 2.00 | 23.7 | 1330 | -.327 | 6.57 | .205 |
| 103 | 2.00 | 24.0 | 1340 | -.333 | 7.56 | .279 | | | | | | | |
| 104 | 2.00 | 23.8 | 1043 | -.341 | 6.73 | .359 | 104 | 2.00 | 24.4 | 1050 | -.359 | 5.88 | .255 |
| 105 | 2.00 | 24.4 | 1042 | -.321 | 5.30 | .238 | | | | | | | |
| 106 | 2.00 | 24.0 | 567 | -.352 | 4.53 | .516 | 106 | 2.00 | 23.9 | 567 | -.334 | 3.51 | .335 |
| 107 | 2.00 | 24.0 | 567 | -.345 | 4.61 | .537 | | | | | | | |
| 108 | 2.00 | 23.5 | 302 | -.335 | 2.88 | .673 | 108 | 2.00 | 23.2 | 302 | -.339 | 2.77 | .634 |
| 109 | 2.00 | 24.9 | 219 | -.409 | 2.16 | .693 | 109 | 2.00 | 24.3 | 219 | -.381 | 2.17 | .701 |

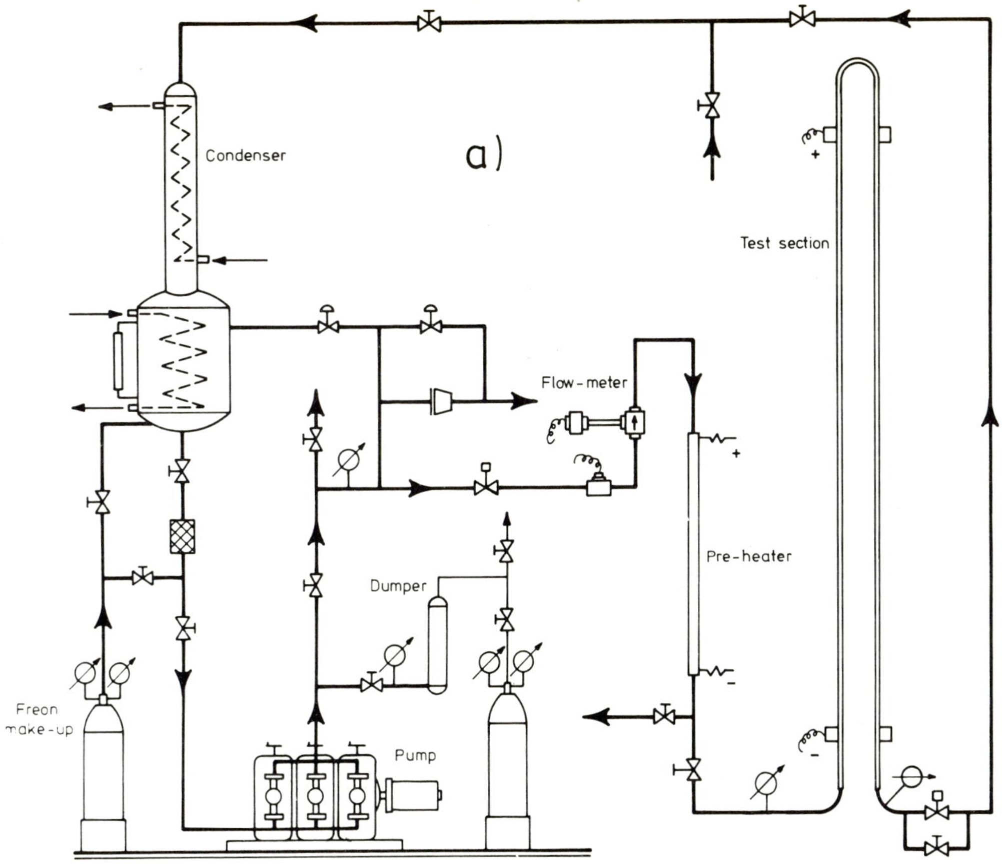


**Fig. 1(a): Flow diagram of the experimental loop**

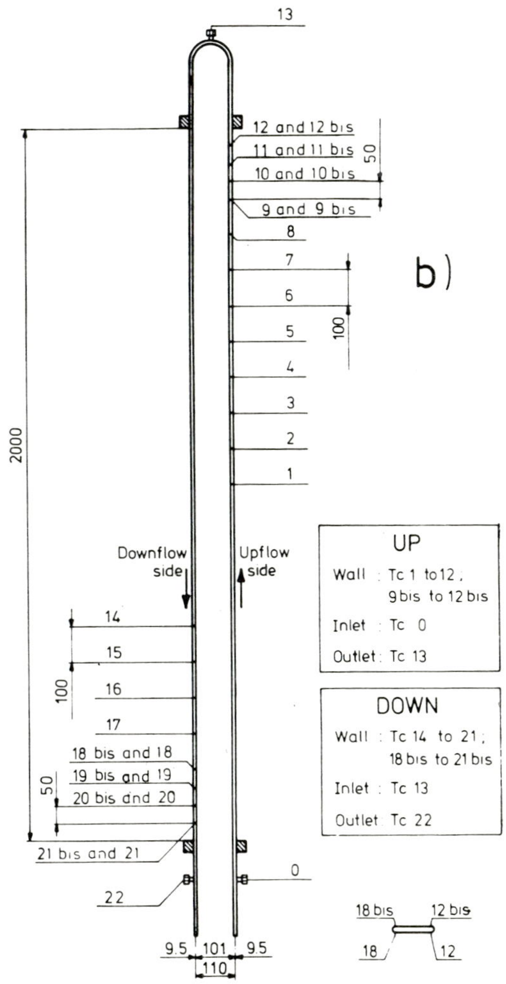


**Fig. 1(b): Test section with thermocouples position**

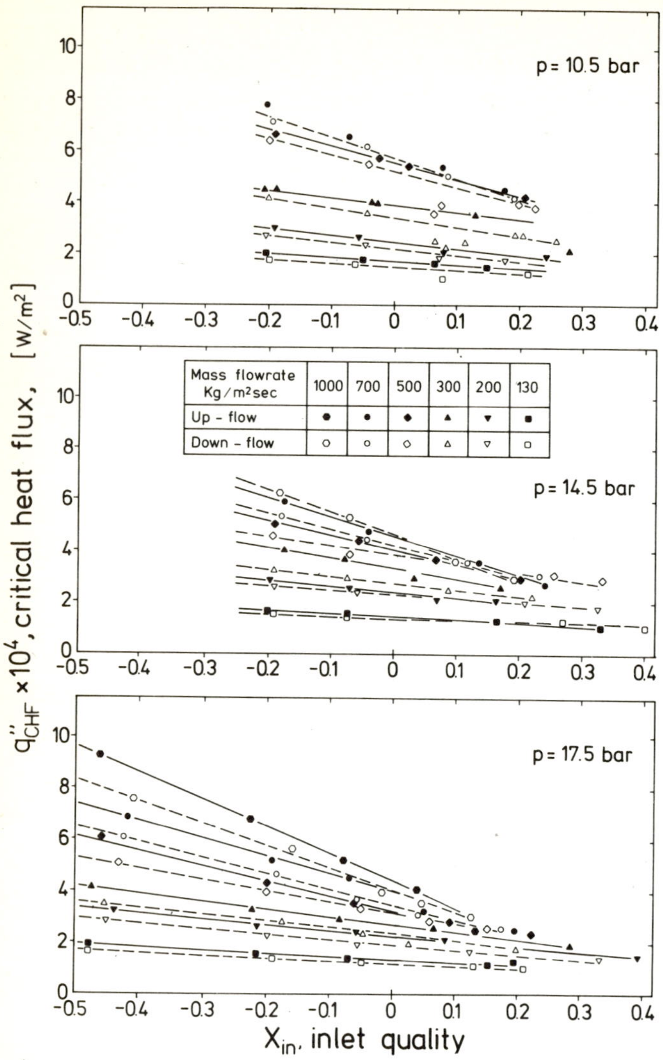


**Fig. 2: Comparison between UP and DOWN FLOW critical heat flux versus inlet quality**

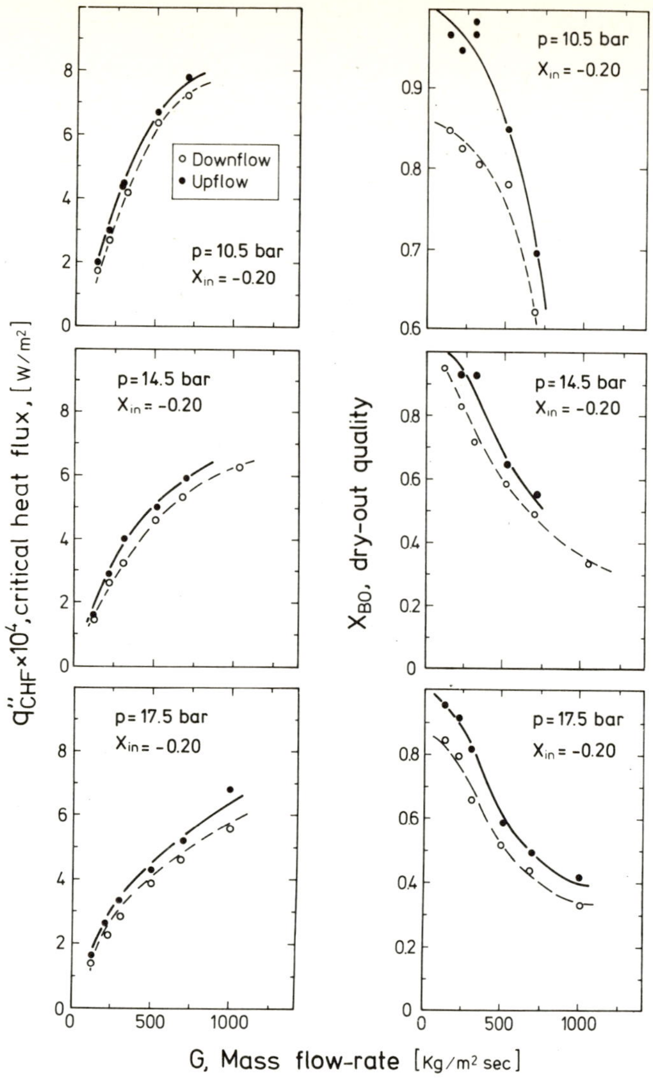


**Fig. 3: Critical heat flux and dry-out quality versus mass flow-rate in two different flow situations**

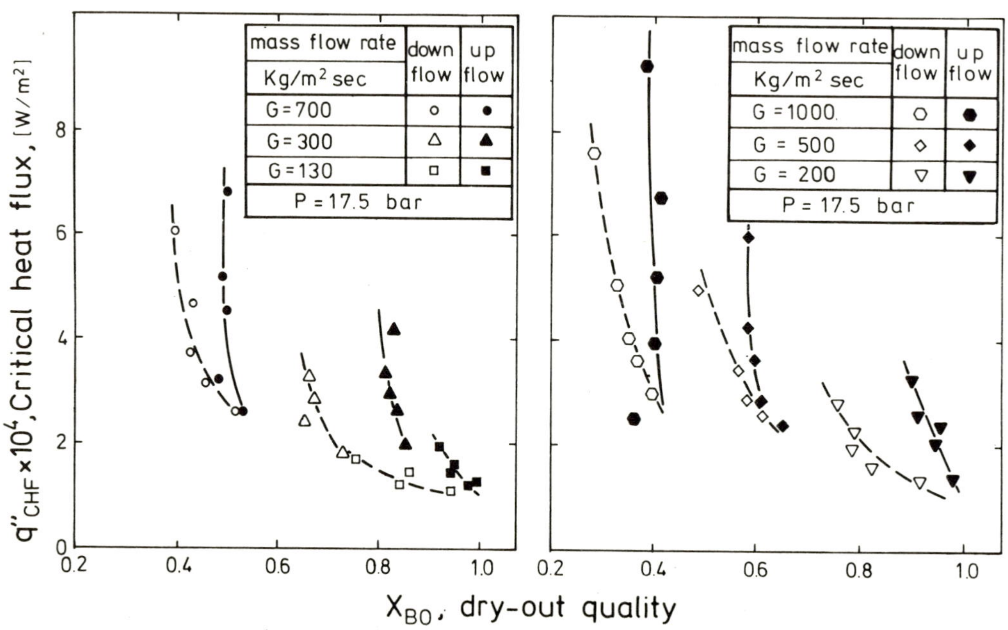


**Fig. 4: Critical heat flux versus dry-out for UP–FLOW and DOWN–FLOW at different mass flow-rates**

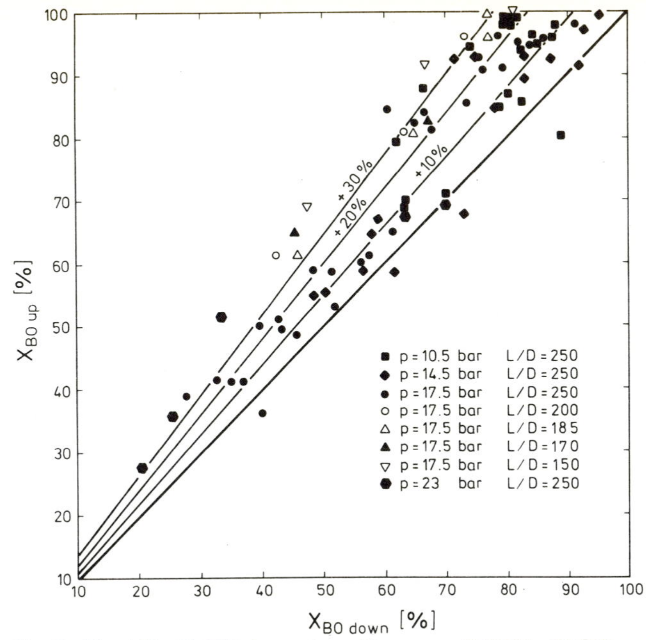


**Fig. 5: The UP–FLOW dry-out quality versus DOWN–FLOW dry-out quality, other parameters being fixed**

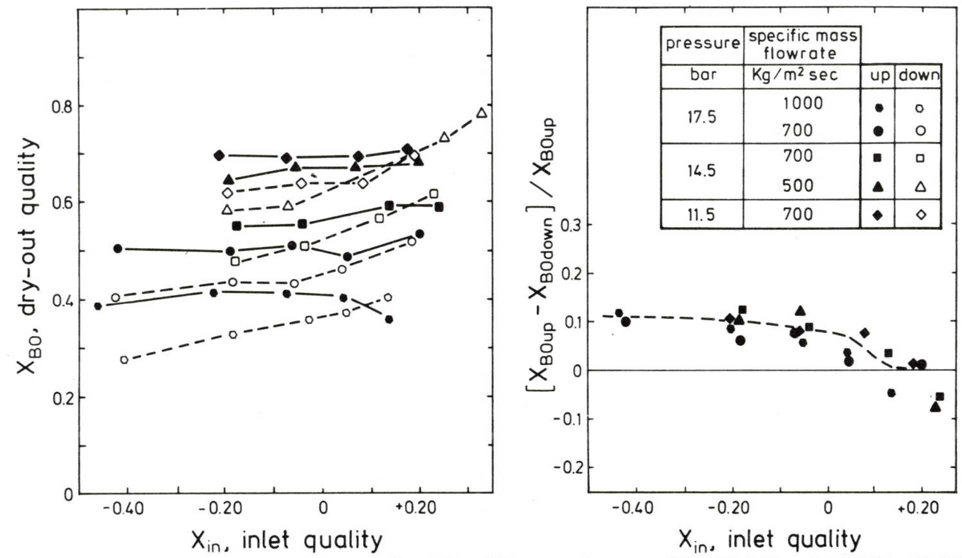


**Fig. 6: Dry-out quality versus inlet quality. The difference between UP–FLOW and DOWN–FLOW dry-out quality decreases as the inlet quality increases and disappears at $x_{in} \simeq 15\%$**

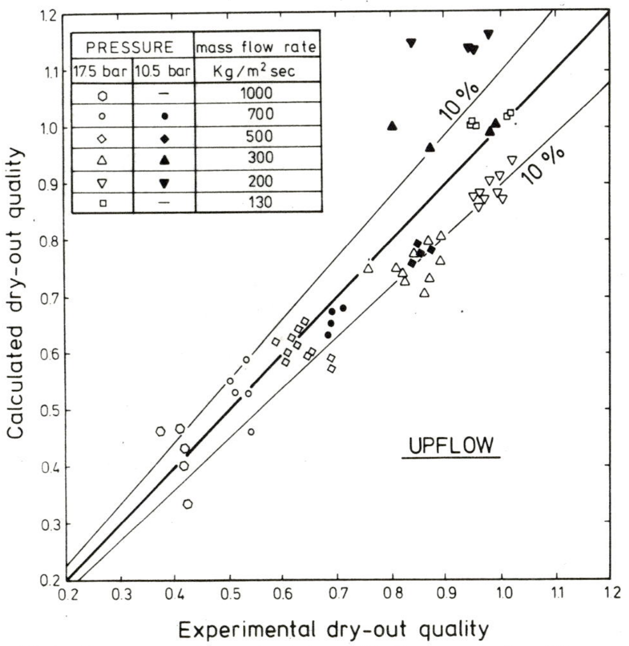


**Fig. 7(a): Dry-out quality calculated with the correlation (2), versus experimental dry-out quality**

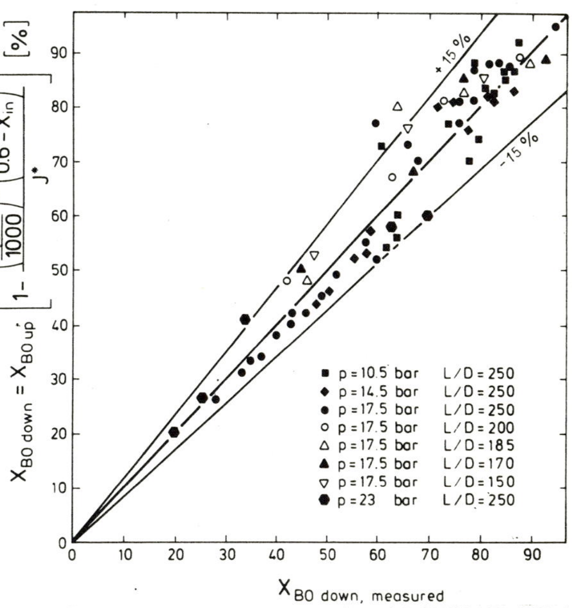


**Fig. 7(b): Comparison between predicted DOWN–FLOW dry-out and experimental data**

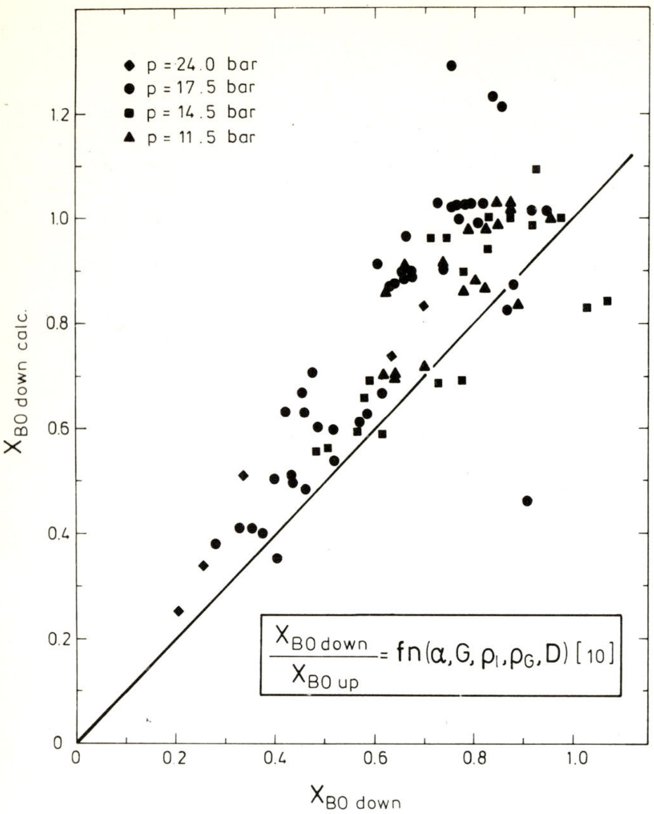


**Fig. 8: Comparison between computed and experimental dry-out in DOWN–FLOW**

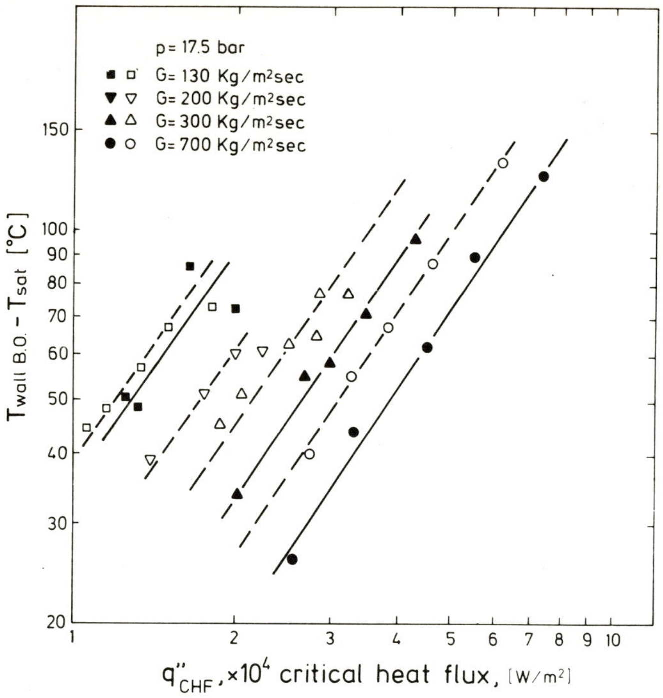


**Fig. 9: Comparison between UP–FLOW and DOWN–FLOW temperature rise in dry-out points**

C219/77

# THE EFFECT OF NON-UNIFORM AXIAL HEAT FLUX ON CRITICAL POWER

R. T. LAHEY, PhD
Rensselaer Polytechnic Institue, Troy, New York, USA
J. M. GONZALEZ – SANTALO, PhD
Universidad Autonoma Metropolitana-Azcapotzalco, Mexico

SYNOPSIS The integral techniques currently used to appraise LWR thermal margins are shown to be uniquely related. Moreover, an important relationship between uniform and non-uniform axial critical heat flux data is derived.

## NOTATION

$D_H$ = Hydraulic diameter

$G$ = Mass flux

$L_{B_c}$ = The critical boiling length

$p$ = System pressure

$q''(z)$ = The local heat flux

$\overline{q}''_{c(Eq)}$ = The equivalent critical heat flux (CHF)

$\langle x_e \rangle$ = Cross-sectional average thermodynamic equilibrium quality

$\lambda$ = Distance from the beginning of the heated length to the position of the boiling boundary

$\Omega$ = Upstream "memory" parameter

## INTRODUCTION

1. Boiling transition (DNB or dry-out) in light water nuclear reactor (LWR) fuel rod bundles is a rather complicated phenomenon. This is due not only to the complex geometry involved, but also the non-uniform axial and local (i.e., rod-to-rod) heat flux profiles which exist. The purpose of this paper is to present a unified treatment of techniques which can be used to quantify the effect of non-uniform axial heat flux on critical power.

2. Collier (Ref. 1) has summarized the various techniques currently used in correlating non-uniform axial heat flux data. These schemes are based either on the "local conditions hypothesis" or on an integral approach. The so-called "local conditions hypothesis" states in essence that it is only the local heat flux and local quality that control the boiling transition; that is, the upstream history is not important. In contrast, the integral concept implies that the upstream history is normally quite important. The integral concept can take many analytic forms, all of which imply that it is not sufficient to just know the system's local thermal-hydraulic parameters, but one must also know how the local situation came about. For example, the local flow quality ($\langle x(z) \rangle$) tells one the liquid content which exists at a particular axial position but yields no information as to how much of this liquid is actually on the heated surface.

3. The effect of axial heat flux profile, sometimes referred to as the "upstream memory effect", depends strongly on the flow regime and thus the quality. In an interesting experiment, De Bortoli (Ref. 2) demonstrated that the "local conditions hypothesis" was only valid for boiling transition (i.e., DNB) at high subcooling.

## DISCUSSION

4. In general, it has been found that the integral concept is much better than the "local conditions hypothesis" for heat flux profiles of practical concern in LWR technology. This can be clearly seen in Fig. 1, which presents FREON-114 boiling transition data taken by Shiralkar (Ref. 3) in an internally heated annulus.

If only the local heat flux and quality were important (as implied by the "local conditions hypothesis"), then the data for the various axial heat flux profiles would coalesce. Obviously they do not; in fact, a strong dependence on axial heat flux profile is evident.

5. There are various integral techniques for introducing the effect of upstream history. One of the most popular and widely used scheme is known as the Tong-F factor (Ref. 4) method. The Tong-F factor method is basically a flux-quality plane scheme in which boiling transition data, taken in a test section having uniform axial heat flux, is divided by the appropriate upstream weighting factor, F, to yield the appropriate thermal limit for the non-uniform axial heat flux profile under consideration. The functional form of the Tong-F factor, widely used in PWR evaluations, is given by Tong (Ref. 4) as,

$$F \triangleq \frac{\overline{q}''_{c(Eq)}}{q''_c(L_{B_c})} = \frac{\Omega}{q''_c(L_{B_c})\left[1 - e^{-\Omega L_{B_c}}\right]} * \int_{\lambda}^{L_{B_c}+\lambda} q''(z)\, e^{-\Omega\left[(L_{B_c}+\lambda)-z\right]}\, dz \qquad (1)$$

In general, it has been found (Ref. 4) that the parameter $\Omega$ decreases as critical quality increases. In order to appreciate the implications of this variation of $\Omega$ with quality, let us consider the integrand of Equation (1). For large values of the parameter $\Omega$, the flux weighting term, $\mathrm{Exp}\{-\Omega[(L_{B_c}+\lambda)-z]\}$, dies out strongly for axial positions less than the critical position, i.e., $z<(L_{B_c}+\lambda)$. Hence, for low quality boiling transition, there is little upstream memory effect and the local heat flux is very important. That is, the "local conditions hypothesis" may be valid for some PWR conditions of interest. In contrast, for higher quality boiling transition typical of BWR technology, $\Omega$ is small, and thus the exponential weighting term does not attenuate the effect of upstream heat flux as strongly. In the limit, as the parameter $\Omega$ goes to zero, there is equal weighting of the local and upstream heat fluxes. Using L'Hospital's rule, Equation (1) yields,

$$\lim_{\Omega\to 0} F \stackrel{\Delta}{=} \frac{\bar{q}''_{c(Eq)}}{q''_c(L_{B_c})} = \frac{1}{q''_c(L_{B_c})L_{B_c}} \int_{\lambda}^{L_{B_c}+\lambda} q''(z)\,dz \quad (2)$$

which implies that, $\bar{q}''_{c(Eq)}$, the equivalent uniform axial critical heat flux (CHF), is given by the non-uniform axial heat flux profile averaged over the critical boiling length. This interesting observation shall prove of value shortly.

6. There is another integral technique which has been developed specifically for boiling transition in the higher quality annular flow regime. This technique was originally introduced into the literature by investigators (Ref. 5) at CISE, but has been widely used by others (Refs. 6,7). The CISE type correlation is basically of the critical quality-boiling length ($\langle x_e\rangle_c - L_{B_c}$) type, where the upstream history is contained implicitly in the critical boiling length.

7. In order to appreciate the $\langle x_e\rangle_c - L_{B_c}$ approach, consider the situation of a heated rod bundle having a chopped cosine axial heat flux profile operating with fixed inlet subcooling, flow rate and system pressure. Moving from right to left in Fig. 2, the heat flux profile is shown with the typical location of a boiling transition indicated; the axial quality distribution is shown with the critical quality ($\langle x_e\rangle_c$) indicated; and, on the left, the boiling boundary ($\lambda$) and critical boiling length ($L_{B_c}$) are indicated. It can be seen that the boiling length is just the length over which bulk boiling occurs. It is, by convention, measured from the boiling boundary ($\lambda$). The critical boiling length is then just the length from the boiling boundary to the point at which a boiling transition is observed.

8. Techniques which assume the validity of the "local conditions hypothesis" attempt to correlate in terms of local values of $q''_c$ and $\langle x_e\rangle_c$. Integral flux-quality techniques (i.e., the Tong-F factor approach) also attempt to correlate in terms of these local parameters through appropriate modification of uniform axial CHF correlations. As seen in Fig. 2, rather than correlating $q''_c$ and $\langle x_e\rangle_c$, one could correlate $\langle x_e\rangle_c$ and $L_{B_c}$, where the boiling length is given implicitly by,

$$\int_{\lambda}^{L_B+\lambda} P_H q''(z)\,dz = GA_{x-s}h_{fg}\langle x_e\rangle \quad (3)$$

and the position of the boiling boundary ($\lambda$) is given by,

$$\int_{o}^{\lambda} P_H q''(z)\,dz = GA_{x-s}\left[h_f - h_i\right] = GA_{x-s}\Delta h_{SUB} \quad (4)$$

The integral nature of the boiling length parameter ($L_B$) is evident in (3) and (4). In order to determine whether this integral weighting of the upstream heat flux history is appropriate, non-uniform axial heat flux data, including the flux-quality data given in Fig. 1, were plotted in Fig. 3 in the critical quality-boiling length plane. It can be seen that this plane does an excellent job of collapsing the axial heat flux effect.

9. It should be evident that, in principle, one could develop a "best fit" line to the data and thus synthesize a critical quality-boiling length correlation for design purposes. The functional form of this correlation is quite arbitrary and depends completely on the ingenuity of the investigator. As an example, the well-known "CISE Correlation" has a functional form given by,

$$\langle x_e\rangle_c = \frac{a(p,G)L_{B_c}}{\left[b(p,G,D_H) + L_{B_c}\right]} \quad (5)$$

although more recent work at CISE (Ref. 8) indicates that a quadratic form of the correlation may yield a better data fit.

10. Either of the integral techniques discussed above can be used for LWR design applications, although due to its relative simplicity, the critical quality-boiling length technique is preferred for BWR evaluations. Let us now investigate further the relationship between these integral flux-quality and the critical-quality-boiling length techniques.

11. It is well known that boiling transition data for uniform axial heat flux profiles is frequently linear in the flux-quality plane. This trend can be seen in the data plotted in Fig. 1. Hence, the appropriate CHF correlation is of the form,

$$\bar{q}''_c = n(G,p) - m(G,p)\langle x_e\rangle_c \quad (6)$$

In order to better understand the significance of the $\langle x_e\rangle_c - L_{B_c}$ correlation given in (5), let us assume, in accordance with Equation (2), that the appropriate heat flux to compare with $q''_c$ is the heat flux averaged over the boiling length. That is, the equivalent uniform axial heat flux is

$$\bar{q}''_{c(Eq)} = \frac{1}{L_{B_c}} \int_{\lambda}^{\lambda+L_{B_c}} q''(z)\,dz \quad (7)$$

Combining (6) and (7) with the identity,

$$\langle x_e\rangle_c = \frac{\int_{\lambda}^{\lambda+L_{B_c}} P_H q''(z)\,dz}{GA_{x-s}h_{fg}} \quad (8)$$

one obtains,

$$\langle x_e \rangle_c = \frac{(n/m)L_{B_c}}{\left[(GA_{x-s}h_{fg})\,/\,(mP_H) + L_{B_c}\right]} \qquad (9)$$

Comparison of (5) and (9) indicates the unique relationship between the "CISE Correlation" and the parameters (n and m) in the uniform axial CHF correlation. This is an interesting and important result, since it implies that, at least in simple geometries, one need only perform uniform axial heat flux experiments to obtain the slope and intercept (m and n); then using Equation (9), one can predict boiling transition for situations having non-uniform axial heat flux.

12. In more general situations, in which the appropriate CHF correlation is no longer linear, as in Equation (6), the same procedure can be followed to obtain the corresponding critical quality-boiling correlation. In this case, however, the correlation will no longer have the same functional form as in Equation (5)

13. It has been found (Ref. 5) that for cases of extreme axial heat flux profiles, the simple flux averaging given by (7), is inadequate. In order to account for this effect, Silvestri (Ref. 9) introduced an axial weighting factor $\zeta(z)$ to obtain the expression for the equivalent uniform critical heat flux,

$$\bar{q}''_{c(Eq)} = \frac{\int_{\lambda}^{\lambda + L_{B_c}} q''(z)\,\zeta(z)dz}{\int_{\lambda}^{\lambda + L_{B_c}} \zeta(z)dz} \qquad (10)$$

where,

$$\zeta(z) = \left[q''(z)\right]^n e^{-\Omega(L_{B_c}-z)} \qquad (11)$$

14. It is interesting to note that for the case in which n=0, (10) and (11) yield,

$$\bar{q}''_{c(Eq)} = \frac{\Omega}{(1-e^{-\Omega L_{B_c}})}\int_{\lambda}^{\lambda+L_{B_c}} q''(z)e^{-\Omega\left[(L_{B_c}+\lambda)-z\right]}dz \qquad (12)$$

Comparing (11) and (12), it is seen that the Tong-F factor approach and the generalized critical quality-boiling length approach are completely equivalent.

15. For conditions of interest in BWR technology, the ratio of peak-to-average axial heat flux is normally less than two, and thus to a good first approximation, standard $\langle x_e \rangle_c - L_{B_c}$ correlations are valid. Indeed, the General Electric Company currently uses a critical quality-boiling length correlation (GEXL) for BWR thermal analysis (Ref. 7).

16. The main difference between the two integral techniques just discussed has to do with how the upstream history is introduced. As shown schematically in Fig. 4, the Tong-F factor approach modifies the uniform axial CHF correlation to yield the appropriate correlation for the existing non-uniform axial profile. In contrast, the critical quality-boiling approach modifies the heat balance, $q''(\langle x_e \rangle)$, rather than the CHF correlation ($\bar{q}''_c$). Both techniques are equivalent and yield the same thermal margin.

## SUMMARY

17. A unique relationship has been demonstrated between the two most widely used LWR integral techniques. It has been shown that, for the same operating conditions, these techniques give the same thermal margin. Moreover, an important, and highly useful, relationship between uniform axial and non-uniform axial boiling transition data has been derived.

## REFERENCES

1. COLLIER, J. G. Convective boiling and condensation. McGraw-Hill, 1972.
2. DE BORTOLI, R. A., et al Forced convection heat transfer burnout studies for water in rectangular channels and round tubes at pressures above 500 psia. WAPD-188, 1958.
3. SHIRALKAR, B. S. Analysis of non-uniform flux CHF data in simple geometries. NEDM-13279, 1972.
4. TONG, L. S. Boiling crisis and critical heat flux. TID-25887, 1972.
5. BERTOLETTI, S., et al Heat transfer crisis with steam-water mixtures. Energia Nucleare, 12, 1965.
6. KEEYS, R. K. F., et al Post burnout heat transfer in high pressure steam-water mixtures in a tube with cosine heat flux distribution. AERE-R6411, 1971.
7. General Electric BWR thermal analysis basis (GETAB): Data, correlation and design application. NEDO-10958, 1973.
8. GASPARI, G. P., et al Dryout onset in flow stoppage, depressurization and power surge transients. Proceedings of CREST Specialists Meeting on LWR ECC, 1972.
9. SILVESTRI, M. On burnout equation and location of burnout points. Energia Nucleare, 13, 1966.

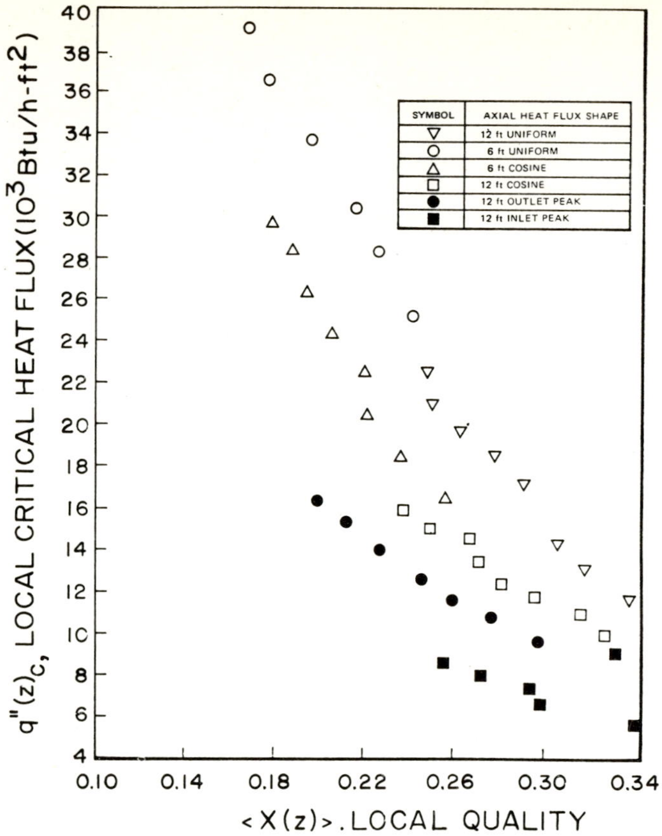


**Fig. 1: Heat — flux vs quality at location of boiling transition, Freon 114 annulus data, $D_1$ = 0.563 in., $D_2$ = 0.875 in., 6- and 12- ft heated length, various axial profiles, P=123 psia, G=0.54 x $10^6$ lb/h — $ft^2$ (Reference — 3)**

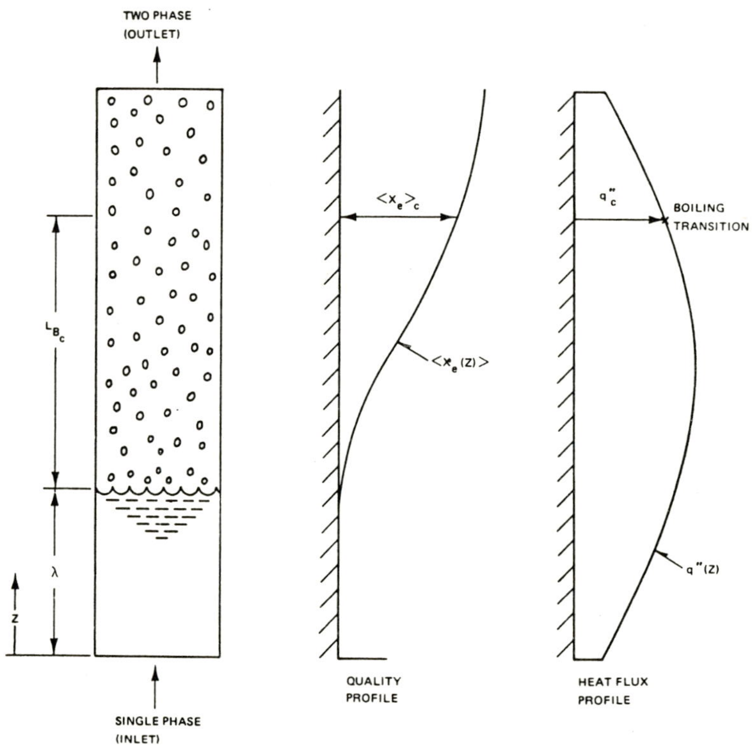


**Fig. 2: Schematic representation of boiling length $L_B$**

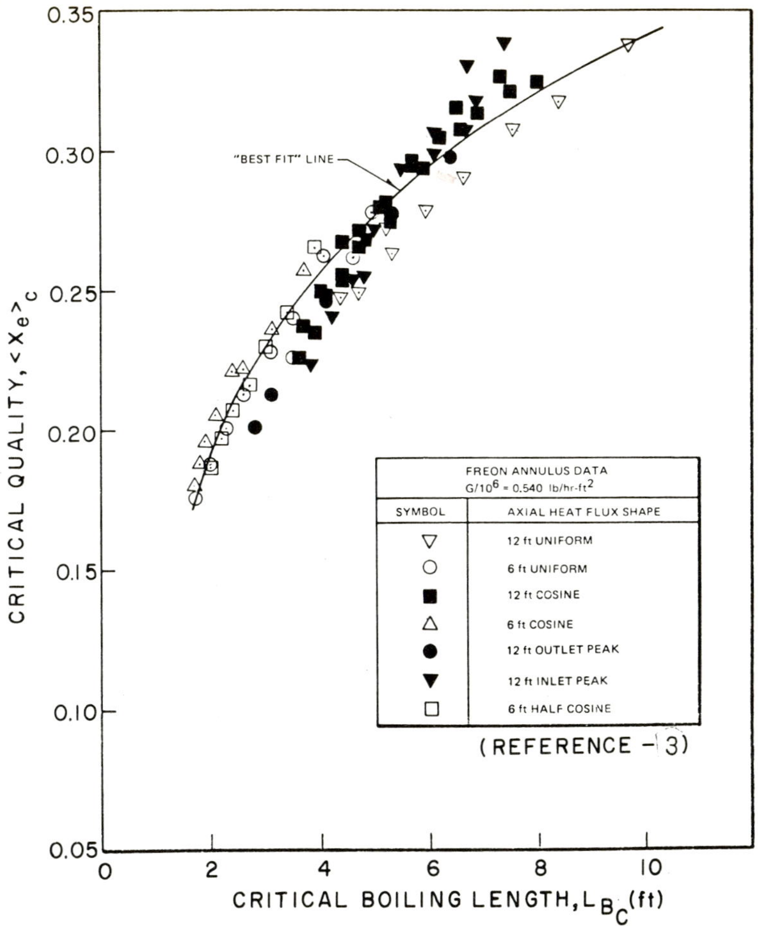


Fig. 3: Critical quality vs boiling length, Freon — 114 annulus data

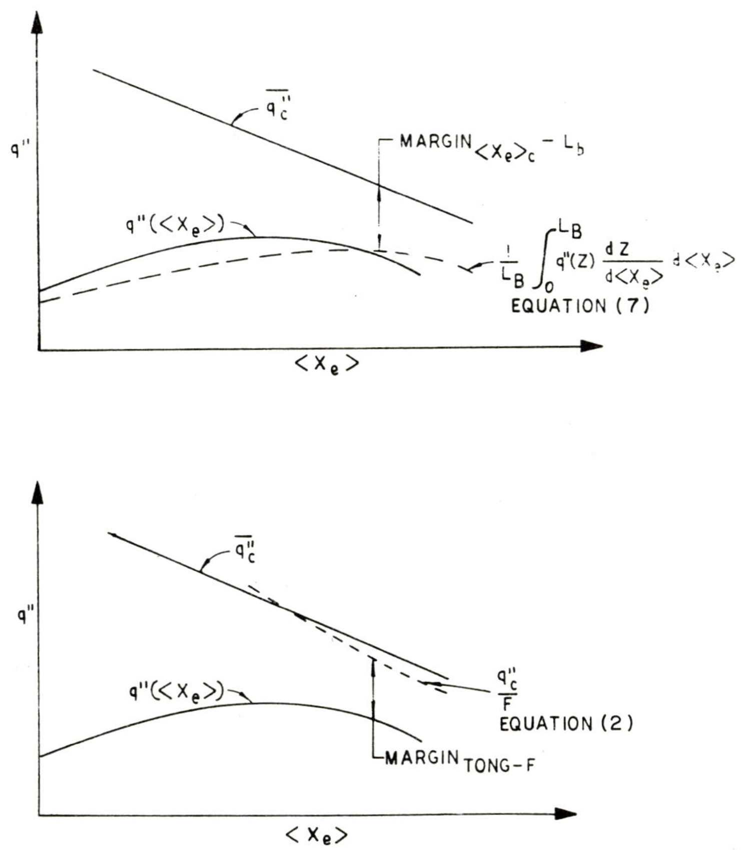


Fig. 4: Comparison of $< X_e >_c$ and Tong — F factor methods in the flux — quality plane

# C220/77

# HEAT TRANSFER ANALYSIS OF EXPERIMENTS SIMULATING A LOSS-OF-COOLANT ACCIDENT

S. MALANG, K. RUST
Institut für Reaktorbauelemente, Kernforschungszentrum Karlsruhe, GFR

SYNOPSIS Thermodynamic out-of-pile experiments simulating a loss-of-coolant accident (LOCA) are performed with electrically heated rods, which are instrumented with internal thermocouples because surface measurements would influence the coolant flow. The data analysis problem is therefore the solution of the nonlinear problem to determine the surface temperature, the surface heat transfer coefficient, and the surface heat flux from internal temperature measurements. A digital computer code was developed for the analysis of the experimental data. The code has different options. The major application of the code is the numerical solution of the inverse heat conduction problem involving temperature dependent material properties and complex multilayer geometries.

## INTRODUCTION

In all experiments for investigating loss-of-coolant accidents (LOCA), the fuel rods are simulated with electrically heated rods. The main purpose of these experiments is to determine the sheath temperature and the heat transfer to the coolant during a LOCA. To measure the sheath temperature, the rods are instrumented with thermocouples which are placed in the sheath wall instead of on the rod surface to avoid any coolant flow disturbance. Therefore they do not show the surface temperature, which is necessary to calculate the heat transfer coefficient.

For the steady-state case the temperature difference between the thermocouple and the surface can be determined by a simple calculation or by a calibration experiment. However, it is much more difficult to calculate the temperature difference for a transient case with fast changes in power and temperatures. Moreover, due to the heat capacity of the rod, the surface heat flux is not only influenced by the instantaneous power, but also by the history of the transient.

To take all these influences into account, the computer code HETRAP (ref. 1) was developed to calculate the time-dependent value of the heat transfer coefficient during a fast transient. This code determines the temperature field in the rod and the heat transfer coefficient using data measured during the experiment. New options have been added to the code and improved calculation procedures are implied (ref. 2).

Special emphasis was placed on the numerical technique to avoid oscillations which are mainly caused by thermocouple noise. The well known method suggested by J.V. Beck (ref. 3) has been modified and is used in the code.

An other problem in analysing experiments performed with electrically heated rods is the difference in the thermal behaviour between a fuel rod and a heater rod due to their different design and structure. If the same power trace is used, the two rods will have generally a different surface temperature, even for identical transients of the heat transfer coefficient. Therefore a code option has been developed to calculate the temperature profiles in both rods in parallel to show the quality of simulation.

The application of this code to the analysis of experiments simulating the reflood phase of a loss-of-coolant accident will be demonstrated.

## PROBLEM DEFINITION

In an experiment simulating a loss-of-coolant accident usually the following time dependent values are measured

- the clad temperature,
- the power supply to the rod,
- the coolant pressure.

The measurement of the local coolant temperature is difficult and usually not very reliable. On the other hand, the transients of the surface heat flux as well as the surface temperature can be calculated without knowledge of the coolant temperature, since they do not depend on the definition of the coolant temperature. For the calculation of the heat transfer coefficient it is more important to have a clearly defined coolant temperature which allows the recalculation of the heat transfer coefficient for any differently defined temperature than a local temperature which is not very reliable. In a blowdown experiment usually the saturation temperature is used, calculated from the measured pressure. This method can result in negative heat transfer coefficients in the later part of a reflood test because the water and the surface temperature of the rod are lower than the saturation temperature. Therefore it is suggested to use the water temperature at the inlet of the bundle, taking into the bargain that the heat transfer coefficient, calculated with the so defined coolant temperature, is lower than the physical one.

The basic problem is to determine the surface

heat flux as a function of time. Because of the heat capacity of the rod, this heat flux will lag the power during a transient and will vary with the cooling condition. Therefore, the heat transfer coefficient at a given moment cannot be calculated using only the momentary values of the sheath temperature and supplied power; the whole history from the beginning of the transient must be taken into account.

## MODEL AND METHOD FOR CALCULATING THE TEMPERATURE FIELD IN A ROD

The rod may consist of an arbitrary number of cylindrical shells with or without heat generation. This heat generation can be a function of either radius for a fuel rod or temperature for an electrically heated rod. In the first case the radial power profile is an input; in the second case either the power or the current may be given. All material properties can be temperature dependent. The material functions of the HETRAP code are revised and physical properties of additional materials are added. These functions are compiled in the subroutine PEW (ref. 4) which can be used independently of the HETRAP code.

Heat is conducted inside the shells, and there can be radiation and/or contact resistance between the shells. The contact resistance can vary from zero to a very high value.

Since a finite-difference method is used, the rod must be divided into an arbitrary number of nodes in which the whole mass is concentrated. The radial location of these nodes is described by input data, and the volumes associated with each node and the thermal resistance between the nodes are calculated.

Two intergration methods are implied in the code. The code itself selects either an explicit or a mixed explicit/implicit method in order to minimize the computing time.

## CALCULATION OF THE HEAT TRANSFER COEFFICIENT AND THE SURFACE HEAT FLUX

One option of the code solves the inverse heat conduction problem determing the surface temperature, the heat transfer coefficient, and the resultant surface heat flux as function of the time by using measured sheath temperatures. That option is similar to the HCODE written by General Electric (ref. 5). A typical problem hereby is the sensitivity to thermocouple noise which is mainly caused by SCR (silicon controlled rectifier) controlled power supply. Any numerical technique that seeks to solve the inverse heat conduction problem exactly is inherently instable since small changes in the measured data can cause large changes in the solution, i.e. the surface temperature. The oscillations are amplified with increasing distance between the thermocouple position and the rod surface. The best known method to dampen numerical oscillations in the surface temperature and consequently in the heat transfer coefficient as well as in the heat flux has been suggested by J.V. Beck. Using Beck's approach for the determination of the heat transfer coefficient for a single time step the heat flux at the surface is kept constant for several advanced time steps. For one given heat flux the sum of the squares of the differences between the measured temperature and the temperature calculated for the position of the thermocouple is determined. This sum is minimized by variation of the heat flux. Since the minimization has been done for a number of time steps, but the so obtained heat flux is applied to one time step only, there will result a difference between the measured and the calculated temperature. To avoid that, Beck's approach has been modified by adding of an integral correction with the effect that the averaged value of the integral

$$\int_0^t \left( T_{measured} - T_{calculated} \right) \cdot dt \qquad (1)$$

becomes zero. That modified approach of Beck has been implemented in our computer code.

## OPTION FOR COMPARING OF THE HEATER ROD AND FUEL ROD

The calculation of heat transfer coefficient has been shown above. In another option that heat transfer coefficient is applied to a nuclear fuel rod, resulting in the surface temperature and the heat flux of this fuel rod. This is a good check, how well the heater rod simulates the fuel rod since in the ideal case the two fluxes are identical.

## APPLICATION OF THE CODE TO A REFLOOD EXPERIMENT

An experiment simulating the reflood phase of a loss-of-coolant accident is performed at the Nuclear Research Centre Karlsruhe in the frame of the Project Nuclear Safety (PNS). The code is used for the analysis of this experiment. In Fig. 1 the surface heat flux is plotted versus the time for one test. Varied is the number of the advanced time steps. It can be seen that the oscillation are considerably reduced by using 6 advanced time steps as illustrated in Fig. 1c. A further improvement can be obtained by applying of an additional smoothing routine. The result is pointed out in Fig. 1d. In this case each point is substituted by the arithmetic mean value of the previous and following 9 points.

The comparison to a nuclear fuel rod is shown in Fig. 2. The cross section of the two rods can be seen from Fig. 2a. For the heater rod the transients of the linear rod power, surface heat flux, and heat transfer coefficient are shown in Fig. 2b. It should be mentioned that the heat transfer coefficient is based on the flooding water temperature instead of the local coolant temperature as mentioned above. In Fig. 2c and Fig. 2d the surface temperatures and the stored energies, respectively are plotted for the two rods. The graph shows that there is a difference in the surface temperature which stems from the different geometrical structure and the materials. The conclusion of Fig. 2c is, that the fuel rod would be quenched earlier in this case. However, it depends of course strongly on the used parameters, especially the gap coefficient between fuel and cladding. The material properties are specified in the subroutine PEW of the modified HETRAP code as functions of the temperature. The following interface coefficients were assumed:

a) heater rod
   h = 3,0 W/(cm K) at all locations

b) fuel rod
   h = 1,0 W/(cm K)

## REFERENCES


1. Malang S.
   "HETRAP: A Heat Transfer Analysis Program"
   ORNL-TM-4555, Sept. 1974

2. Malang S., K. Rust
   "HETRAP II: A Heat Transfer Analysis Program"
   KFK-Report will be published 1977

3. Beck J.V.
   "Nonlinear Estimation Applied to the Nonlinear Inverse Heat Conduction Problem"
   Int. J. Heat Mass Transfer. Vol. 13, pp. 703-716, 1970

4. Rust K., Malang S., Götzmann W.
   "PEW - Ein FORTRAN IV-Rechenprogramm zur Bereitstellung physikalischer Eigenschaften von Werkstoffen für LWR-Brennstäbe und deren Simulatoren"
   KFK-Ext. 7/76-1, Dec. 1976

5. Mussy R.J., Avila J.H., Root D.E.
   "Determination of Transient Heat Transfer Coefficients and the Resultant Surface Heat Flux from Internal Temperature Measurements"
   GEAP-20731, Jan. 1975


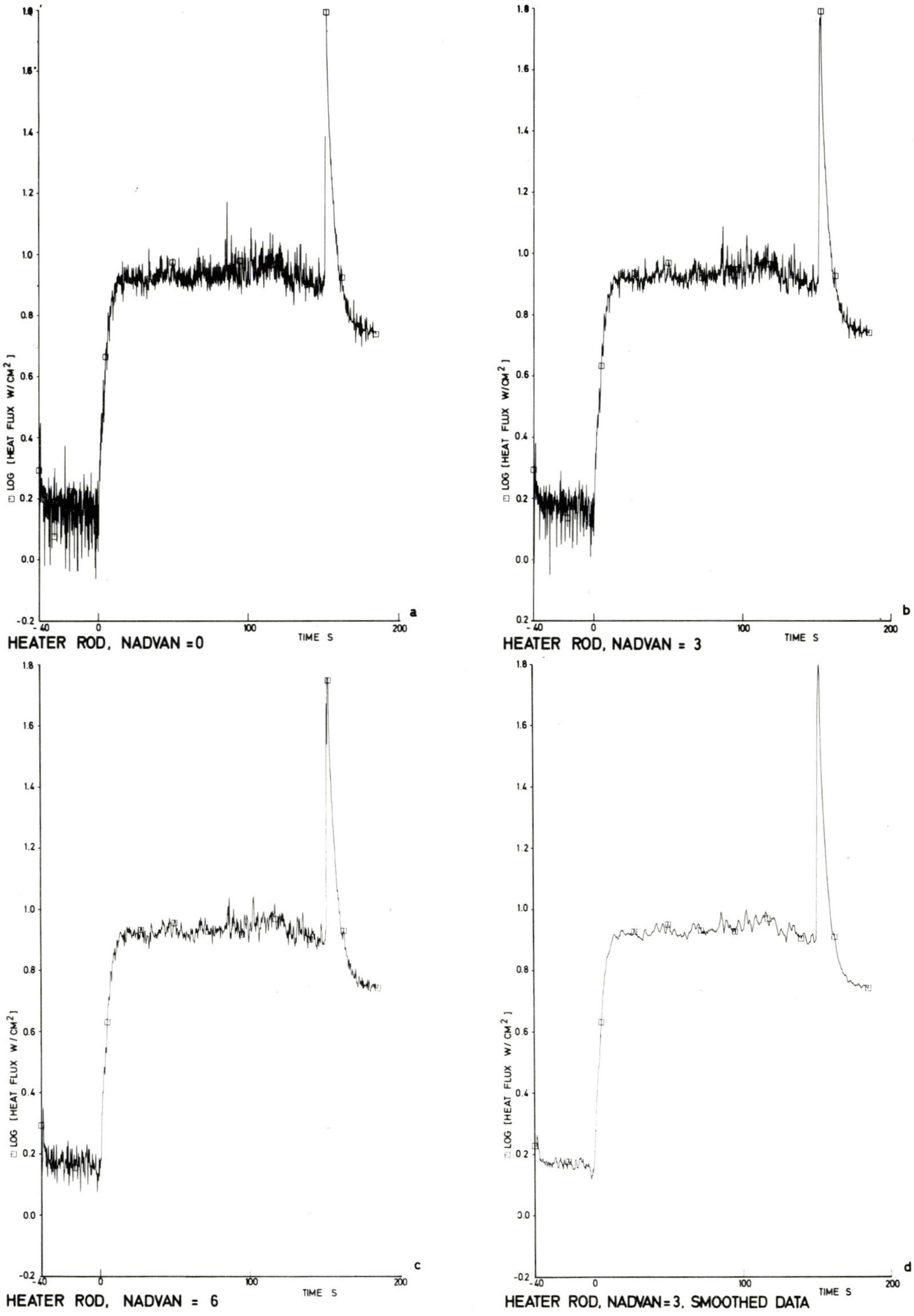


Fig. 1: Influence of the 'advanced time steps' on the heat flux

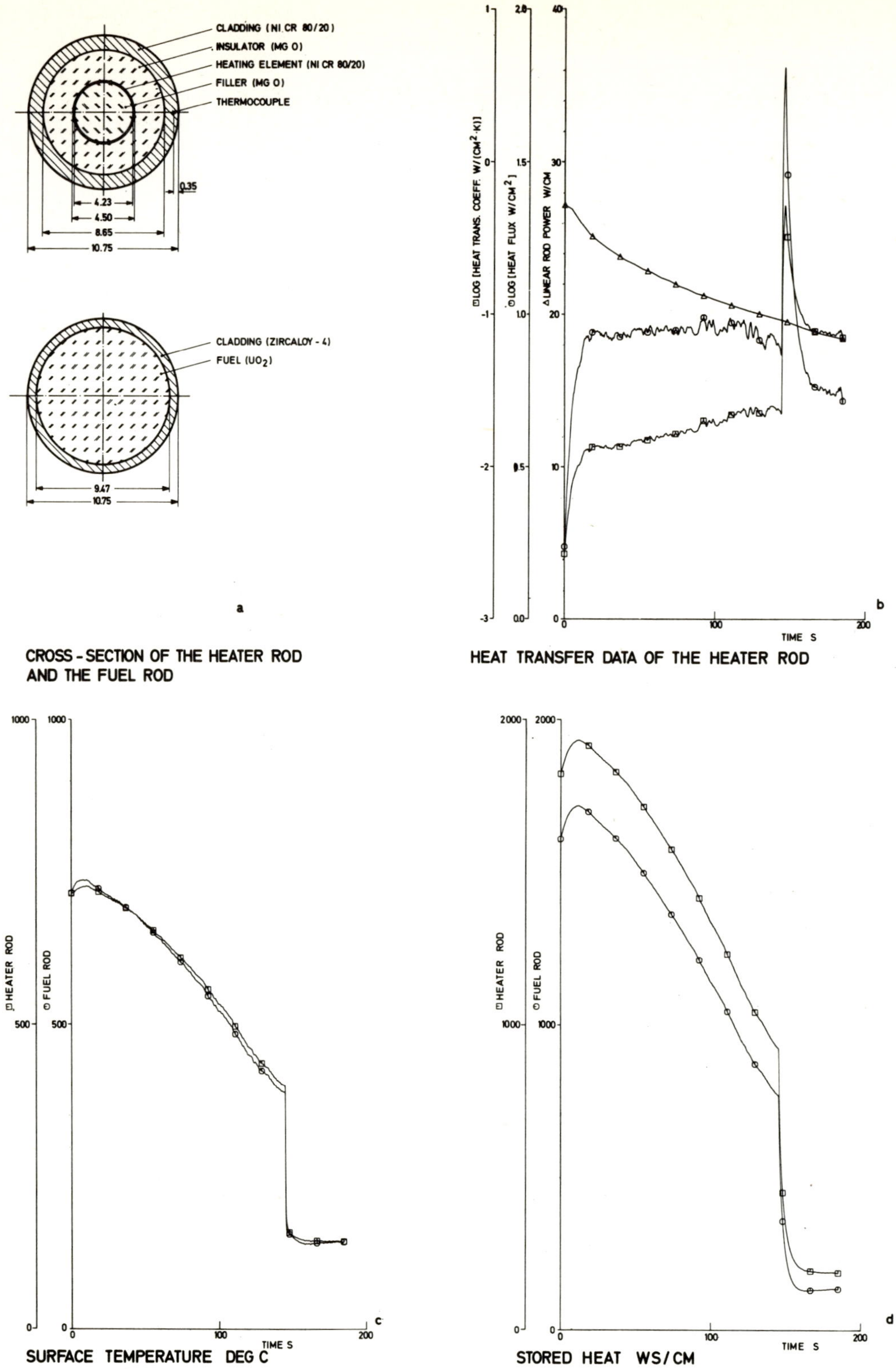


Fig. 2: Heat transfer analysis for heater rod and fuel rod

C221/77

# A REVIEW OF DRYOUT DATA IN SGHWR FUEL BUNDLE SIMULATIONS

N. A. BAILEY, MA, PhD and D. H. LEE, BSc, MSc, MIMechE
UKAEA, Reactor Development Division, Winfrith

SYNOPSIS This paper reviews the large number of experiments which have been carried out at AEE Winfrith investigating dryout in electrically heated simulations of SGHWR fuel clusters. The effect of flow conditions is examined in the context of a set of reference experiments and the effect of varying geometrical factors is discussed in detail, the large number of experiments allowing the effect of single parameter variation to be explored.

NOMENCLATURE

1. G Mass velocity ($kg/m^2s$)
   P Pressure (bars)
   Qp Dryout Power at P bars (MW)
   $\Delta H_{in}$ Inlet Subcooling (KJ/kg)

## INTRODUCTION

2. The SGHWR (Steam Generating Heavy Water Reactor) is designed so that the fuel pins do not suffer dryout under normal conditions although it may occur during severe transients. If dryout occurs in such a reactor and the surface of the fuel pins is no longer cooled by an evaporating film of water, significant surface temperature excursions are experienced which could affect the integrity of the fuel. Information is therefore required so that the margin to dryout can be assessed over a wide range of reactor conditions.

3. Dryout in a fuel cluster is complex and representative experimental investigations are of paramount importance. This paper summarises the results of a large number of tests which have been carried out on the 9 MW Rig in the Heat Transfer Laboratory at Winfrith. The tests reported here are for 36 pin fuel clusters and are discussed in more detail elsewhere (Ref 1)

## EXPERIMENTAL TECHNIQUES

4. As the SGHWR is a pressure tube reactor with a channel power of about 4.5 MW it has been possible to model accurately a complete flow channel with a full length fuel cluster. The fuel pins in these tests are represented by hollow stainless steel tubes heated by passing an electrical current along them and supported in representative grids. Because of the need for electrical connections at each end of the pins some small differences between reactor and rig are inevitable. However as the electrical connections are well away from the region where dryout is found to occur it is assumed that their effect on dryout is negligible.

5. The radial heat flux profile found in the reactor can be simulated by using different wall thicknesses for each ring of pins and heat flux tilt across an individual pin can be simulated by machining the tube wall eccentrically. In the experiments reported here pin tilt has been restricted to the outer row of pins where it is most marked in the reactor. The axial heat flux profile, which takes the form of chopped cosine in the reactor, can be simulated by having tubes drawn specially to give a tube bore varying with length. However such a process is expensive to carry out and cannot be combined with pin tilt so a substantial fraction of the tests carried out have been with axially uniform heat flux profiles.

6. As the test cluster is electrically heated the pressure tube has to be represented by a ceramic liner inside a pressure vessel and the emergency spray cooling sparge pipe by an insulated rod. The tests reported here were carried out without any spray flow.

7. The experimental techniques is to set up constant inlet conditions of flow, pressure and temperature and to increase the power by small steps of about 1% until a sudden change in wall temperature is detected. For uniform axial flux profiles where dryout always occurs at outlet the temperature rise is detected by a single 4 thermocouple thermopile in each pin 12 mm from the end of the heated length. With non uniform axial flux profiles where the location of dryout varies widely six 300 mm long resistance coils are placed inside each tube over the upper half of its length and their resistances are monitored on 3 Wheatstone Bridges. Local overheating causes one of the bridges to go out of balance and it is therefore possible to determine the approximate location of dryout in addition to the dryout power.

8. A detailed assessment of the accuracy of the measured data and the influence of an error in any parameter on the dryout power has been carried out. This showed that the data for dryout power should lie within an r.m.s. error band of $\pm$ 2.4% and that the electrically heated tests, compared with the reactor, would be biased on the low side by up to 1.5%.

THE REFERENCE EXPERIMENT

9. A wide range of arrangements have been tested to investigate the effect of bundle eccentricity, radial and axial heat flux distributions and heat flux tilt and this has allowed the effect of varying one parameter at a time to be investigated. One series of tests has been selected as the reference and this represented a differentially enriched concentric bundle with pin heat flux tilt on the outer ring of pins and with an axially uniform heat flux profile. The relative power levels of the three rings of pins were 100, 100, 80 for outer, middle and inner rings respectively and the heat flux tilt resulted in the heat flux on the outward facing side of the pin being 15% above the mean and 15% below on the opposite side.

10. These reference tests covered a wide range of inlet subcoolings from 20 to 430 kJ/kg, mass velocities from 680 to 2370 kg/m$^2$sec and pressure of 44.8, 58.6 and 68.3 bars. Dryout was found to occur on the middle ring of pins and the results for dryout power are shown in (Fig 1) where it can be seen that dryout power increases with increasing inlet subcooling, increasing mass velocity and, marginally, with decreasing pressure. The results for a pressure of 68.3 bars have been correlated on a least squares basis by

$$Q_{68.3} = 0.28G^{0.416} + (1.21 + 0.0034G)\Delta Hin/1000 \qquad (1)$$

and similar expressions were obtained for the other pressures. These correlations fit the data with an r.m.s. scatter of $\pm$ 1.8% which compares favourably with the scatter band predicted. More sophisticated correlations (Ref 2) are required to cover the whole range of parameters of interest but increased scatter must be expected.

11. Examination of results from other tests shows that although dryout power does increase marginally as pressure is reduced from 68.3 to 44.8 bars this trend does not continue to lower pressures. A further decrease in pressure below 40 bars leads to a decreasing dryout power and it has been found for a wide range of cluster tests that the pressure effect can be normalised to give

$$Q_p = Q_{68.3}\left(1 + 0.36(1 - P/68.3) - 0.93(1 - P/68.3)^{3.5}\right) \qquad (2)$$

for the pressure range 16-70 bars and mass velocities of 1000 kg/m$^2$sec and above. Equation 2 is based on a constant inlet subcooling. For the lower mass velocities the pressure effect is even smaller.

THE EFFECT OF VARYING PARAMETERS

12. A number of factors are considered to influence the onset of dryout and these effects are considered in turn.

SINGLE VERSUS DIFFERENTIAL ENRICHMENT

13. Two sets of tests have been carried out on bundles with uniform axial heat flux profiles without pin tilt. One corresponded to a differentially enriched situation with relative pin power ratings of 100, 100, 80 on the outer, middle and inner rings of pins while the other corresponded to a single enrichment situation with relative power ratings of 100, 66, 57. The results of these two sets of results are shown in (Fig 2) where it can be seen that the dryout power is about 6½% higher for the differentially enriched situation. In the differentially enriched cluster dryout occurred on the middle ring of pins but in the single enrichment case it occurred on the outer, most highly rated, ring of pins. Thus although the single enrichment cluster dried out at a lower power and lower exit steam quality the local heat flux at dryout was about 12% higher than in the differentially enriched case leading to a larger temperature excursion. Similar results have been found on clusters with non-uniform axial heat flux profiles. The overall effect of having differential enrichment is therefore to increase dryout power and reduce temperature excursions at dryout which is a very desirable trend.

AXIAL HEAT FLUX PROFILE

14. Two sets of tests have been carried out on single enrichment clusters without pin tilt. One had a uniform axial heat flux profile and the other a chopped cosine profile with a peak to average heat flux ratio of 1.38. The results of these tests are shown (Fig 3) where it can be seen that there is no significant difference at the lower mass velocities but an increase of about 6% in dryout power is observed for mass velocities above 2000 kg/m$^2$sec with the cosine heat flux profile. For such a profile the axial dryout position moved upstream with increasing mass velocity to the region where the local heat flux would be higher than the average value and the steam quality lower than at outlet. The temperature excurions with a cosine heat flux profile at high mass velocity would thus be expected to be higher than with a uniform heat flux profile.

BUNDLE CONCENTRICITY AND PIN TILT

15. Several tests using axially uniform heat flux profiles have been carried out with the bundle concentric or eccentric and with or without pin tilt. Both differentially enriched and single enrichment representations have been used. The eccentricity (0.6 mm compared with a maximum possible 0.85 mm) was induced by fixing thicker pads on one side of the spacer grids.

16. For the differentially enriched cluster pin tilt and bundle eccentricity had negligible effects and all the data could be correlated within an r.m.s. scatter of about $\pm$ 2.7%. This is not surprising as dryout occurred on the middle ring of pins and might not 'sense' either a movement of the cluster towards the pressure tube or pin tilt on the outer ring of pins.

17. As expected the effect of pin tilt and eccentricity was much more marked for the single enrichment clusters where dryout normally occurred on the outer ring of pins and examples of the results are shown (Fig 4). For a bundle with pin tilt eccentricity induced a reduction

in dryout power of about 12% which is due to the reduction in subchannel flow area adjacent to the pin undergoing dryout while for a concentric cluster the omission of pin tilt involved a marginally significant reduction in dryout power of 3%.

PIN SPACING

18. Pin spacing tests on a 7 pin cluster using Freon-12 as the working fluid (Ref 3) have shown that dryout power is not very sensitive to pin spacing except when the pins are very close together or when the outer pins are very close to the shroud tube and this is consistent with the eccentricity effect observed for the single enrichment cluster. As the variation in individual pin spacing in the SGHWR will be very small the effect of such variation on dryout can be neglected.

PIN BOWING

19. Pin bowing tests on a single cosine flux profile pin in an annulus showed negligible differences between a straight pin and one bowed into contact with the wall under the vast majority of conditions. Dryout normally occurred remote from the point of contact and only occurred at that point when the local heat flux was above 160 $W/Cm^2$, a value far in excess of SGHWR ratings.

CREEP AFFECTED PRESSURE TUBE

20. Tests on a 19 rod bundle with Freon-12 have shown that an increase in pressure tube diameter increases the dryout power for a given total flow rate for situations, such as single enrichment, where dryout occurs on the outer pins. Even so 3% creep in the pressure tube would only yield an increase of about 1½% in dryout power. For a differentially enriched cluster the effect will be even smaller and can be neglected.

FOULING

21. The 9 MW Rig tests have been carried out 'clean'. Single pin tests (Ref 4) have shown a decline in dryout power with increasing fouling thickness. For a mass velocity of 2700 $kg/m^2sec$ and a fouling thickness of 0.03 mm (typical of the reactor) the decline was negligible but the same thickness induced a 5% reduction at 1350 $kg/m^2sec$. Surface coatings up to 0.15 mm thick reduced the dryout power by up to 10%.

COMPARISON WITH REACTOR

22. Measurements of dryout power have also been made in the prototype SGHWR (Ref 5). When allowances were made for the small differences between the reactor and the 9 MW Rig the inpile results are found to agree satisfactorily with the rig data with some eccentricity of the cluster. The reactor data suggested that the eccentricity might by varying with flow rate.

CONCLUSIONS

23. The following conclusions may be drawn from this paper.

(a) a large number of experiments have been carried out on 36 pin clusters over a wide range of conditions and the experimental scatter is small.

(b) the effects of flow conditions and geometrical variations on dryout power have been investigated and quantified. Bundle eccentricity and pin tilt only influence dryout power for single enrichment.

(c) comparison of 9 MW Rig and reactor data shows no significant deviations and validates the concept of "out-of-pile" testing.

(d) Correlations are given for the reference experiment dryout power at 68.3 bars and a pressure correction correlation for the pressure range 16-68 bars.

(e) In terms of dryout characteristics a differentially enrichment cluster has several advantages over a single enrichment cluster.

REFERENCES

1. BAILEY N A and LEE D H. A review of dryout data in SGHWR fuel bundle simulations. AEEW - R 1001, 1976

2. BOWRING R W. A new mixed flow cluster dryout correlation for pressures in the range 0.6 - 15.5 $MN/m^2$ - For use in a transient blowdown code. Paper no 71. I. Mech. E Conference on Heat and Fluid Flow in Water Reactor Safety.

3. HERON R A et. al. Burn-out power and pressure drop measurements on 12 ft 7-rod clusters cooled by Freon-12 at 155 psia. AEEW - R655, 1969

4. MACBETH R V et. al. An investigation into the effects of 'Crud' deposits on surface temperature, dryout and pressure drop with forced convective boiling of water at 69 bars in an annular test section. AEEW - R 705, 1971

5. REDPATH W. Winfrith SGHWR in-reactor dryout tests. Jour. BNES. Vol 13. No 1, 1974.

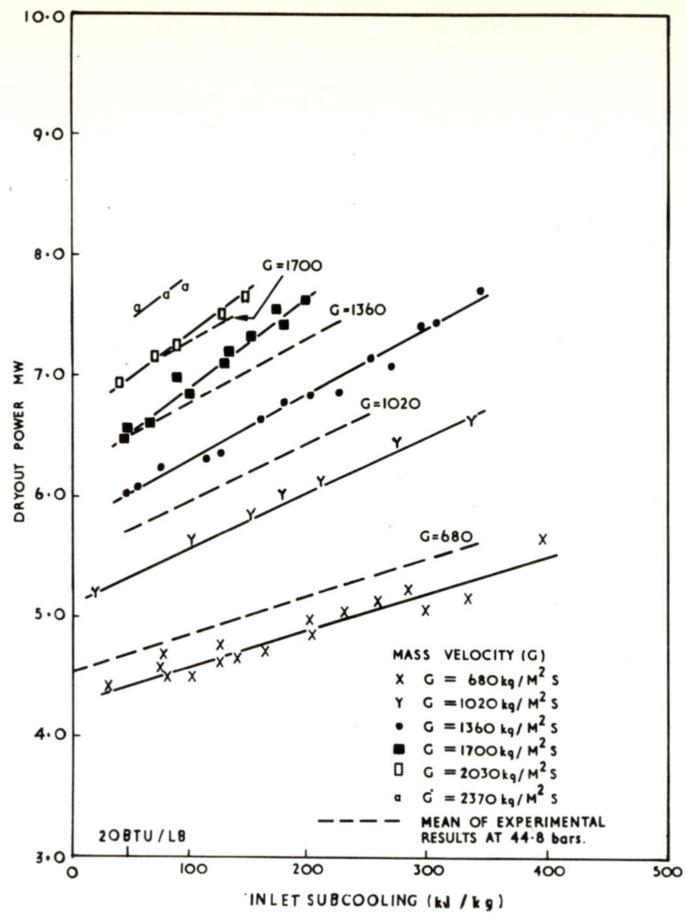
DRYOUT POWER MW
INLET SUBCOOLING (kJ /kg)
G=1700
G=1360
G=1020
G=680
MASS VELOCITY (G)
X G = 680 kg/M² S
Y G = 1020 kg/M² S
• G = 1360 kg/M² S
■ G = 1700 kg/M² S
□ G = 2030 kg/M² S
a G = 2370 kg/M² S
MEAN OF EXPERIMENTAL RESULTS AT 44·8 bars.
20BTU/LB

Fig. 1:

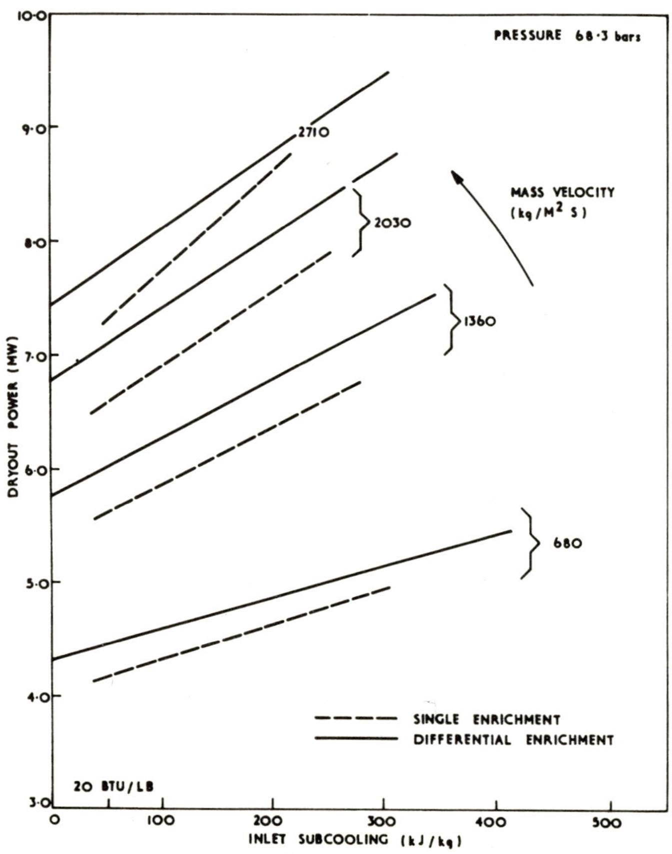
PRESSURE 68·3 bars
DRYOUT POWER (MW)
INLET SUBCOOLING (kJ/kg)
2710
2030
1360
680
MASS VELOCITY (kg/M² S)
SINGLE ENRICHMENT
DIFFERENTIAL ENRICHMENT
20 BTU/LB

Fig. 2:

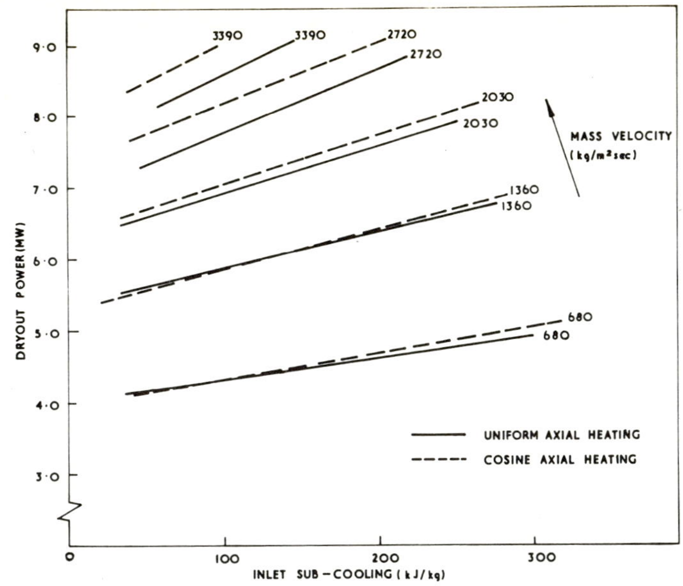

3390
3390
2720
2720
2030
2030
1360
1360
680
680
MASS VELOCITY (kg/m²sec)
UNIFORM AXIAL HEATING
COSINE AXIAL HEATING
DRYOUT POWER (MW)
9·0
8·0
7·0
6·0
5·0
4·0
3·0
0
100
200
300
INLET SUB-COOLING (kJ/kg)


Fig. 3:

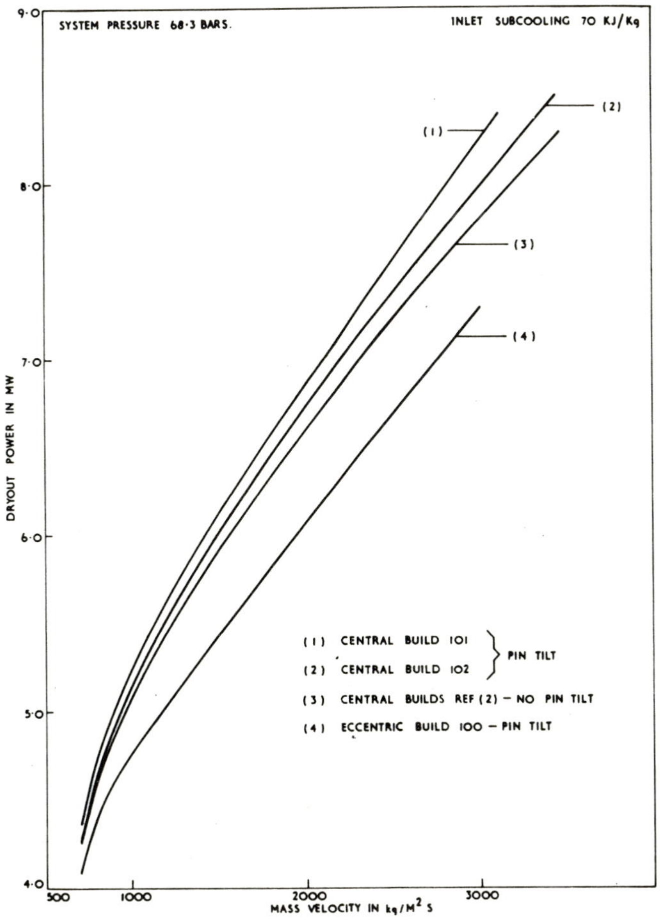

SYSTEM PRESSURE 68·3 BARS.
INLET SUBCOOLING 70 KJ/Kg
(1)
(2)
(3)
(4)
(1) CENTRAL BUILD 101
(2) CENTRAL BUILD 102
PIN TILT
(3) CENTRAL BUILDS REF (2) – NO PIN TILT
(4) ECCENTRIC BUILD 100 – PIN TILT
DRYOUT POWER IN MW
9·0
8·0
7·0
6·0
5·0
4·0
500
1000
2000
3000
MASS VELOCITY IN kg/M² S


Fig. 4:

C253/77

# A REVIEW OF SOME ASPECTS OF THE VALIDATION OF THE RELAP-UK CODE FOR PRESSURE TUBE REACTOR APPLICATIONS

F.J. FAYERS, BSc, PhD, I. BRITTAIN, BA, J.P. HARDIMAN, BSc
United Kingdom Atomic Energy Establishment, AEE Winfrith, Dorchester, Dorset

SYNOPSIS This paper briefly outlines the type of safety arguments that have been made in the UK for analysing the consequences of postulated loss-of-coolant accidents in Steam Generating Heavy Water Reactors. The role of the RELAP-UK code is identified and the type of noding arrangement adopted for a typical case is indicated. The results of three series of large experiments specific to the geometric configuration of pressure tube reactors are discussed as illustrations of code validation studies. The three experimental series are:

(i) CISE blowdown experiments in the CIRCE Rig investigating the occurrence of core stagnation following a small feeder breach.

(ii) The Japanese LIP experiments at the O'Arai facility which entail blowdown of a three-channel configuration.

(iii) The heat transfer measurements made under full power reactor conditions in the Cluster Loop of the Winfrith Prototype SGHWR.

It is shown that the RELAP-UK calculations are reasonably successful in representing the primary experimental phenomena, although the need to provide an adequate rewetting model is evident from the results.

## INTRODUCTION

1. A separate line of development of the RELAP family of codes has been undertaken in the UK, primarily intended to meet the needs of the safety studies for the Steam Generating Heavy Water Reactor project. The code has also been used in LOCA assessment studies for PWRs. The latest developments released in the Mark IV version of RELAP-UK have been described by Brittain et al (Ref 1). Two of the important features include an improved 'Drift Flux' model (Holmes (Ref 2)) and a new dryout correlation (Bowring (Ref 3)). The latter has been developed to fit a wide range of dryout experiments, including Winfrith dryout measurements on full sized clusters covering a range of pressures and flow rates. Other features of RELAP-UK which differ from RELAP4, include provision of a two-phase slip correlation which ensures a hyperbolic system for the equations of motion (Bryce (Ref 4)), and inclusion of a bubble rise treatment which operates successfully for a multi-node vertical stack.

2. A continuing task in association with the code development activity is the validation of the code capabilities against experimental data. One valuable aspect of following the RELAP line of development is the experience gained in the solution of the standard international benchmark series of problems which have been set up by the Committee for Safety of Nuclear Installations. RELAP-UK solutions have been presented to CSNI for the Edwards' blowdown pipe (Benchmark 1), the LOFT semi-scale unheated experiment (Benchmark 2), the IETI Rig experiment (Benchmark 3), and the LOFT semi-scale heated experiment (Benchmark 4). Improvements in the analysis and interpretation in terms of two-phase models and heat transfer in these experiments are continuing, and inter-comparison with the American RELAP4 experience is helpful.

3. The present paper examines some aspects of comparisons with three series of experiments which have been selected because of their particular relevance to pressure tube reactor geometries. A schematic diagram of one loop of a typical SGHWR circuit is shown in Figure 1, with a typical noding arrangement which might be used in an accident study. One important aspect of such a circuit arrangement is the multiple parallel paths which result from the use of an individual pressure tube for each fuel channel with its associated feeder and riser pipes being connected to the bottom and top headers of the reactor. The behaviour of the steam drum and of the primary circulators also have a significant influence on the flow transients during an accident. The type of safety case which is being developed for SGHWRs is briefly described in the next section. The experiments which are of direct value in developing this case are:

(i) The Italian/AECL blowdown experiments (Azzalini et al (Ref 5)) in the CIRCE rig at CISE which experimentally demonstrate the occurrence of flow stagnation resulting from a feeder breach in the single channel within a loop representing a pressure tube reactor.

(ii) The Japanese LIP series of blowdown experiments (Tokumitsu (Ref 6)) on a simulated feeder breach for a loop with three parallel channels, which were conducted by PNC at O'Arai, thus furnishing evidence on flow separation between intact and broken flow paths.

(iii) A series of in-reactor heat transfer experiments (Redpath and Read (Ref 7)) which were performed in a specially instrumented channel in the SGHW Prototype reactor at Winfrith during the Autumn of 1976. These experiments have furnished direct information on dryout, post dryout heat transfer and rewetting under realistic reactor conditions for a range of flow rates and pressures.

## THE NATURE OF SAFETY ARGUMENTS FOR SGHWRs

4. Safety arguments for loss-of-coolant accidents in SGHWRs have placed a large degree of emphasis on the performance of a specially designed spray cooling system, which delivers spray in a central sparge tube in each cylindrical fuel assembly in this reactor type. It has been argued that the worst reactor conditions which might ensue in a hypothetical accident would involve complete stagnation of the core flow under high steam voidage conditions, and for the period of a transient where this state persists, the blowdown cooling would be almost negligible in the absence of the specially provided spray system. There are no clear arguments which indicate that the likelihood or persistence of stagnation in SGHWR core flow under accident conditions should be worse than in other water cooled reactors; nevertheless, it has been considered expedient to develop a safety philosophy against such a limiting situation, since this essentially simplifies the type of safety case which has to be proved with adequate confidence.

5. The properties of spray cooling heat transfer under stagnated core flow conditions have been measured in an extensive programme of large scale rig measurements at AEE Winfrith. These measurements have indicated that the dry surface heat transfer is a strong function of pressure, as well as varying significantly as a function of position in the cluster geometry fuel element. Low values of heat transfer are observed to occur at low pressures at locations remote from the spray injection levels. Apart from quantifying the heat transfer achievable under the most pessimistic conditions, the rig measurements have demonstrated a number of important features of the spray cooling system, namely:

(i) A relatively cool wet pressure tube is maintained as an important radiation sink.

(ii) Thermal equilibrium between steam and water is promoted, thus avoiding high steam superheat conditions in an accident.

(iii) Rewetting can occur which is induced by falling films on the rod surface, initiated at the spray and sometimes the cluster grid levels.

6. The measured changes in dry surface heat transfer from spray which occur under decreasing pressure conditions imply a need for a demonstrated ability to calculate the reactor pressure transient during an accident. In particular, it is necessary to be able to predict the critical discharge rate which will induce low flow or near stagnation conditions in the core, and thence to compute the core pressure history associated with this critical rate. Intermediate breach sizes, rather than the maximum credible severance, are found to be in this category, and therefore such breaches normally constitute the 'design basis accident'. A modest breach upstream of the pump typifies the worst accident case. The UK programme of work has therefore undertaken the development of the computer program RELAP-UK whose functions in order of importance are as follows:

(i) Compute pressure histories for accidents which cause the poorest core flow situation, ie. near stagnation. These pressure histories are then utilised in conjunction with a pessimistic spray cooling heat transfer correlation to predict clad temperature transients.

(ii) Compute core flow rates and appropriate levels of pessimistic blowdown cooling for accidents involving either large breaks or small breaks which lead to well defined flow patterns.

(iii) Use a RELAP-UK best estimate prediction to quantify the duration of core stagnation in a 'design basis accident' and to quantify the safety margin which is associated with the additional blowdown cooling derived from low flows in the non-stagnated period.

7. The Italian CIRCE experiments provide direct evidence on item (i) above, the Japanese LIP experiments give data useful in demonstrating item (ii) where multiplicity of flow paths may be a factor, and the Winfrith in-reactor measurements give vital heat transfer data under low flow conditions to support item (iii).

## ANALYSIS OF THE CISE/AECL BLOWDOWN EXPERIMENTS IN THE CIRCE LOOP

8. The CIRCE loop is illustrated schematically in Figure 2 where the RELAP-UK nodal arrangement adopted in most of the simulation studies is indicated. The test section was an electrically heated 19-rod fuel cluster 4 metres in length. The experimental procedure involved inducing various sizes of simulated breach in the feeder pipe, and thence to observe circuit pressures and temperature transients in the wall of the heater. The total rig power of 1.3MW was held constant for approximately 7.5 seconds in each transient, and then sharply cut back to 20% of this value. The pump used to circulate coolant during the run-up to the steady condition preceeding the breach was rapidly valved out by a bypass line at the start of a transient. A variety of transients were studied experimentally ranging in breach size from 0% to 100% of the feeder pipe area of 19.6sq cm. It was observed that the worst temperature conditions were associated with breach areas in the range of 10% to 12.5% of the feeder area. RELAP-UK simulation of the loop indicates that near stagnation conditions occur for this range of breach sizes. An important feature of the CISE experimental arrangement was provision of fast acting valves at selected places in the circuit which allow fluid mass holdup measurements to be made over various circuit sections. Mass holdup measurements were made at a number of time intervals for the 10% breach case and we therefore select this experiment to illustrate the nature of predictions and measurements for pseudo-stagnated flows.

9. The calculations were undertaken with the Mark IV version of RELAP-UK using the Holmes drift flux model and the Bryce slip correlation.

The steam drum was represented as a single 'bubble rise' node in which the bubble rise velocity was determined by the code using a Kutateladze number of 3.2 and the bubble gradient constant was fixed at 0.8. The critical discharge was calculated using thermal equilibrium and a Moody slip ratio with a vena contracta for the discharge orifice of 0.8. (Experience indicates that this choice of discharge model is reasonably satisfactory for critical flow from small breaches). The predicted channel flows at inlet and outlet of the test section are shown in Figure 3, which illustrates the timing of the flow reversal and subsequent low flow oscillations. The channel flows eventually subside into a low reverse flow condition which typifies the situation for 'underside' breaches in an SGHWR circuit.

10. The predicted pressure history for the steam drum is compared with the measured pressure in Figure 4a, and the steam drum level, as inferred from a $\Delta P$ tapping is compared with the calculation in Figure 4b. The choice of the equilibrium discharge model with Moody slip has given a quite good result, although the 'knee' seen in the pressure measurements is not predicted by RELAP-UK. The clad temperatures are compared with thermocouple measurements at three levels in Figure 5 where we note that dryout and the ensuing temperature rise are well predicted up to about 7.5 seconds. The delayed rewetting following the power setback, which is observed in the measurements, is not predicted by the code. The failure to predict rewetting subsequently leads to inadequate representation of the evaporation of the remaining water in the channel, which in the experiment then causes a second dryout and a rising clad transient associated with superheated steam. The calculated channel voidage fractions are close to unity at the end of the transient and steam superheating is predicted thereafter. Some comparisons between the measured and predicted mass holdups for this experiment are indicated in Figure 6, where good agreement is seen, except for the channel during the period of significant changes in heat transfer. In particular, the calculated increase in channel mass in the period adjacent to the mass measurement at 20 sec occurs because of the absence of rewetting in this interval, with a corresponding absence of steam evolution. The calculated mass history in the channel section is also sensitive to the correct prediction of dryout and is improved by use of a drift flux model in preference to a co-current flow model.

11. We conclude from this experiment that the RELAP-UK capabilities in predicting the discharge rates which will produce near stagnation conditions, and the associated rate of pressure decline, are reasonably well confirmed. However, the prediction of heat transfer and rewetting under near stagnation conditions is very complicated because of the interaction between flow and heat transfer phenomena. This experiment illustrates that under conditions of near stagnated flow deliberate biassing of a code to discourage rewetting may not always be pessimistic.

## ANALYSIS OF THE JAPANESE LIP BLOWDOWN EXPERIMENT IN THE O'ARAI RIG

12. The loop configuration used in the LIP-12 case of this series of feeder breach experiments is illustrated schematically in Figure 7 with the adopted noding arrangement indicated. The three parallel channels each contained 28-pin electrically heated clusters, 3.7 metres in length. The breach in one of the feeder pipes was induced via a T-off into a rather complicated blowdown line some 4.3 metres in length. The experimental arrangement leads to a significant reverse flow during the transient on the 'breached' channel, but rather slow forward flows in the two 'intact' channels. The steam drum again plays a rather sensitive role in the modelling of this case, accentuated by the fact that the characteristics of the turbo-separators present near the riser entries to the drum influence the flow division between the circuit paths. We have dealt with this problem in an 'ad hoc' manner by adopting a compromise two-node model of the drum. These nodes, representing the separator belt and the main body of the drum were both given a bubble rise treatment. A further complication in the modelling is caused by the presence of the lower header which feeds the three channels from the downcomer pipe. It was assumed that conditions here are sufficiently turbulent for a homogeneous assumption to apply. The experimental procedure again involves by-passing the pump at the start of the transient. The breach size for this particular case involved the full pipe area of 30.5 sq cm. It was considered appropriate to use the homogeneous equilibrium discharge model for these conditions with a vena contracta ratio of 1.0. We generally find that the HEM model gives good results for larger breach sizes when analysing large rig experiments. Clad temperature measurements were taken at five levels in both the 'breached' channel and on the 'intact' channels. Pressure measurements are also available, but in contrast to the Italian experiments, measurements of mass distribution were not made.

13. The calculated steam drum pressure history is compared with the measurements in Figure 8a and the drum levels are compared in Figure 8b; although there are some minor departures in detail the overall results for both pressure and drum mass are encouraging and lie within the probable experimental accuracies. The measured and calculated fuel clad temperatures are shown in Figure 9a for the breached channel, and Figure 9b for the intact channel. The calculation for the breached channel indicates some delay in dryout, but this does not appear to be very significant in this slowly varying transient. The dryout predicted on the intact channel occurs close to the correct time and also in the correct positional sequence. The post-dryout rates of temperature change are also predicted quite well, although there is some evidence of overprediction of heat transfer to superheated steam in the intact channel when the power is cut on that channel.

14. In spite of some limitations in detail we conclude that the main aspects of the experimental behaviour are calculated rather well. The timing and positional sequence of dryout in the breached and unbreached channels has been adequately predicted, and the post dryout heat transfer is also adequately represented. The adduced success of the Bowring dryout correlation is believed to confirm its use under conditions of low flow and reduced pressure, and we note that the inlet quality correction influences the prediction for this experiment.

ANALYSIS OF THE IN-PILE HEAT TRANSFER EXPERIMENTS IN THE WINFRITH SGHWR

15. These experiments were conducted under genuine reactor conditions for a range of powers, flows and pressures in the Cluster Loop of the Winfrith Prototype reactor which contained an instrumented SGHWR 36-pin fuel element, each pin being 365 cm in length and 1.54 cm O.D. The loop is provided with its own coolant flow circuit which can be valved out from the main reactor circuit. A separate heat rejection system allows an independent estimate to be made of channel power, which was backed by measurements obtained from a carefully checked and calibrated quality meter. The flow in the loop was varied during the transients using a control valve provided in the feeder pipe and a calibrated venturi was used to measure this flow. The experimental procedures entailed ramping down the flow until some selected limiting peak temperature was registered on one or more of the 50 thermocouples attached to the fuel pins and thence holding flow constant until asymptotic temperatures were attained, the transient then being terminated by a ramped increase in flow. Rewetting of the region of the cluster in dry-out was observed to occur quite quickly during the ramp increase in flow. Experiments were performed at two loop pressures, namely 6.0MPa and 2.0MPa, and with spray cooling on or off as a further variation. We illustrate here the results of three RELAP-UK simulations of this large range of experiments. The main features of the experiments selected are summarised below:

the measured and calculation points refer to heights above the base of the fuel. The input flow variation is also plotted against the right hand axis of the figure. The calculated dryout is slightly early, and occurs first at the highest level which is contrary to the measurements. The use of the Dougall-Rohsenow correlation for post dryout heat transfer is seen to be slightly pessimistic in the calculation. The failure of the RELAP-UK model to achieve rewetting is clearly indicated when the flow is stepped up again. Rather similar results are obtained for Experiment (ii), as shown in Figure 11, although the time of dryout is now very slightly late.

18. The simulation of Experiment (iii) which provides the high temperatures in the presence of spray cooling is again broadly successful. The early onset of dryout now observed at the top of channel is somewhat delayed in the calculation, and the subsequent temperature rise is not as large as seen in the measurement. A sensitivity study indicates that increasing the overall channel power by only 1% (well within the estimated experimental uncertainty of ±3%) would cause the top node to dryout at the correct time. The integrated power in the top node is very sensitive to the precise nature of the rapidly varying power profile in this region (which changes markedly with small variations of the $D_2O$ moderator height), thus the rating used on the calculational slave pin may well be somewhat inconsistent with the local pin rating, thus causing the steeper rise in the measured temperature transient. The most severe temperatures

Table 1 Summary of Experimental Conditions in Cluster Loop Experiments

| Experiment | Channel power MW | Drum pressure MPa | Spray Flow Kg/sec | Initial Flow rate Kg/sec | Minimum Flow rate Kg/sec | Initial outlet quality % | Maximum outlet quality % | Peak clad tempera- °C |
|---|---|---|---|---|---|---|---|---|
| (i) | 4.96 | 2.0 | 0 | 5.9 | 2.9 | 42 | 87 | 660 |
| (ii) | 2.04 | 6.0 | 0 | 1.5 | 1.1 | 82 | 100 | 580 |
| (iii) | 1.98 | 6.0 | 0.6 | 2.0 | 0.3 | 49 | 100 | 900 |

Experiment (iii) is of particular interest since this was nearly the last in the series and a peak clad temperature of 900°C was attained. In this case it was not practical to achieve an asymptotic state in post dryout and we are thus very dependent on the dynamic aspects of the RELAP modelling capability. The characteristics of the pin thermal model are important in this regard.

16. The RELAP simulation was achieved by using the measured flow and enthalpy variation at channel inlet as one boundary condition, and the measured pressure at the loop steam drum as the second condition. An axial power profile which was deduced from axial scans with copper wires for several channel qualities, has been assigned to the node arrays. 10 nodes were used in the channel heated length, and measured temperatures on the pins are compared with the RELAP slave and driver pin predictions at similar levels.

17. Some results for Experiment (i) are illustrated in Figure 10, in which the elevations of

are observed on the thermo-couples in the lower half of the core, but still in the highly rated region of the axial power profile. The conditions here indicate flows through the channel of less than 0.6Kg/sec with fairly high steam voidage (ie. in excess of 50%). In these conditions the RELAP-UK calculation selects the rather low heat transfer associated with a spray cooling correlation, and this choice is confirmed by the observed rates of increase in the measured temperature profiles. For points higher in the channel, the steam flow rates became large enough to cause selection of the Dougall-Rohsenow correlation, and this again performs quite well in matching the temperature observations. It is noticeable that when the flow is increased the worst of the clad temperature predictions does not turn over, which appears to be due to use of the Dougall-Rohsenow correlation at the lower qualities which prevail with increased flow.

19. We conclude that the dryout and post-dryout heat transfer correlations utilised in RELAP-UK

are reasonably satisfactory under specified flow conditions. The need for the introduction of an adequate rewetting model in the code is underlined by these experiments, and the first results obtained from a conduction controlled rewetting model (Blackburn and Pearson (Ref 8) recently incorporated into RELAP-UK are encouraging.

ACKNOWLEDGEMENT

The authors gratefully acknowledge the provision of experimental information from CISE Italy, and PNC Japan which has been utilised in this paper. Thanks are due to W Redpath and R Read for discussions on the Winfrith Cluster Loop measurements and to C Green who has undertaken a substantial portion of the RELAP-UK code development.

REFERENCES

1. BRITTAIN I et al. The status of RELAP-UK Mark IV at August 1977. In preparation.

2. HOLMES J. Description of the drift flux model in the LOCA Code RELAP-UK. Conference on Heat and Fluid Flow in Water Reactor Safety. Manchester 1977.

3. BOWRING R W. A new mixed flow cluster dryout correlation for pressures in the range 0.6 - 15.5 MN/m$^2$ (90-2250 psia) - for use in a transient blowdown code. Conference on Heat and Fluid Flow in Water Reactor Safety. Manchester 1977.

4. BRYCE W M. A new flow-dependent slip correlation which gives hyperbolic steam-water mixture flow equations. May 1977. AEEW - R 1099.

5. AZZALINI A et al. An experimental investigation of blowdown in pressure tube reactor conditions. November 1973. CISE-R-342.

6. TOKUMITSU M. Full scale safety experiments of FUGEN. Paper N341 72.12 CREST Specialist Meeting on Emergency Core Cooling for Light Water Reactors. Munich October 1972.

7. REDPATH W and READ R. In-reactor measurements of CHF, post-CHF heat transfer and rewetting in a 36-pin fuel element in the Winfrith SGHWR 100 MW(e) Power Station. To be published.

8. BLACKBURN D and PEARSON K G. Analytical approximation to rewetting rate of overheated composite fuel pins,under low pressure, high temperature conditions. Conference on Heat and Fluid Flow in Water Reactor Safety. Manchester 1977.

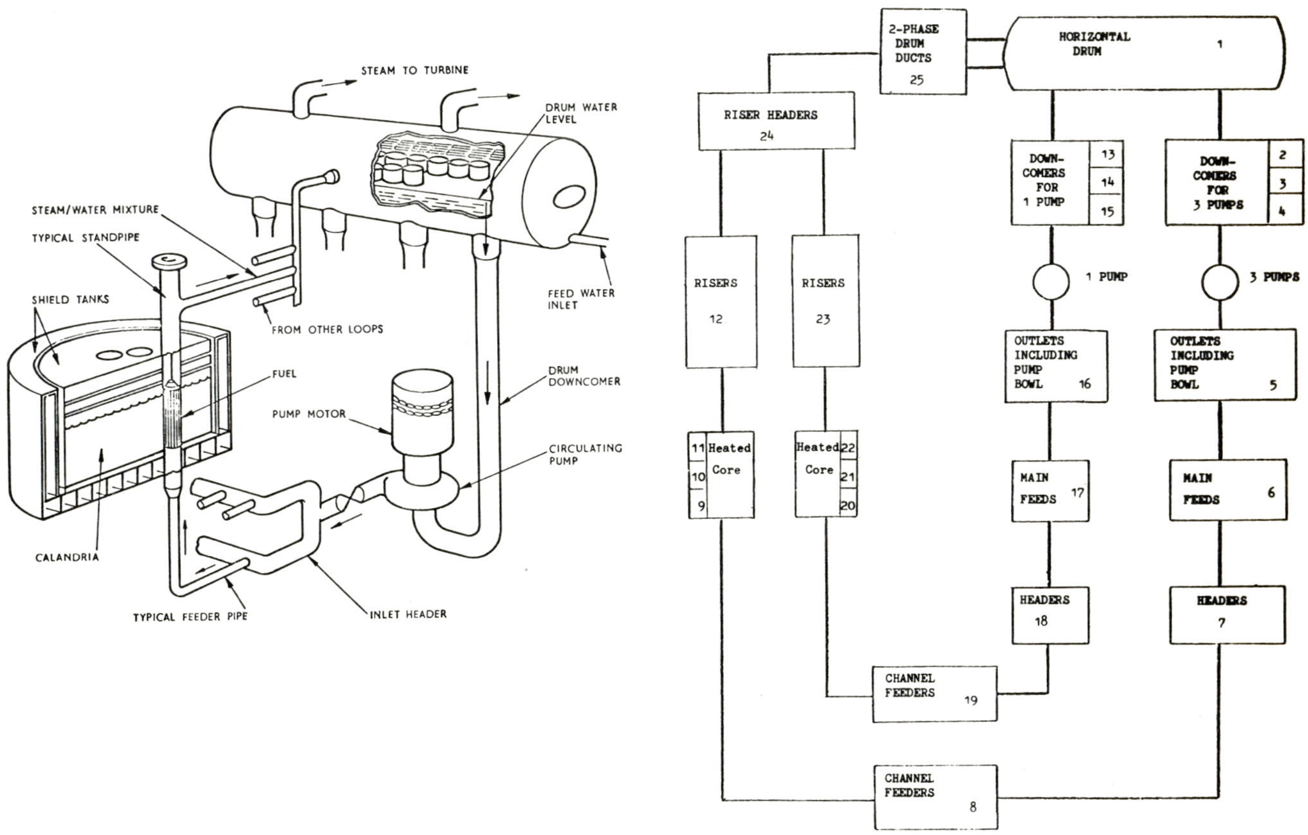


Fig. 1: SGHWR circuit schematic and noding arrangement for typical LOCA studies

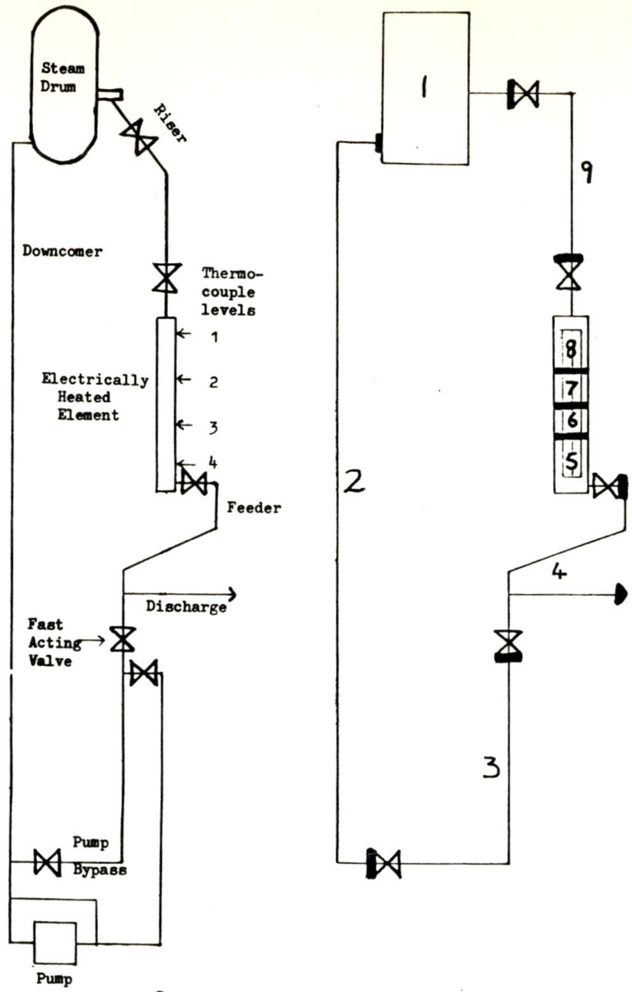

Fig. 2: CIRCE loop schematic and noding arrangement

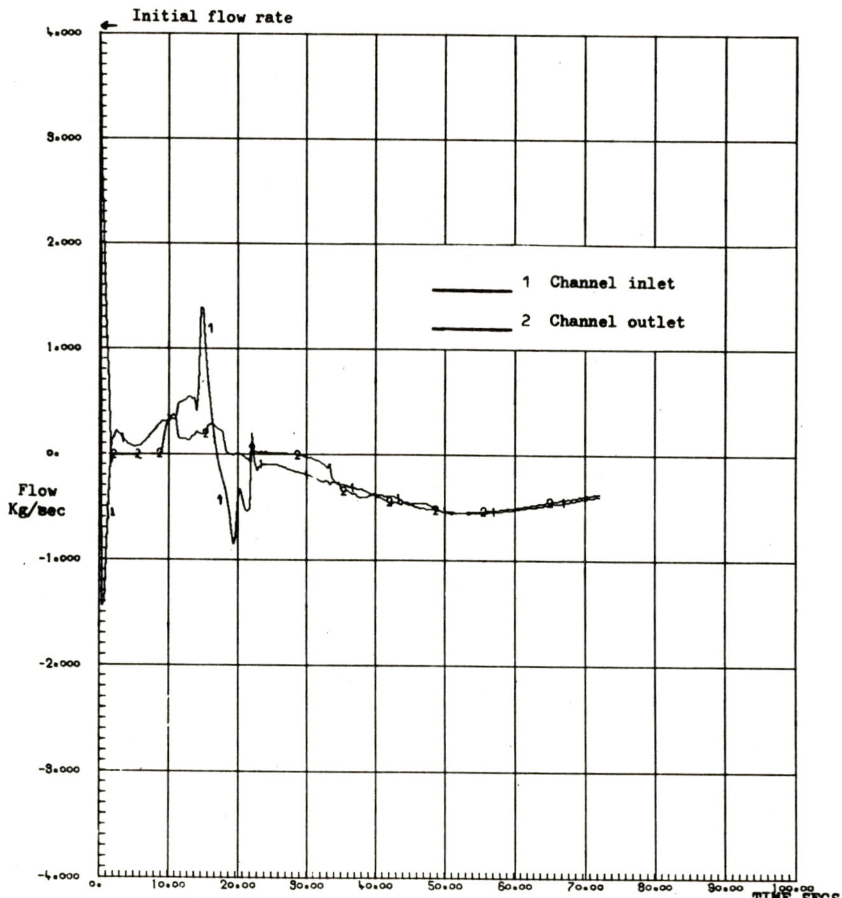

Fig. 3: Channel inlet and outlet flow rates for CIRCE 10% feeder breach

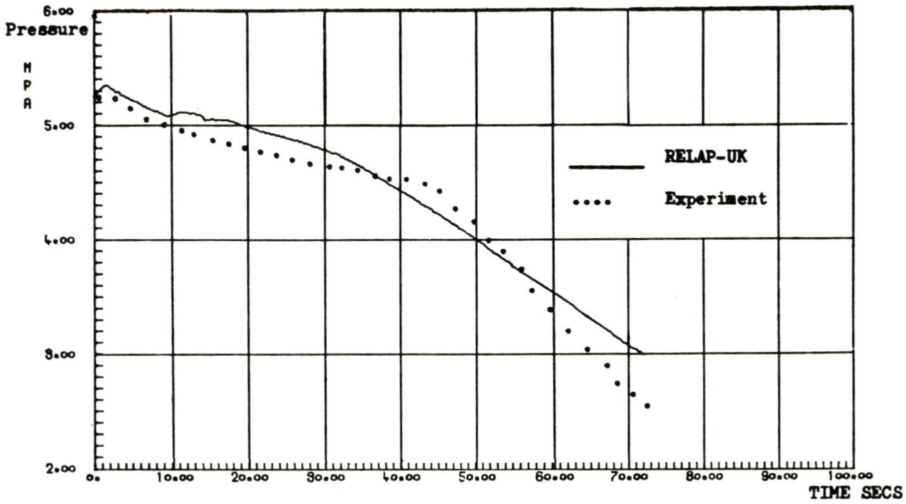


Fig. 4(a): Loop steam drum pressure history for CIRCE 10% feeder breach

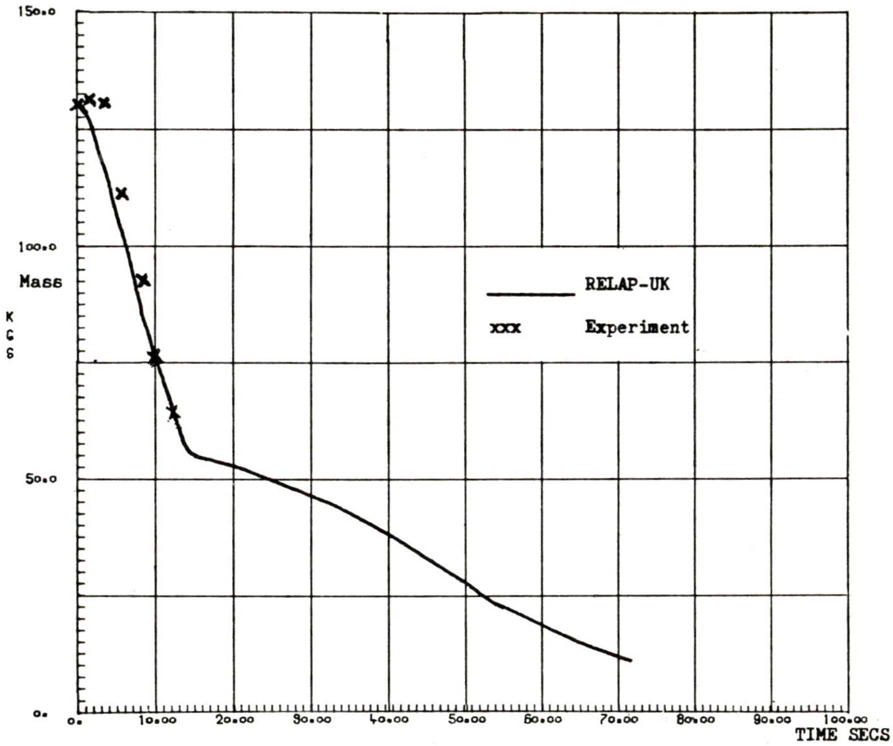


Fig. 4(b): Loop steam drum mass history for CIRCE 10% feeder breach

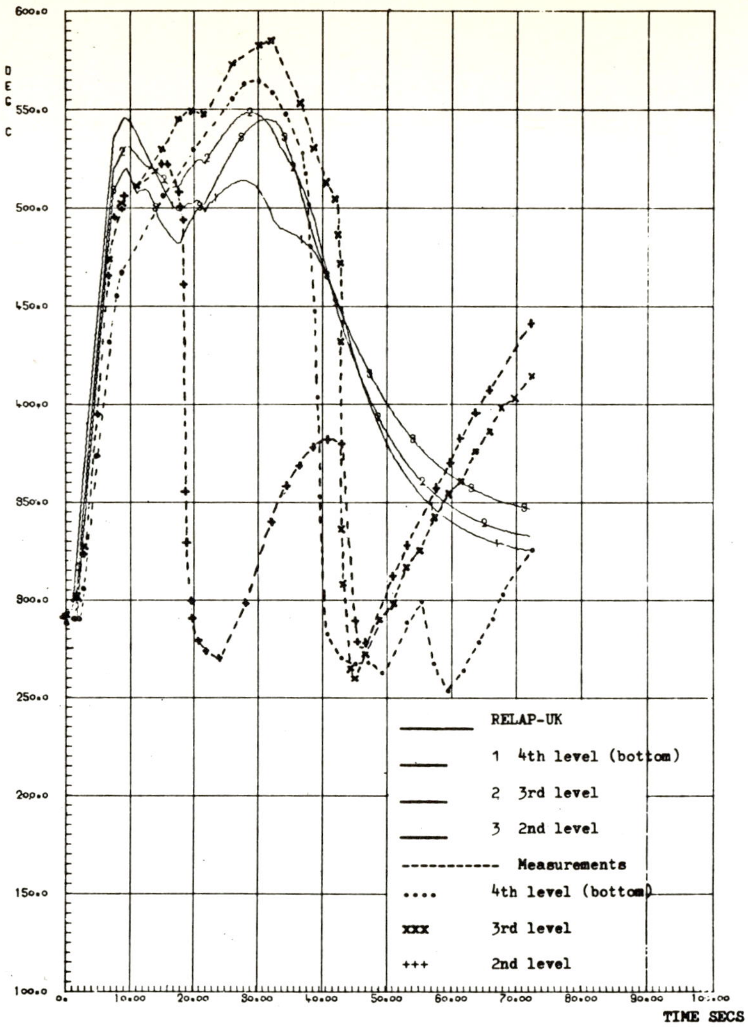


Fig. 5: Clad temperatures for CIRCE 10% feeder breach

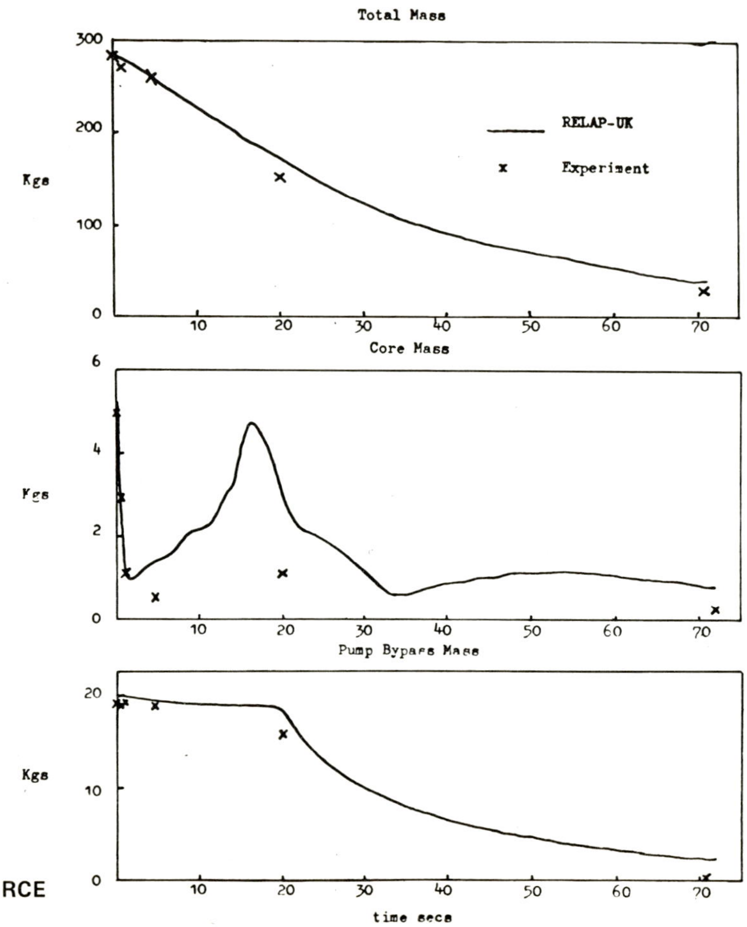


Fig. 6: Mass holdups for CIRCE 10% feeder breach

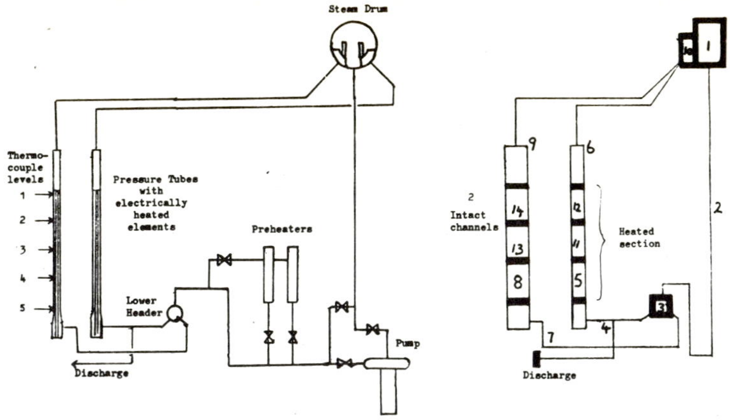

Fig. 7: Loop schematic and noding arrangement for O'Arai LIP experiment

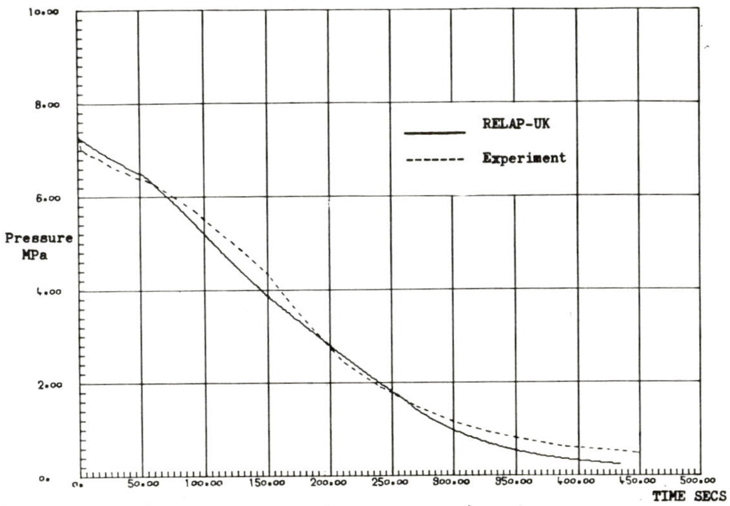

Fig. 8(a): Steam drum pressure for O'Arai LIP experiment

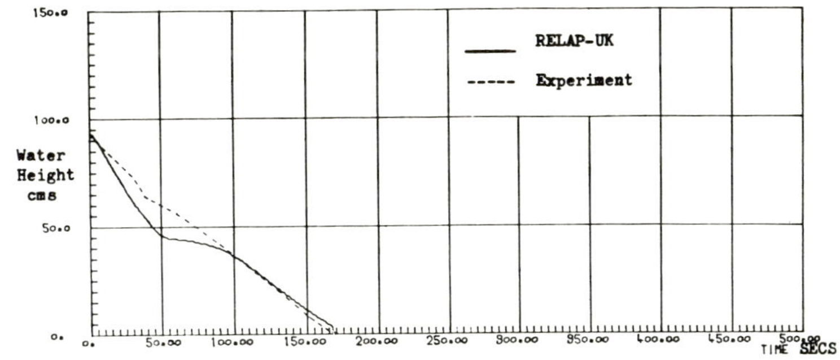

Fig. 8(b): Steam drum water level for O'Arai LIP experiment

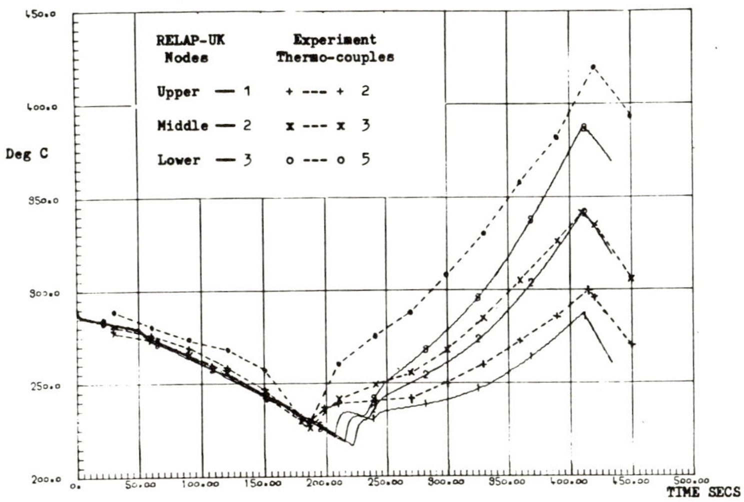

Fig. 9(a): Clad temperatures for breached loop of O'Arai LIP experiment

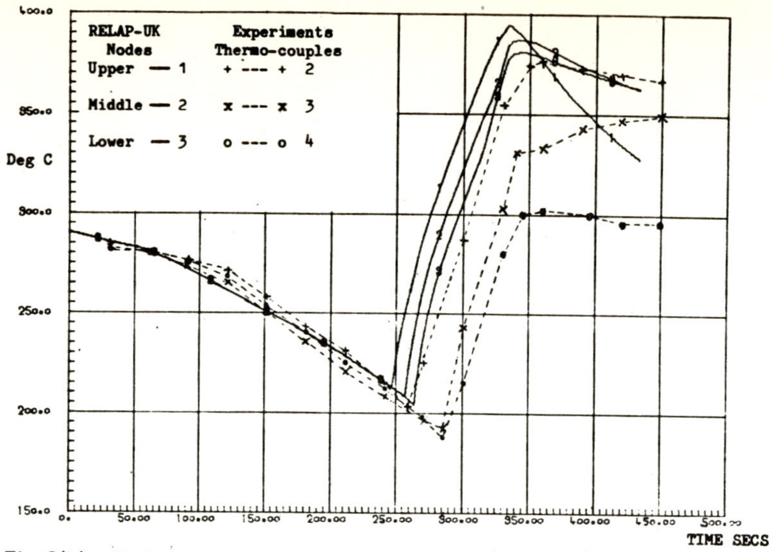

Fig. 9(b): Clad temperatures for intact loop of O'Arai LIP experiment

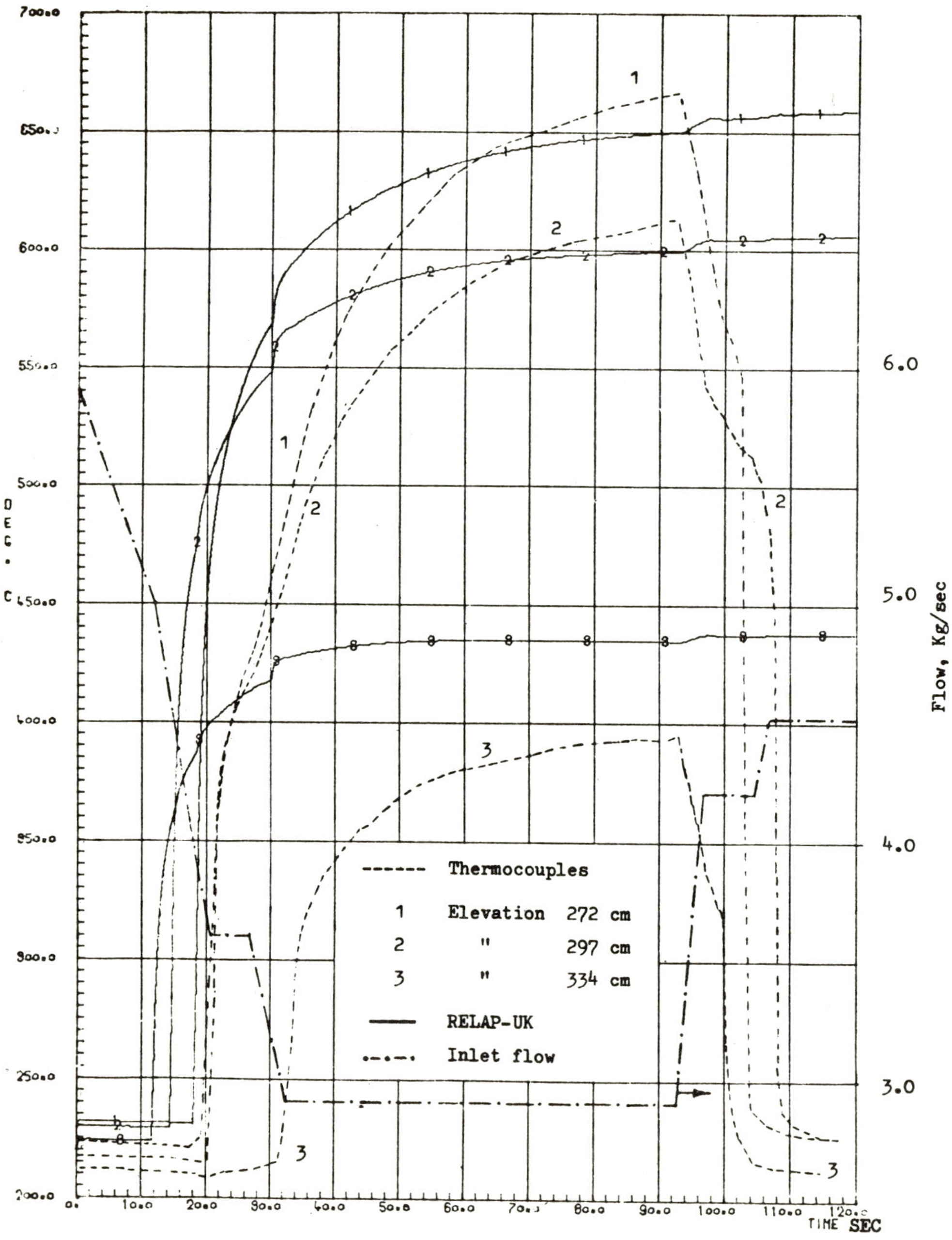

Fig. 10: Clad temperature transients for SGHWR cluster loop experiment (i)

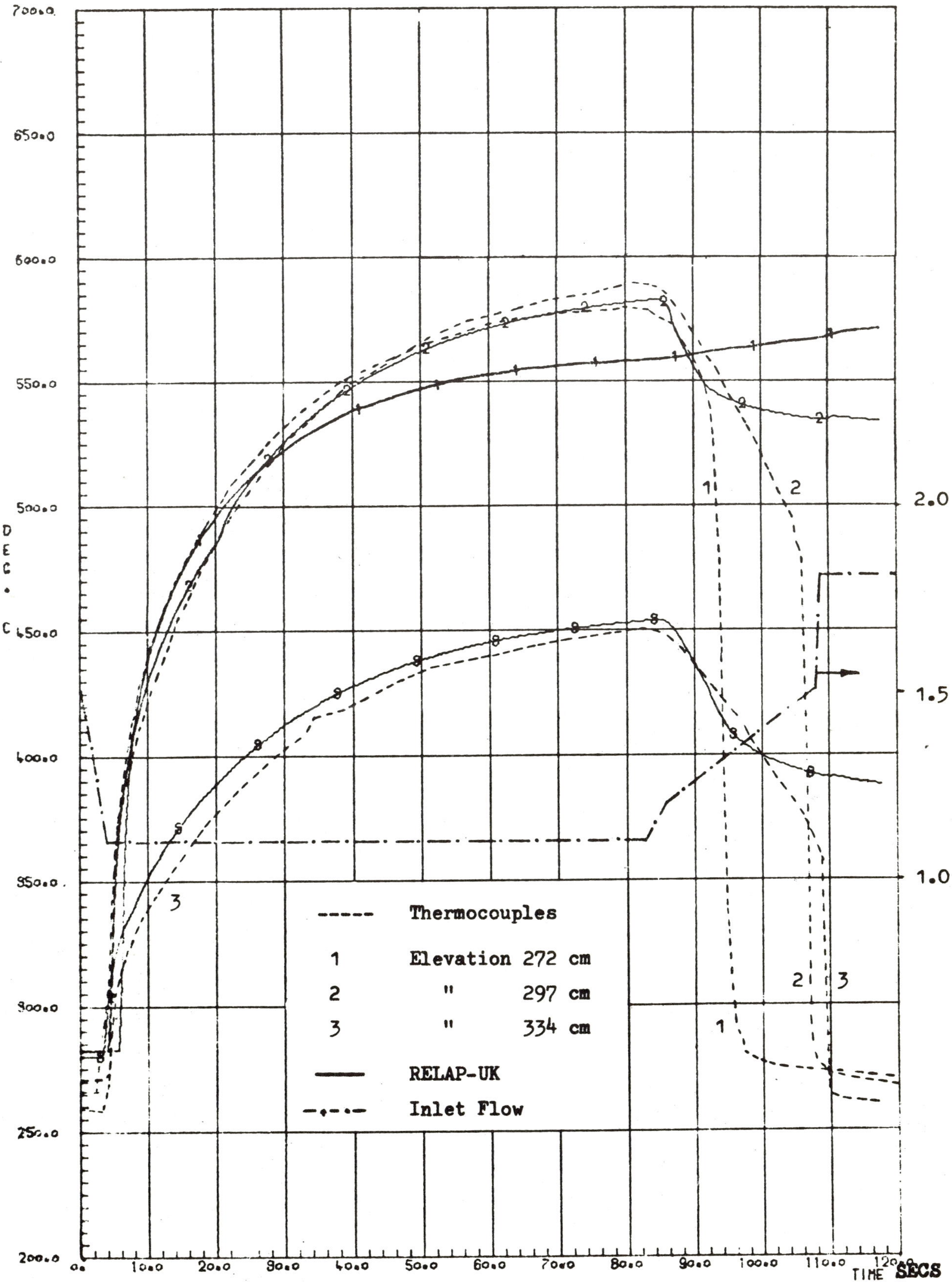


Fig. 11: Clad temperature transients for SGHWR cluster loop experiment (ii)

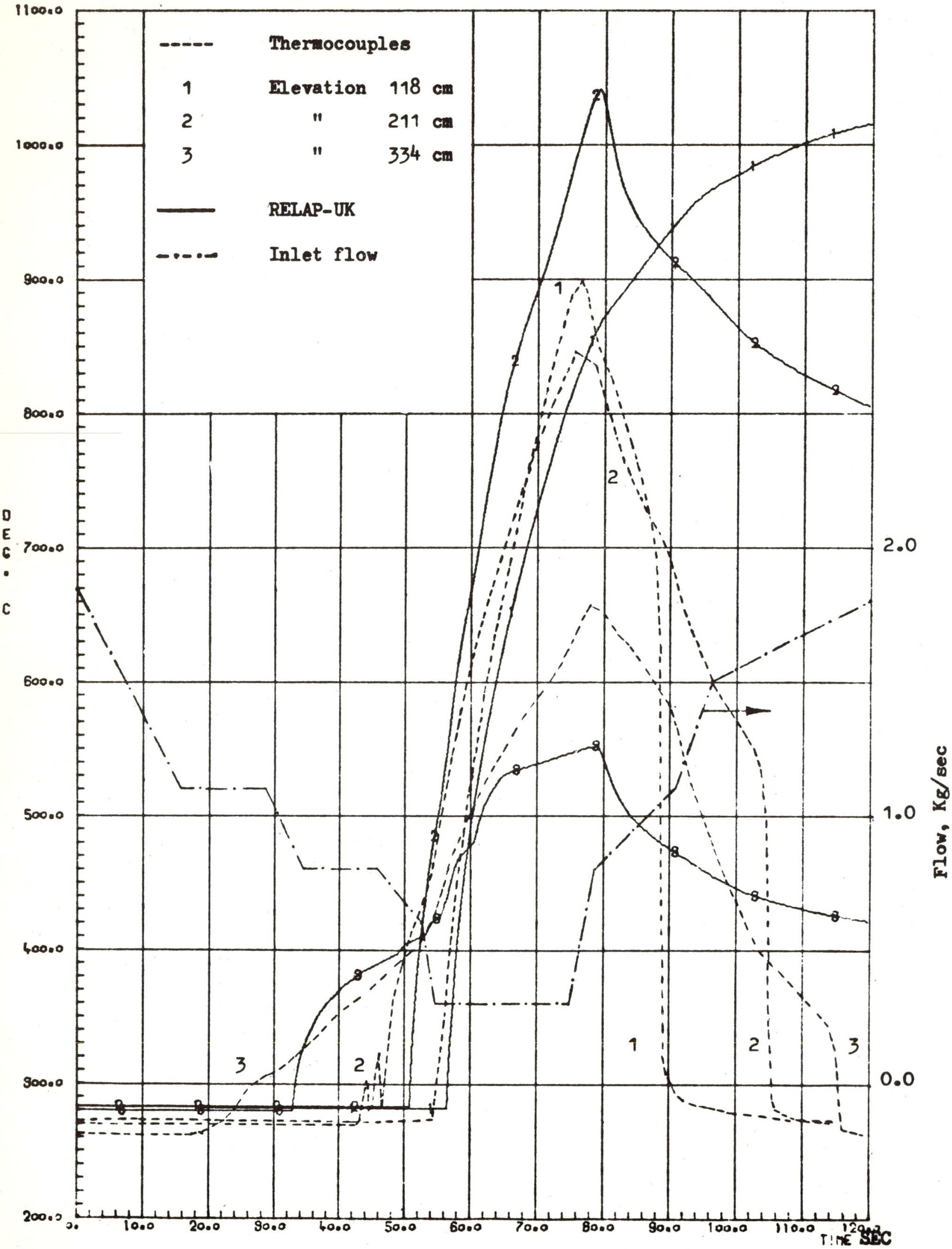


Fig. 12: Clad temperature transients for SGHWR cluster loop experiment (iii)